Molecular Gerontology
Research Status and Strategies

Molecular Gerontology
Research Status and Strategies

Edited by

Suresh I. S. Rattan

University of Aarhus
Aarhus, Denmark

and

Olivier Toussaint

The University of Namur
Namur, Belgium

Springer Science+Business Media, LLC

Library of Congress Cataloging-in-Publication Data

Molecular gerontology : research status and strategies / edited by
  Suresh I.S. Rattan and Olivier Toussaint.
      p.   cm.
  Includes bibliographical references and index.
  ISBN 978-1-4613-7701-6   ISBN 978-1-4615-5889-7 (eBook)
  DOI 10.1007/978-1-4615-5889-7
  1. Aging--Molecular aspects.   I. Rattan, Suresh I. S.
II. Toussaint, Olivier.
QP86.M692 1996
612.6'7--dc21   96-37284
                                                                CIP

ISBN 978-1-4613-7701-6

© 1996 Springer Science+Business Media New York
Originally published by Plenum Press, New York in 1996
Softcover reprint of the hardcover 1st edition 1996

PREFACE

Ageing is an important biological, social and emotional issue that affects almost all aspects of our lives. In recent times, there has been a significant increase in interest in research on the molecular and biochemical basis of ageing. This is not only due to a rapidly changing demography throughout the world in which the proportion of the elderly in the population is increasing progressively, but also due to the realisation that the process of ageing is the fundamental reason behind most of the major diseases, such as cardiovascular and cerebrovascular diseases, cancer, cataracts, insulin-independent diabetes, muscular atrophy, osteoarthritis, osteoporosis, senile dementia (for example, Alzheimer's disease), renal failure and retinopathy.

The aim of gerontological research is to achieve the ultimate goal of healthy old age and increased years of independence for older people. It is clear that if we want to know the aetiology of an age-related disease and if we want to find out preventive and curative measures to counter that disease, then we need to study the process of ageing itself. Furthermore, unravelling the fundamental mechanisms of ageing which ultimately lead to the development of a plethora of functional impairments and diseases is a pre-requisite for developing appropriate means of increasing mobility, activity, creativity, and independence in the elderly. Therefore, there is not only a need of undertaking more basic research in ageing, but also it is essential that biogerontology is properly taught to graduate and postgraduate students of biology, biotechnology, and medicine.

The aim of this book is to give an overview of molecular gerontological research in terms of its present status and future research strategies. The molecular gerontological perspectives covered in this book represent the latest and most topical areas of research, such as the stability of the nuclear and mitochondrial genomes, regulation of gene expression from DNA to RNA to functional proteins, molecular similarities and differences between ageing and cancer, and origin of various age-associated diseases. The issue of the genetic regulation of ageing and longevity is discussed in the framework of premature ageing syndromes and centenarians. The failure of homeostasis as the underlying principle of ageing is the basis of all discussion about various mechanisms of defence and repair, including altered responsiveness and constant remodelling of the immune system. Furthermore, advantages and disadvantages of some new technologies, such as transgenic organisms, and the use of various nutritional and chemical modulators of ageing and lifespan are evaluated. Suggestions are made for future research strategies in the field of molecular gerontology.

This book is the outcome of the collective effort of the participating groups in the European Commission's BioMed and Health Research Concerted Action Programme titled "Molecular gerontology: the identification of links between ageing and the onset of age-related diseases" (BMH1 CT94 1710). The book is aimed at students and researchers as a text book on molecular aspects of ageing to be followed during any courses on bio-gerontology being offered in the universities and medical colleges throughout the world.

Suresh I.S. Rattan   Olivier Toussaint
Aarhus, Denmark   *Namur, Belgium*

CONTENTS

GENETICS OF AGEING AND MULTIFACTORIAL DISEASES

P. E. Slagboom and D. L. Knook

Gaubius Laboratory, TNO-PG
Department of Vascular and Connective Tissue Research
P.O. Box 430, 2300 AK, Leiden, The Netherlands

1. INTRODUCTION

The ageing process of an organism may be defined as the, mainly intrinsic, progressive accumulation of changes with time, associated with or responsible for the increasing risk of dying. Only some of these changes are clearly associated with disease. Ageing studies of many different species have revealed that at time points at which age-related changes become detectable, they can be found at all levels of organismic organization. These involve changes in hormone status, response mechanisms to stress, structural components, such as the intra- and extracellular matrix and chromatin, the primary DNA sequence, the protein synthesis machinery, etc.

Ageing research in the past decades has been dedicated to the search for common factors and mechanisms determining the rate of ageing in different species. There are four characteristics by which ageing in different species is usually described and compared: the maximum life span (which depends on a single last survivor), the 50% survival age (chosen because the occurrence of multiple pathology is generally significantly less in those dying before that age), the initial mortality rate (IMR, determined by the main cause of death at a relatively young age), and the mortality rate doubling time (MRDT, expressing the acceleration of mortality).

The maximum life span and the MRDT phase appear to be specific for the species, even if diverged populations are compared, which indicates that these characteristics are determined by a significant genetic component[1,2]. Life span is correlated with species characteristics such as body size and temperature, metabolic rate, blood glucose levels, etc. The IMR, in contrast, can vary extensively among different populations of the same species, which may be caused both by environmental influences (if, for example, a population has many predators) and genetic ones (if a strong genetic predisposition for a disease causing an early death prevails in a population). The fact that the largest differences in life span and MRDT are those between species (40-fold between mammalian species) suggests a genetic basis for the interspecies variations in ageing rate. A useful basis for the

Molecular Gerontology, edited by Rattan and Toussaint
Plenum Press, New York, 1996

discussion of how ageing may be caused was provided by three theories hypothesizing the evolution of ageing[3,4]:

i) programme theory of ageing, ii) pleiotropic gene theory of ageing and iii) disposable soma theory of ageing. Predictions from these theories explain ageing by:

  i. programmed events and processes
  ii. pleiotropic effects mediated by alterations in gene expression and late-acting deleterious effects of germ-line mutations;
  iii. accumulation of somatic defects, at a rate determined by networks of longevity assurance (somatic maintenance) genes.

(i) The length of the maturational phase of life correlates with the length of the period of ageing across many mammalian species (despite vast differences in total life span). Hence, similarly programmed genetic processes may control maturational development and ageing[5]. Many hormones governing circadian rhythms, development and seasonal sexual cycling (such as pituitary, thyroid, adrenocorticoid, ovarian, and testicular hormones) have major effects on the length of the reproductive period and on median and maximal survival in nematodes, fruit-flies, rodents and primates[6]. Hypophysectomy, ovariectomy and castration delay age-related physiological changes and ageing in mice and rats[7]. In old rats the activity of the hypothalamo-pituitary-adreno cortical axis is increased. This is associated with an age-related derangement of thyroid function. The beneficial effect of hypophysectomy in rodents may involve enhancement of thyroid hormonal action. Other hormones such as growth hormone[8] and melatonin, the pineal hormone, have many rejuvenating and longevity-prolonging effects when administered to old mice[9,10]. A reduction in pineal melatonin levels, especially during darkness, is associated with ageing of rats and humans[10].

These observations could indicate the widespread existence of ageing programmes. Hormone-driven programmed processes indeed affect tissues of iteroparous species (menopause, thymic involution), and the age of onset of age-related diseases. There is no evidence, however, that these programmed processes determine the functional decline of the whole organism. Several lines of evidence indicate a general inability of higher organisms to maintain the homeostatic balance of the endocrine and nervous system at later ages. This may result from late-life alterations in the level of key hormones, neurotransmitters and peptides. Such alterations may either be programmed or due to somatic defects such as a random loss of transcriptional control of the gene systems involved in neuroendocrine functions.

(ii) The genome of species with a post-reproductive period may harbour mutations which are advantageous and/or neutral to reproduction with late-acting deleterious effects[11,12]. Late-acting deleterious effects may arise from genes with 'mild' mutations, the declined quality of their products becoming a hazard gradually. Such mild mutations may accumulate in gene regions coding for protein domains on which selection is not too stringent.

(iii) Ageing may have evolved as a consequence of a trade-off between early fecundity and late-survival[13]. The disposable soma theory states that the optimal investment in somatic maintenance is always below the hypothetical threshold necessary for indefinite survival. The theory predicts a central role of energy metabolism (and therefore of mitochondria in eukaryotes) and somatic maintenance functions in determining lifespan. In comparative studies, relations were found between life span and species-specific characteristics such as metabolic rate, growth rate, etc. These relations are compatible with the disposable soma theory.

Two major lines of research have dominated the experimental investigations into the genetics of ageing. One line was focused on identification of candidate longevity genes; another line on monitoring the accumulation of somatic defects. In spite of common elements in the ageing process of different species, the most critical life span-limiting function may vary among species. Genetic control of longevity in lower species has been demonstrated in selection experiments and by construction of transgenic animals carrying candidate longevity genes[14]. Mutants of increased longevity have been obtained for Drosophila[15] and the nematode Caenorhabditis elegans[16,17] but not for mammals. It is thus far not clear whether the findings in lower species can be extrapolated to gradually ageing species such as humans, with MRDT phase determined by multi-organ pathology. Except for some life-shortening genotypes, no genes or gene clusters have yet been shown to be directly involved in the control of the maximum life span of higher species.

2. AGEING POPULATIONS

In an ageing population, different subgroups may be discriminated for which the nature of the factors determining mortality varies. The survival curve of gradually ageing species can hypothetically be divided into three parts[18] (Figure 1). The 'A' part would be dominated by the initial mortality rate (IMR), as a result of trauma, early onset diseases caused by severe genetic, physiological or developmental defects. Short-lived genotypes of mice (and inbred offspring) are frequently associated with early onset diseases, which may resemble late onset diseases, but do not otherwise generally accelerate senescence. These shorter-lived genotypes may ultimately reveal candidate genes for age-related diseases, if late and early onset versions of the disease have a common basis. Determinants of a disease may be found more easily in short-lived genotypes than in longer-lived genotypes where the disease occurs among many other age-changes.

The 'B' part of the curve is determined by the MRDT; mortality in this age-group is brought about by common and individual-specific susceptibilities and disease patterns (usual ageing). The type of disease and the rate and age of onset at which a disease becomes clinically expressed, are influenced by the individual's genetic background, life style (life events) and environment. This creates an increasing heterogeneity between individuals of the same chronological age, which is reflected in data on many age-related parameters. Following the three concepts of ageing mentioned above (1.1), three types of genes are expected to be involved in determining the MRDT:

1. gene variants with programmed deleterious effects in cells and tissues
2. gene variants which have unprogrammed late-acting deleterious effects,
3. networks of genes determining the accumulation rate of somatic defects exerting their effect on longevity from the moment of conception.

The 'C' part of the curve is determined by relatively long-lived individuals (successful ageing). The process of ageing in these individuals may occur more slowly than in other subjects, not dominated by diseases arising in a single tissue or organ. Many physiological age-changes are indeed retarded in these individuals, although a general decline of organ reserve and impairment of homeostasis occurs, which probably results in a diverse pathology at death[19]. The variability of overall health status, disease patterns, pathological and molecular age-changes in part 'B' and 'C' increases. This even occurs in inbred laboratory animals[20]. Studies aimed at the identification of determinants of disease patterns should actually be performed with subsets of individuals with comparable bio-

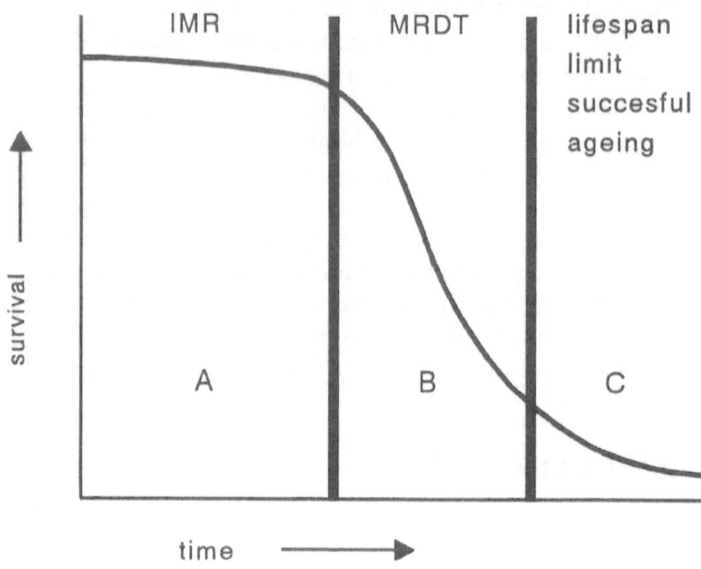

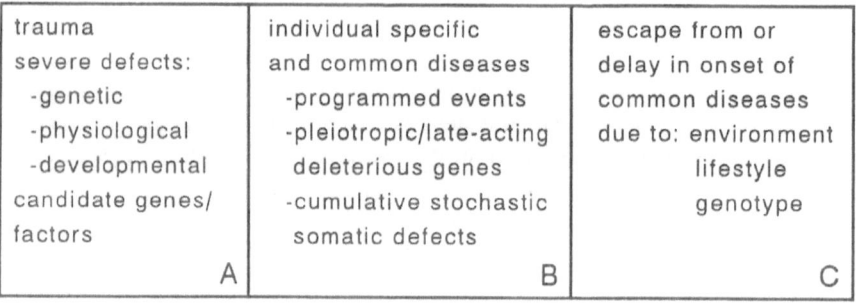

Figure 1. Hypothetical separation of subpopulations in a schematic survival curve of a gradually ageing species. Text in boxes relate to factors that determine mortality in phase A, B and C.

logical age (concerning overall health status within an age group) and not only on the basis of chronological age. The study of specific subpopulations in the age-groups of section 'B' may lead to identification of genes determining the MRDT, which may represent functions involved in developmental programmes that set the homeostatic capacity, genes involved in somatic maintenance and risk alleles for late-onset diseases (late-acting deleterious genes). The study of exceptionally healthy subpopulations in section 'C' (the oldest old) may reveal loci involved in positive selection for survival. Alleles predisposing to common disease patterns are expected to have lower frequencies in these populations. Higher frequencies are expected from alleles involved in protection from common disease patterns. Except for specific loci determining common diseases occurring in section B, somatic maintenance loci may be identified in populations of the oldest old involved in functions such as DNA repair/proofreading, accuracy of protein synthesis, protein turnover, oxygen radical scavenging and acute phase response. These genes are expected to be organized as networks, the quality and synchronization of which may vary greatly from lower to higher species.

3. MOLECULAR AGE-CHANGES IN SOMATIC CELLS

A diverse spectrum of alterations is observed at the cellular level as a function of age which may be the consequence of programmed, pleiotropic and stochastic events. A schematic overview of the nature of the molecular age-changes[18] is presented in Figure 2. Naturally, most of these changes occur through inter-related mechanisms. Stochastic damage at the DNA level may induce cellular defense mechanisms (DNA repair, heat shock

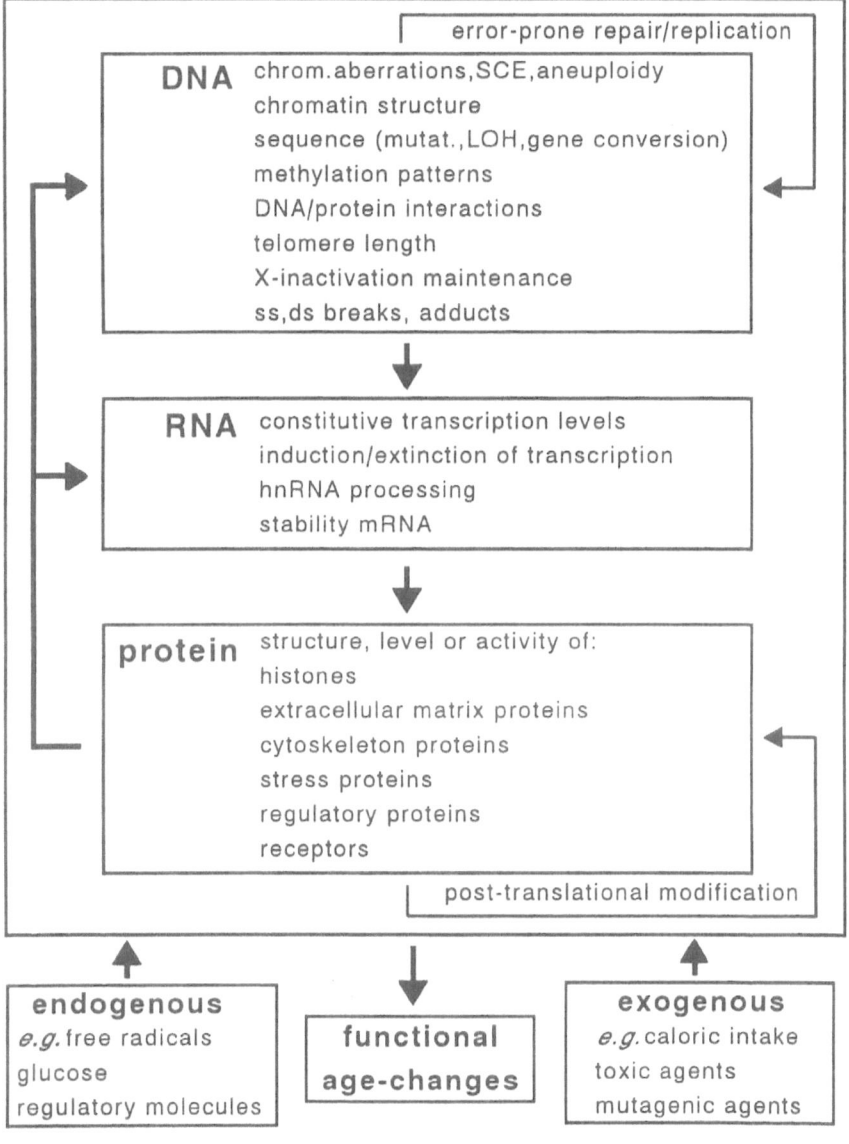

Figure 2. Schematic overview of the nature of molecular age-changes observed in somatic cells of gradually ageing species.

response etc.) and qualitative and quantitative changes in specific gene products. Stochastic damage to proteins may affect translation, transcription and may also induce cellular defense mechanisms.

Diverse somatic instabilities (both programmed and stochastic) of nuclear and mitochondrial DNA increase with age. However, the frequency of mutations is generally low[18,19]. Both in the nuclear and mitochondrial genome of mitotic and post-mitotic cells, a diverse spectrum of DNA defects accumulates as a function of age. Amongst a stochastic accumulation some site-specificity is observed at hot spots of instability corresponding to sites of instability in the germ line. The effect of randomly occurring mutations on ageing and disease is, if present, not clear. No direct causal relationship has thus far been established between accumulation of spontaneous genomic instability and longevity. An interesting well-controlled phenomenon of genetic instability is the reduction of human telomeres in mitotic cells as a function of donor age. This is explained as a separate subject below.

Intrinsic age-associated alterations in gene expression occur widespread. The alterations are generated by changes in membrane parameters, chromatin structure, function and levels of transcription factors, hormones and hormone receptors[21–25]. The data obtained thus far in this field, also include alterations in RNA processing and stability. Intrinsic age-associated alterations at the protein expression level occur with age in most of the tissues and cells of the species studied thus far. The constitutive expression levels of housekeeping, tissue-specific and inducible genes becomes affected with age. In addition, the inducibility and extinction of the latter type of genes are delayed and/or decreased with age. Some of these effects involve changes at the level of gene transcription which may, among others, be caused by alterations in DNA methylation, DNA-protein interactions and availability of hormones, transcription factors and/or their receptors. Other effects at the protein level involve post-transcriptional changes and post-translational modifications.

These molecular alterations generated as a function of age may provide the basis for a declining organismal organization of ageing animals and humans. One of the most widespread problems in ageing research, however, is to uncover direct causal relationships between molecular age-changes and decreased organismal functions. Apart from the individual variation in gradually ageing populations, this is also due to the fact that among many age-related changes that occur, some are deleterious and irreversible, whereas others are without any effect or even compensatory. Molecular age-changes can be observed at ages when diseases do not yet become clinically manifest. The lack of markers indicative for biological age as opposed to chronological age hampers the causal interpretation of molecular age-changes. One of the processes that has been suggested a potential marker of biological age is the shortening of human telomeres.

3.1 Telomere Shortening in Human Mitotic Cells

The highest concentration of genes and transcriptional activity in the human genome is present at the telomeric genome regions[26]. The ends of human chromosomes consist of tandem arrays of the (TTAGGG)n repetitive motif[27] which stabilize chromosomes during replication. The length of these telomeric repeat fragments (TRFs) can be determined by Southern hybridization analysis of digested genomic DNA. Since the TRF length in a cell population varies extensively, only the average TRF length of a cell population can be determined. In human somatic cells, the telomeric repeat arrays undergo a length reduction as a function of donor age. The average length of telomeric restriction fragments (TRFs) in white blood cells and colorectal mucosa was found to be shorter in old than in young human individuals, corresponding with a rate of telomere loss of 33 bp per year *in vivo*[28]. The mean TRF length in individuals of ages between 4 and 95 years of age decreases in a range from 10 to 5 kb[29]. *In vitro* telomere reduction occurs progressively with serial passage of cultured cells ($\approx$50 bp per population doubling for human fibroblasts). The initial telomere length of cells in culture predicts their replicative capacity[30,31].

Telomeric repeats are lost during each cell cycle at the discontinuously replicating DNA strand. Removal of the RNA primer at the end of this strand results in a 3' terminal overhang; thus, the DNA molecule will lose sequences from the telomeric end with each round of replication. In germ cells, the loss of telomeric DNA is compensated by the activity of the telomerase enzyme, which repairs telomeric repeat gaps[32]. Telomerase activity in somatic cells, replicating at a normal rate, is usually absent or low[33]. The loss of telomeric DNA at each cell cycle in normal somatic cells, is considered a recording system of the number of replication rounds. It was postulated that the sequential loss of telomeric DNA from the ends of human chromosomes with each somatic cell division eventually reaches a critical point that triggers cellular senescence[34]. Beyond a critical length expression of adjacent genes may be affected in such a way that cellular senescence is induced[35,36]. Further telomere shorting may result in chromosomal abnormalities. Immortalized cells in culture which do not express telomerase activity accumulate dicentric chromosomes and undergo cell death when a critical minimum telomere length of 1.5 Kb is passed[37]. Many immortalized cells in culture on the other hand and tumour cells of a wide range of tumour types express telomerase activity[38]. Synthesis of DNA at chromosome ends by telomerase to stabilize the ends may be necessary for indefinite proliferation of human cells.

Reduction of telomere length occurs in human peripheral blood lymphocytes as a function of donor age. A considerable TRF length variation is present among subjects of the same age at all ages. This variation was found to be strongly influenced by genetic factors (heritabilities 60–85%), as demonstrated in a human twin study[29]. Telomere length variation between individuals of the same age may be explained in four ways: 1) variation in initial telomere length in the germ line; 2) variation in incidental decrease of telomeric repeats in somatic cells by, for example free radical induced single strand breaks[39]; 3) variation in the number of telomeric repeats lost at each cell division, or 4) variation in turn over rate of cells at a constant loss of telomeric repeats per cell division. Another question is whether TRF length in blood lymphocytes reflects loss of telomeric DNA in other cell types. The genetic basis for individual variation may be found in the presence of telomerase gene variants with variable activity in germ line and/or stem cell pools or in a genetically determined variation in cell turnover between individuals.

If telomere length reduction through chromosomal damage or cellular senescence of mitotic cells is causally involved in ageing, TRF length might be a marker of the biological age of an individual. A causal relation between telomere shortening and organismal decline has not been established, however. To test the possibility that telomere size represents a biomarker of human ageing, age-structures and disease patterns may be compared in individuals or families with long versus short telomeres as compared to their age group. Telomere shortening could be a marker of human pathology if a high rate of telomere loss is indicative for high cell turnover which may relate to susceptibility to cancer. Alternatively, short telomeres may be indicative for cells to reach the senescent state at a high rate. Telomere related cellular senescence was suggested to be involved in atherosclerosis[40]. Telomere loss in vivo was more prominent in tissues involved in atherosclerosis (iliac arteries) than in tissues known to be free of atherosclerosis even among the elderly (internal thoracic arteries). It was suggested that early replicative exhaustion of endothelial cells from iliac arteries is involved in atherosclerotic plaque formation and thrombosis because of a reduced ability to form a continuous monolayer at sites of continuous damage to the arterial wall. If cellular senescence is involved in the etiology of aging[41] and/or disease, then TRF size may be a biological marker of ageing.

3.2 Potential Causes of Molecular Age-Changes

How are molecular age-changes caused? In the first place an accumulation of stochastic defects may occur, generated by endogenous mechanisms such as oxidation by free radicals of lipids and DNA, glycation and oxidation of long-lived slowly replaced molecules (DNA, collagen) and accumulation of other Advanced Glycation End (AGE) products. Whereas intermediates of these mechanisms are reversible, end products such as lipid peroxides, crosslinks and DNA mutations are not. The rate at which the defective molecules accumulate depends on the rate at which they are generated (depending on glucose levels, free radical concentrations, etc.) and at which they are removed (depending on levels and quality of proteases, antioxidants, DNA repair enzymes, etc.). For AGE products, this rate is subject, for example, to local levels of sugar substrates, metals and metabolites that influence degradation of glycated intermediates. It is therefore expected that tissue areas that accumulate metals, such as zinc in the hippocampus, may accumulate glycated proteins[41,42]. Somatic defects need not accumulate completely stochastic. The functional importance of somatic defects may be found in accumulation at specific areas of tissues, specific cell types, specific sites in the cell and in the genome. Another example of such area specificity is the accumulation of mitochondrial DNA (mtDNA) deletions in the cortex of the human brain[43]. The mechanism of accumulation of mtDNA defects affecting energy metabolism would fit the predictions of the disposable soma theory. The frequency at which these deletions have thus far been detected is very low, however.

A second pathway through which age-related alterations in gene expression are induced, is by changes in the homeostatic network, involving systemic levels of hormones and growth- and transcription factors etc. (Figure 3).

These levels are influenced by lifestyle and environmental factors. Many of the small changes in hormone levels and temporal patterns of secretion during non-pathological ageing are expected to alter gene activity in target cells. Example of systemic age-related alterations in the levels of human hormones and growth factors[1]:

- sex hormones: estrogen (ovarium), decreases
- melatonin night peak levels (pineal gland), decreases
- parathyroid hormone (parathyroid gland), increases
- growth hormone, nocturnal secretion, decreases
- luteinizing hormone (hypothalamus), testosterone, pulse frequency, decreases
- corticosteroids (adrenal gland), increases
- dopaminergic-2 receptor (neurostriatum), decreases
- blood glucose level, decreases
- growth factors: IGF-1 decreases

Apart from programmed switches in the level of specific hormones, more diverse systemic alterations may result from endogenous accumulation of somatic defects in cells from the neuroendocrine and endocrine system. These may affect patterns of gene expression in target cells distributed over the whole organism.

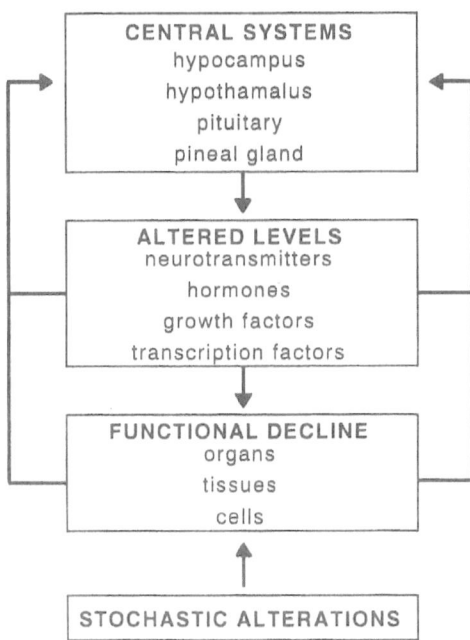

Figure 3. Schematic overview of a pathway through which systemic and cellular age-changes may accumulate and contribute to organismal ageing.

4. HUMAN AGEING AND DISEASE

Human ageing is associated with a spectrum of chronic diseases and non-pathologic changes that may be precursors of diseases such as age-effects in hearing, vision, renal function, glucose tolerance, systolic blood pressure, bone density, pulmonary function, immune function and sympathetic neuron system activity. Twin and family studies have demonstrated a considerable influence of genetic factors for some of these traits and for most of the common diseases of old age such as osteoporosis, osteoarthritis, diabetes type II, dementia, atherosclerosis, coronary artery disease and cancer. This is also the case for conditions such as obesity and hypertension. Gerontologists have long speculated about the existence of genetic influences on human longevity. Pearl's[44] classic study of five large New England families indicated that parental age at death had minimal albeit non zero prognostic significance for offspring longevity. This indicated that genetic influences on longevity cannot be detected as concordant age at death among first degree relatives. A more suitable design to investigate genetics of longevity, however, would involve following a large genetically informative cohort from birth to death[45]. This was accomplished in a Danish study of 218 pairs MZ twins and 382 pairs like-sex DZ twins[46]. The highest heritability of age at death (estimated at 20–35%) was found at premature death before the age of 60 and at extreme old age. Except for genetic factors there is a large impact of environmental factors and lifestyle habits in the etiology of ageing and disease.

The genetic influence on human disease patterns can be discussed following the hypothetical division of phases in the survival curve depicted in Figure 1. Examples of human premature 'aging' genotypes (part A in Figure 1) are given in Table 1. Some of these

Table 1. Human progeroid syndromes

Disease	Life span (years)	Genotype
Hutchinson-Guilford	10–20	autosomal recessive
Werner's	50–60	autosomal recessive
Huntington's	40–70	autosomal dominant
Down's	50–70	trisomy 21
Cockayne	50	autosomal recessive
Ataxia telangiectasia	40	autosomal recessive

diseases have an early-onset clinical expression of various ageing phenomena. The study of these syndromes may reveal genes involved in these segmental ageing phenomena. For example, in Down's syndrome patients with partial trisomy of chromosome 21, gene clusters and aspects of the total progeroid phenotype may eventually be dissociated and causally related. In addition to the progeroid syndromes, there are a number of early-onset heritable diseases associated with atherosclerosis, dementia, diabetes (type I), cancer and osteoarthritis. These heritable disorders resemble (multifactorial) disease patterns of aged individuals and prematurely represent aspects of senescence, although none of these disorders shows a uniform acceleration or intensification of the usual spectrum of age-related degenerative changes (except for Werner's diseases which is associated with a diverse spectrum of pathologies). Genes identified in these early onset diseases are currently tested as candidate loci for common diseases at later ages of onset.

One of the most stable components of the MRDT (part 'B', Figure 1) in different human populations is atherosclerosis. Heritable risk factors for vascular diseases are associated with defects in lipid transport proteins, of which low density lipoprotein (LDL) and high density lipoprotein (HDL) are important, as well as their receptors. These defects may lead to increases in the LDL/HDL ratio[47,48]. LDL is the major carrier of cholesterol; HDL protects against atherosclerosis by the reverse cholesterol transport from peripheral tissues to the liver, where it may be recycled. The common basis for early and late onset atherosclerosis may be found in the damaging effect of oxidized lipids/LDL particles. Any dysfunction leading to high levels of blood LDL, which is under polygenic control, may result in accumulation of oxidized and glycated LDL. The importance of lipid peroxidation for atherosclerosis is confirmed by the fact that the disease in humans[49] and Watanabe rabbits which have a defect in the LDL-receptor gene, is retarded by treatment with the antioxidant probucol[50]. It can be understood that the age of onset of atherosclerosis may be severely influenced by age-related dysfunctions in the metabolic homeostatic systems regulating the accumulation of free oxygen radicals and glycation products. The pattern of human age-related diseases that dominate the MRDT, varies among different populations. Programmed and pleiotropic events, late-acting deleterious germ-line mutations and accumulation of somatic defects all seem to influence the age of onset of human diseases occurring in subpopulations determining the MRDT and longevity.

Programmed events, such as the menopause, and switches in hormones connected to this process, have diverse effects on the onset of human diseases. Switches in levels of sex steroids during menopause are protective for atherosclerosis, and cancer of the mammary and female reproductive tract; decreased estrogen secretion is associated with increased osteoporosis and loss of bone, which can be partly prevented by steroid supplements. Other programmed events that may contribute to ageing are cellular senescence (cell cycle exit) and apoptosis. There is no direct evidence, however, that exhaustion of cell replication or apoptosis contribute to disease.

Table 2. Association between common diseases, related traits and alleles
from polymorphic loci

Disease	Trait	Polymophic locus
Osteoporosis[51]	peak bone density	Vitamin D receptor
Osteoarthritis[52]	cartilage integrity	Collagen type II
Myocardial infarction[53]	fibrinolysis	Plasminogen activator inhibitor1
Dementia[54]		Apolipoprotein E

Improved statistical methodology and the availability of a dense map of polymorphic markers in the human genome have increased the chance of identifying late-acting deleterious alleles that may be involved in common age-related diseases. Genetic association studies and linkage studies in affected relative pairs are currently being performed to identify genes of which specific (sometimes rather common) alleles increase the risk for the disease. Examples of candidate genes which have been associated with various of these diseases by polymorphic markers within or flanking the gene are listed in table 2.

Accumulation of stochastic somatic defects are likely to be involved in carcinogenesis: somatic mutations, genomic instabilities leading to the revelation of recessive germline mutations and methylation changes. Glycation of osteocalcin may be involved in osteoporosis; collagen cross-links in osteoarthritis; in diabetes the glycation of hemoglobins and collagens is expected to be important[55,56]. Accumulation of lipid peroxides is involved in atherosclerosis, whereas cross-link formation may increase trapping of plasma proteins that could lead to aggregation of blood platelets, important in ischaemic heart disease and stroke. A role for nuclear and mtDNA damage and mutations may be expected especially in diseases which involve functional changes in post-mitotic cells (neurons, heart muscle cells) such as neuro-degenerative diseases and ischaemic heart disease.

In centenarians (part C, Figure 1) death, in the absence of substantive organ pathology, appears to occur mainly as a result of impairment of homeostatic functions. This is indicated by the low organ reserve and the "cascade breakdown"[19] frequently occurring once one disease is treated. In these individuals, stochastic somatic defects might be the main source of deleterious age-effects, possibly accumulating at a rate slower than the population average. The accumulation of collagen crosslinks by non-enzymatic glycation, for example, is a common characteristic of the aged. There is, however, individual variation in the aged to such extent that the level of collagen crosslinks in the skin of some 80-year-old individuals resemble average levels of 40-year-old individuals[57].

The oldest 10% of the human population in industrialized societies consist of survivors from common diseases (85+ fraction) and centenarians, extreme old individuals. Some age-effects associated with disease in middle aged individuals are now being found in centenarians in the absence of disease (hypertension; highly increased levels of coagulation factors[58]). Many disease risk alleles for common complex diseases will be identified in the coming years. By studying octo-, nona- and centenarians the importance of such alleles to human mortality of the past generations may be revealed. In addition to studying the extremes, gene frequencies must be studied in a gradient of younger age groups from the 'B' part (Figure 1). Alleles associated with mortality are expected to be decreased in the survivors. Alternatively, some specific genotypes increase life span above the average for the species by lowering incidence of specific diseases. Examples of this are the rare heritable hypo-β or hyper-a lipoproteinemias in humans, which are associated with life spans 5 to 10 years above the general population[59–61]. These genotypes have two to threefold less LDL levels than the general population, which may reduce the risk for myocardial infarction and ischaemic heart disease. These alleles enhance longevity by protecting against atherosclerosis and may resemble mouse Ath-1r and Ath-2r alleles which confer resistance to atherogenic diets[62,63].

By studying the healthy old, potential factors in environment, lifestyle and genotype may be identified that minimise the generation of deleterious changes over lifetime. Early diagnosis of age-related diseases may become possible if genotypic risk factors become identified and understood in relation to modifying factors in lifestyle and environment of the phenotypic expression. Treatment could be directed to compensation of relevant age-effects, for example by retardation of oxidation (antioxidants) and glycation in relevant tissues. It is not unlikely that also in humans a general decline of homeostasis and the occurrence of diseases can be retarded by caloric restriction[64]. It was suggested that reduction in some age-related diseases such as cancer, Alzheimer's, Parkinson's and cardiovascular disease, is associated with dietary antioxidant supplementation (vitamin E and β-carotene) and reduced food intake[65]. The occurrence of age-related diseases seems inevitable in the light of all molecular changes discussed. The age of onset and progression, however, may be delayed and modifiable, at least for some groups of patients. Since the average maximum life span in industrialized countries approaches the maximum life span of the human species, elimination of one common cause of death will establish a new equilibrium between the remaining causes, thereby hardly leading to extension of the average life expectancy[66]. Subpopulations in part 'B' (Fig. 1) may however be identified with an increased environmental and/or genetic risk for common diseases to occur at earlier ages. Insight into genetic factors involved in multifactorial age-related diseases (late-acting deleterious alleles), will be obtained in the years to come, by complementary genetic association and pedigree pair studies and from the study of the oldest old.

REFERENCES

1. Finch CE. (1990) Longevity, senescence and the genome. University of Chicago Press, Chicago, IL.
2. Finch CE. (1992) Mechanisms in senescence: some thoughts in April 1990. Exp Gerontol 27: 7–16.
3. Kirkwood TBL. Comparative evolutionary aspects of longevity. 1985. In: Handbook of the Biology of Aging, 2nd ed. Finch CE, Schneider EL, eds. Van Nostrand, New York, NY, pp 27–45.
4. Kirkwood, T.B.L. (1991) Genetic basis of limited cell proliferation. Mutat Res 256: 323–328.
5. Russel ES. (1978) Genetic origins and some research uses of C57BL/6, DBA/2 and B6D2F1 mice.In: Development of the rodent as a model system of aging. Gibson DC, Adelman RC, Finch CE, eds. NIH Publication DNEW, Bethesda, MD, pp 37–44.
6. Everitt A, Meites J. (1989) Aging and anti-aging effects of hormones. J Gerontol 44: B139-B141.
7. Regelson W. 1983. The evidence for pituitary and thyroid control of aging: is aging reversal, a myth or a reality? The search for a "death hormone". In: Interventions in the Aging Process. Regelson W, Sinex FM, eds. Alan R. Liss, New York pp 3–52.
8. Gresik EW, Wenk-Salamone K, Onetti-Muda A, Gubits RM, Shaw PA. 1986. Effect of advanced age on the induction by androgen or thyroid hormone of epidermal growth factor and epidermal growth factor mRNA in the submandibular glands of C57Bl/6 male mice. Mech Ageing Dev 34: 175–189.
9. Reiter RJ. 1987. The melatonin message: duration versus coincidence hypothesis. Life Sci 40: 2119–2131.
10. Pierpaoli W, Dall'Ara A, Pedrinis E, Regelson W. 1991. The pineal control of aging. The effects of melatonin and pineal grafting on the survival of older mice. Ann N Y Acad Sci 621: 291–313.
11. Medawar PB. 1952. An unsolved problem of biology. H.K. Lewis, London, 1952.
12. Williams GC. 1957. Pleiotropy, natural selection, and the evolution of aging. Evolution 11: 398–411.
13. Kirkwood TBL. 1977. Evolution of aging. Nature 270: 301–304.
14. Shepherd JCW, Walldorf U, Hug P, Gehring WJ. 1989. Fruit flies with additional expression of the elongation factor EF-1α live longer. Proc Natl Acad Sci USA 86: 7520–7521.

15. Luckinbill LS, Graves JL, Reed AH, Koetsawang S. 1988. Localizing genes that defer senescence in Drosophila melanogaster. Heredity 60: 367–374.
16. Friedman DB, Johnson TE. 1988. A mutation in the age-1 gene in Caenorhabditis elegans lengthens life and reduces hermaphrodite fertility. Genetics 118: 75–86.
17. Friedman DB, Johnson TE. 1988. Three mutants that extend both mean and maximum lifespan of the nematode, Caenorhabditis elegans, define the age-1 gene. J Gerontol 43: B102-B109.
18. Slagboom PE. 1993. Genomic instability and aging. Thesis. Haveka, Alblasserdam.
19. Eulderink F, Heeren TJ, Knook DL, Ligthart GJ. 1993. Inleiding gerontologie en geriatrie. Bohn, Stafleu, Van loghum, Houten/Zaventem.
20. Zurcher C, Slagboom PE. Basic Mechanisms of aging. 1994. In: Jones TC, Mohr U, Hunt, RD (eds.). Pathology of aging animals. ILSI Monographs on Pathology of Laboratory Animals. Springer Verlag Berlin Heidelberg New York.
21. Slagboom PE, Vijg J. 1989. Genetic instability and aging: theories, facts and future perspectives. Genome 31: 373–385.
22. Slagboom PE, De Leeuw WJF, Vijg J. 1990. Messenger RNA levels and methylation patterns of GAPDH and β-actin genes in rat liver, spleen and brain in relation to aging. Mech Ageing Dev 53: 243–257.
23. Slagboom PE, De Leeuw WJF, Vijg J. 1990. mRNA levels and methylation patterns of the tyrosine aminotransferase gene in aging inbred rats. FEBS Lett 269: 128–130.
24. Murty CVR, Mancini MA, Chatterjee B, Roy A. 1988. Changes in transcriptional activity and matrix association of $\alpha_{2\mu}$-globulin gene family in the rat liver during maturation and aging. Biochim Biophys Acta 949: 27–34.
25. Post DJ, Carter KC, Papaconstantinou J. 1991. The effect of aging on constitutive mRNA levels and lipopolysacharide inducibility of acute phase genes. Ann N Y Acad Sci 1991; 621: 66–77.
26. Saccone S, De Sario A, Della Valle G, Bernardi G. 1992. The highest gene concentrations in the human genome are in telomeric bands of metaphase chromosomes. Proc Natl Acad Sci 89: 4913–4917.
27. Moyzis RK, Buckingham JM, Scott Cram L, Dani M, Deaven LL, Jones MD, Meyne J, Ratcliff RL, Wu J. 1988. A highly conserved repetitive DNA sequence, (TTAGGG)n, present at the telomeres of human chromosomes. Proc Natl Acad Sci USA 85: 6622–6626.
28. Hastie ND, Dempster M, Dunlop MG, Thompson AM, Green DK, Allshire RC. 1990. Telomere reduction in human colorectal carcinoma and with ageing. Nature 346: 866–868.
29. Slagboom P.E., S. Droog and D. Boomsma. 1994. Genetic determination of telomere size in humans: a twin study of three different age groups. Am J Hum Genet 55: 876–882.
30. Harley CB, Futcher AB, Greider CW. 1990. Telomeres shorten during aging of human fibroblasts. Nature 345: 458–460.
31. Allsopp RC, Vaziri H, Patterson C, Goldstein S, Younglai EV, Futcher AB, Greider CW, Harley CB. 1992. Telomerelength predicts replicative capacity of human fibroblasts. Proc Natl Acad Sci 89: 10114–10118.
32. Blackburn EH. 1991. Structure and function of telomeres. Nature 350: 569–572.
33. Broccoli D, Young JW, de Lange T. 1995. Telomerase activity in normal and malignant hematopoietic cells. Proc Natl Acad Sci USA. 92: 9082–9086.
34. Harley CB. 1991. Telomereloss: mitotic clock or genetic time bomb? Mutat Res 256: 271–282.
35. Broccoli D, Cooke H. 1993. Aging, healing and the metabolism of telomeres. Am J Hum Genet 52: 657–660.
36. Biessmann H, Mason JM. 1992. Genetics and molecular biology of telomeres. Adv Genet 30: 185–233.
37. Counter CM, Avilion AA, LeFeuvre CE, Stewart NG, Greider CW, Harley CB, Bachetti S. 1992. Telomere shortening associated with chromosome instability is arrested in immortal cells which express teolerase activity. Embo J 11: 1921–1929.
38. Counter CM. 1994. Telomerase activity in human ovarian carcinoma. Proc Natl Acad Sci USA 91: 2882–2885.
39. Zglinicki von T, Saretzki G, Docke W, Lotze C. 1995. Mild hyperoxia shortens telomeres and inhibits proliferation of fibroblasts: a model for senescence ? Exp Cell Res 220: 186–93.
40. Chang E, Harley CB. 1995. Telomere length and replicative aging in human vascular tissues. Proc Natl Acad Sci USA. 92: 11190–11194.
41. Hayflick L. 1991. Aging under glass. Mutat Res 256: 69–80.
42. Massie HR, Aiello VR, Iodice J. 1979. Changes with age in copper and superoxide dismutase levels in brains of C57BL/6J mice. Mech Age Dev 10: 93–99.
43. Corral-Debrinsky M, Horton T, Lott MT, Shoffner JM, Flint Beal M, Wallace DC. 1992. Mitochondrial DNA deletions in human brain: regional variability and increase with advancing age. Nature Genet 2: 324–329.

44. Linnane AW, Ozawa T, Marzuki S, Tanaka M. 1989. Mitochondrial DNA mutations as an important contributor to ageing and degenerative diseases. Lancet 25: 642–645.
45. Pearl R. 1931. Studies on human longevity. IV. The inheritance of longevity. Preliminary report. Human Biol 3: 245–269.
46. Hrubec Z, Neel JV. 1978. Familial factors in early deaths: Twins followed 30 years to ages 51–61. Hum Genet 59: 39–46.
47. McGue M, Vaupel J W, Holm N, Harvald B. 1993. Longevity is moderately heritable in a sample of Danish twins born 1870–1880. J Geront 48: B237-B244.
48. Ross R. 1986. The pathogenesis of atherosclerosis: An update. N Engl J Med 314: 488–500.
49. Sing CF, Boerwinkle E, Moll PP, Davignon J. 1985. Apolipoproteins and cardiovascular risk: Genetics and epidemiology. Ann Biol Clin 43: 407–417.
50. Lanlois S, Kastelein JJP, Hayden MR. 1988. Characterization of six deletions in the low-density lipoprotein (LDL) receptor gene causing familial hypercholesterolemia (FH). Am J Hum Genet 43: 60–68.
51. Kita T, Nagano Y, Yokode M, Ishii K, Kume N, Ooshima A, Yoshida H, Kawai C. 1987. Probucol prevents the progression of atherosclerosis in Watanabe heritable hyperlipidemic rabbit, an animal model for familial hypercholesterolemia. Proc Natl Acad Sci 84: 5928–5931.
52. Morison NA, Cheng Qi J, Tokita A, Kelly PJ, Crofts L, Nguyen TV, Sambrook PN, Eisman JA. 1994. Prediction of bone density from vitamin D receptor alleles. Nature 367: 284–287.
53. Bijkerk C, Meulenbelt I, Odding E, Valkenburg HA, Miedema H, Breedveld FC,Hofman A, te Koppele JM, Pols HAP, Van Duijn CM, Slagboom PE Association of two alleles of the COL2A1 3' variable region with radiological osteoarthritis in a population based study: The Rotterdam Study. submitted.
54. Eriksson P, Kallin B, Hooft F, Båvenholm P, Hamsten A. 1995. Allele-specific increase in basal transcription of the plasminogen-activator inhibitor 1 gene is associated with myocardial infarction. Proc Natl Acad Sci USA 92: 1851–1855.
55. Pericak-Vance M, Haines JL. 1995. Genetic susceptibility to Alzheime diseases. Tends in Genet 11: 504–508.
56. Bunn HF. 1981. Nonenzymatic glycosylation of protein: Relevance to diabetes. Am J Med 70: 325–330.
57. Garlich RL, Bunn HF, Spiro RG. 1988. Nonenzymatic glycation of basement membranes from human glomeruli and bovine sources: Effect of diabetes and age. Diabetes 37: 1144–1155.
58. Sell DR, Monnier VM. 1990. Structure elucidation of a senescence cross-link from human extracellular matrix. J Biol Chem 266: 21597–21602.
59. Saito F. 1984. A pedigree of homozygous familial hyperalphalipoproteinemia. Metabolism 33: 629–633.
60. Glueck CJ, Garside PS, Fallat RW, Sielski J, Steiner PM. 1976.Longevity syndromes: Familial hypoβ and familial hypera lipoproteinemia. J Lab Clin Med 88: 941–957.
61. Glueck CJ, Garside PS, Mellies MJ. Steiner PM. 1977. Familial hypobetalipoproteinemia. Studies in 13 kindreds. Clin Res 25: 517.
62. Paigen B, Nesbitt MN, Mitchell D, Albee D, LeBoeuf RC. 1989. *Ath-2*, a second gene determining atherosclerosis susceptibility and high density lipoprotein levels in mice. Genetics 122: 163–168.
63. Paigen B, Ishida BY, Verstuyft J, Waters RB, Albee D. 1990. Arteriosclerosis susceptibility differences among progenitors of recombinant inbred strains of mice. Arteriosclerosis 10: 316–326.
64. Masoro IJ, Shimokawa I, Yu BP. 1991. Retardation of the aging process in rats by food restriction. Ann N Y Acad Sci 621: 337–352.
65. Slater TF, Block G. 1991. Antioxidant vitamins and β carotene in disease prevention Am J Clin Nutr 53: 189S-196S.
66. Lohman PHM, Sankaranarayanan K, Ashby J. 1992. Choosing the limits to life. Nature 357: 185–186.

2

AGEING AND CANCER

A Struggle of Tendencies

Anastassia Derventzi, Efstathios S. Gonos, and Suresh I. S. Rattan

National Hellenic Research Foundation
Institute of Biological Research and Biotechnology
48 Vas. Constantinou Ave., Athens 11635, Greece and
Laboratory of Cellular Ageing
Department of Molecular and Structural Biology
Aarhus University
DK-8000 Aarhus - C, Denmark.

1. INTRODUCTION

At the cellular level, ageing and cancer appear to be dialectically opposed phenomena. Whereas a normal diploid cell has a limited capacity to divide in vivo and in vitro, a cell that has been genetically transformed either spontaneously or by various physical, chemical or biological agents has escaped the regulatory mechanisms of limited growth and has a tendency to divide indefinitely and form a clone of cancerous cells. While cellular mortality is characterised by a progressive cessation of cell growth manifested in cell culture by senescence, cellular immortalisation is the escape from senescence as a result of multiple mechanisms involving genetic and epigenetic changes[1-3]. Cell division is regulated by both positive and negative signals. The regulation of cell proliferation involves activation of growth factor receptors, signal transduction pathways, as well as transcriptional and post-transcriptional events[4].

Cell proliferation in response to appropriate and adequate stimuli is considered as the ultimate manifestation of homeostatic balance at the cellular level. However, homeostasis is also achieved and sustained by balancing the rates of cell death and cell proliferation. In this context, apoptosis is a major mechanism of maintenance of homeostasis, by controlling and counteracting cell proliferation[5]. Any loss or deviation of homeostasis can lead to inefficient or defective control of cell proliferation. In a system of normal dividing cells the cessation of proliferation and growth is the most significant manifestation of the declining homeostatic capacity which typifies the process of ageing. In contrast, the control of proliferation in immortal cells is deranged. Immortality is considered as an homeostatic disorder where failure of proliferation control occurs when series of heritable changes (mutations and epimutations) give rise to a clone which grows without proper restrain. The unlimited proliferative capacity demonstrated in immortal and cancerous cells is not a manifestation of just another level of the pre-existing homeostasis (a state of "hyper-homeostasis" where proliferation is uninhibited), but it represents a "new state of being", resulting from the violation and escape from the normal homeostasis.

The causes which initiate and propagate the processes that either result in ageing and mortality or confer unlimited proliferative lifespan are still elusive. However, there is much evidence indicating that the mechanisms governing both mortality and immortality share common elements and are fundamentally linked. Although cellular ageing and cancer are two opposing tendencies in terms of limited and unlimited cell proliferation, respectively, the eventual breakdown of the normal controls of cell phenotype and the failure of maintenance and homeostasis are the common bases for both these phenomena[3].

2. CELL PROLIFERATION

Normal cells have multiple and independent growth regulatory mechanisms. There are several events needed to override these control elements and to induce the numerous changes in cellular structure and functions that characterise the transformed phenotype. It is thought that while senescence is the result of gradual and cumulative changes in the balance of expression of the positive and negative regulatory genes, oncogenesis is the result of such genetic alterations that lead either to the inactivation of the mechanisms restraining cellular proliferation or to the induction of abnormal positive signals for cell proliferation.

During serial subculturing, normal diploid cells undergo a period of rapid proliferation followed by a progressive loss of replicative capacity, manifested by slower population growth rates. The cell population in culture finally reaches a state of growth arrest where almost all cells fail to divide. During this period, whose duration in terms of cumulative population doublings (CPD) depends upon the cell type and its origin, cells undergo a multitude of age-related changes culminating in the complete inhibition of cellular proliferation (Table 1).

Late passage (senescent) non-proliferating cells are characterised by several, physiological, morphological and biochemical alterations, described elsewhere[3,6]. In contrast, immortalised cells escape from normal proliferation control, have altered gene expression, continue to cycle, and their morphology and size do not change after each division[2].

2.1. Cellular Responsiveness and Signal Transduction

It is evident that the disruption of coordination between a cell and its local environment typifies immortality. During immortalisation, the relationship of cells to the local environment is disturbed, the affected cells become unresponsive to negative regulators of cell division leading to a preferential replication of the altered cell, which in turn generates an altered local environment affecting other adjacent cells[7].

The progression of growth of mammalian cells depends on several growth factors which initiate a cascade of events culminating in the transition of cells into the S phase. The cellular potential and capacity to respond to these positive effectors underlines both the processes of mortality and immortality and disorders of responsiveness to endogenous and exogenous stimuli can be considered as causative factors for these processes. In a normal cell population a successful homeostatic control results in a dynamic balance between

Table 1. Phenotype of the senescent cell in vitro

Physiology

Reduced response to growth factors and other mitogens, increased sensitivity to toxins and drugs, altered calcium flux and pH, altered viscosity, altered membrane characteristics and differential regulation of signal transduction pathways.

Morphology

Large size, loss of fingerprint-like arrangement in parallel arrays, irregular shape, increased granularity, increased number of vacuoles, increased number of dense lysosomal granules with high autofluorescence, multinucleation and heteroploidy, cytoskeletal actin filaments highly polymerised, and disorganised microtubules.

Cell cycle characteristics

Cessation of DNA synthesis, cells arrested in G_1 unable to pass the restriction point, over-expression of tumour suppressor genes, accumulation of inactive complexes of some cyclins, reduced levels of cyclin-dependent kinases (cdk) and high levels of cdk-inhibitors.

Biochemical and molecular characteristics

Reduced activity, specificity and fidelity of various enzymes, accumulation of altered and abnormal proteins, reduced rates of DNA and protein synthesis, alterd profile of transcription factors, increased DNA damage, altered pattern of post-translational modifications, reduced rate of protein degradation, altered expression of several genes, loss of cytosine methylation in DNA, reduced length of telomeres.

---

cell proliferation and cell death. Such homeostasis depends on adaptive responsiveness to both the positive and the negative growth regulators. In an ageing cell culture the progressive cessation of cell growth is correlated directly with a declining responsiveness to growth stimulators and an increased sensitivity to hazardous environmental factors[8]. For example, it has been observed that senescent human fibroblasts exhibit increased response to tumour promoters, such as phorbol esters, while they fail to respond to growth stimulators like serum. Since these mitogens convey their message in the cell by activating different signal transduction pathways, a model of differential regulation of the different mitogen-activated signal transduction mechanisms in senescent cells can serve as a likely explanation for the age-related alterations of cellular responsiveness[9-11]. Furthermore, the enhanced cellular sensitivity of aged cells to toxic agents may lead to the propagation of genetic instability to levels beyond the capacity of the in built repair mechanisms and eventually to homeostatic failure. This can lead either to the irreversible cessation of cell proliferation or to unregulated and unlimited cell proliferation resulting in the formation of a tumour.

Late passage cells are irreversibly blocked in the G_1/S and possibly in the G_2/M boundaries of the cell cycle, as has been shown in studies on the nuclear fluorescence patterns, DNA content and cytometric measurements[12]. However, the oncogenic products of DNA tumour viruses, such as the T antigen of the SV40 and polyoma viruses, the E1A oncoprotein of some adenoviruses and the E7 oncoprotein of types -16 and -18 of the human papilloma viruses (HPV), as well as some retroviral gene products and their activated cellular homologues (e.g. v- or c-myc, v- or c-fos) can overcome and cancel the otherwise irreversible proliferation block of senescent cells in several mammalian cell types. Furthermore, the serum response factor (SRF), which recognises the essential component for the induction of c-fos serum response element (SRE), was found hyper-phosphorylated in senescent populations, while there were no major differences in its expression between early and late passage cells[13].

2. 2. Transcription Factors and Tumour Suppressor Genes

There is evidence that altered transcription factors profile specifies the senescent phenotype of human fibroblasts[14]. Furthermore, it is thought that there must be common elements, such as a cAMP regulatory element, in the promoter environment of the senescent specific genes serving as targets for transcriptional regulation of these genes by transcription factors like the CREB[15]. However, and by analogy to other biological processes, as there is not a single transcription factor reported yet which regulates ageing uniformly, it is feasible that different transcription factors regulate set of genes in different tissues.

Transcription factors and their possibly differential regulation in ageing versus immortal cellular systems has received much attention and was highlighted ever since the discovery of the so-called tumour suppressor genes which act as inhibitors of cell cycle progression in normal cells and they play a key role in preventing cancer progression[16]. These genes have been found overexpressed and stabilised in their anti-proliferative forms in non-dividing cells, while in immortal cells conformational or genetic changes inactivate their inhibitory forms, permitting cellular proliferation. Moreover, it has been suggested that whereas tumour suppressor genes play an essential role during development, they eventually become upgraded or aberrantly expressed thus blocking the process of cell division[17].

The p53 gene is a transcription factor characterised by tumour suppressing and anti-proliferative activities. The p53 gene product trans-activates a set of genes, like p21, gadd 45, mdm 2 and bax, while it trans-represses others, such as the bcl-2 and c-myc. The p53 gene serves as a cellular target for the gene products of many DNA tumour viruses and its aberrant expression is considered to be one of the most common genetic feature in a wide range of cancers. Mutations at the normal p53 locus have been associated with approximately half of the human cancers and they are recessive to the wild type p53 allele, contributing only to tumorigenesis when the wild type allele is inactivated. Overexpression of wild type p53 results in growth arrest at the G_1/S and at the G_2/M boundaries of the cell cycle and the suppression of the neoplastic phenotype in malignant osteosarcoma cells, while expression of a mutant form of p53 gene induces escape from senescence in HDF[1,16,17].

2. 3. Cyclin-Dependent Kinases and Their Inhibitors

A lot of attention has recently been given to the function of the genes that act as cyclin-dependent kinase (cdk) inhibitors, such as the WAF-1 (p21), p15, p16, p27 and p57 genes. These cdk inhibitors are blocking the function of cyclin-cdk complexes and they have been found frequently overexpressed in senescent and other terminally differentiated cells. In addition, their overexpression induces growth arrest in proliferating cells while some of them are deleted or mutated in cancer cells, indicating them as tumour suppressor candidates[18,19].

While most of the known cdk inhibitors have been found overexpressed in senescent cells, earlier studies have shown that aged cells express minimal levels of most of the cdks and cyclins examined so far. Senescent human, rat and hamster cells show a down regulation of the expression of the cdc2, cdk2, cdc25, cyclin A and cyclin B genes, as well as inability of serum stimulation of these genes. In contrast, the G_1 cyclins C, D1 and E have retained their ability of becoming stimulated by serum even in senescent cells, although at reduced levels compared to young cells, favouring the possibility that the G_2-block may be predominant and irreversible in cellular senescence[8]. However, overexpression of cdc2, or cdc25, or cyclin A or combinations of these genes into senescent human or rat fibroblasts has not resulted in escape from senescence, suggesting that the molecules which regulate cellular ageing must act upstream the examined cell cycle control genes[20,21].

2. 4. Cell Cycle Regulators

One of the best studied examples of cell cycle regulators is the retinoblastoma (RB) gene which encodes a nuclear phosphoprotein with DNA binding activity. Abnormalities and inactivation of the gene have been associated with the retinoblastoma disease itself, but also with several types of human cancers, implying a vital role for the RB gene in the initiation and/or progression of the disease. Several oncogene products bind and inactivate the RB protein[17]. However, re-introduction of the RB gene into RB-negative tumour cells resulted in suppression of the neoplastic phenotype in some, but not all tested tumours[22]. The activity of RB protein is regulated through the cell cycle by phosphorylation. The hypophosphorylated form of RB is active and blocks cellular proliferation and phosphorylation of RB, by agents such as some cyclins, result in its functional inactivation while TGF-β, or cAMP are blocking RB phosphorylation, by modulating the activities of the cdks that are responsible for RB phosphorylation, thereby preventing cell's advance into late G_1. The levels of hypophosphorylated RB increase during the early and mid- G_1 phase of the cell cycle, while senescent cells fail to phosphorylate RB in response to mitogens[16, 23]. It has also been found that most oncoproteins can disrupt the interaction between the hypophosphorylated RB and the E2F-cyclins/cdks complex which result in RB activation, thus inactivating RB's normal function[24].

3. GENOMIC STABILITY AND REPAIR MECHANISMS

Mutations and several kinds of genetic alterations are considered to be a triggering mechanism observed in cellular immortalisation, while the accumulation of DNA damage has been implicated only in cellular senescence[25]. Single-strand breaks, cross-links, polyploidy, aneuploidy, loss of centromeric tandem, loss of telomeres and a gradual loss of methylated-cytosines (5-metC) do occur as a function of population doubling level in culture. There is evidence that the frequency of cells carrying DNA damages increases as a function of the donor age and the population doubling level in culture. In addition, some age-related changes of the physical and biochemical properties of the chromatin have been reported, the frequency of chromosomal aberrations increases during the serial passaging of cells, while cultured cells from patients with premature ageing syndromes exhibit elevated levels of spontaneous chromosomal breaks[6,25].

Other studies have focused on the in built capacity of normal cells to repair DNA. At present there is minimal evidence for an age-related loss of the overall DNA repair capacity in older cells and tissues, but it has been shown that cells from patients with premature ageing syndromes do indeed have reduced ability to repair DNA[26]. However, a positive correlation between the lifespan of a species and its capacity to repair certain kinds of DNA damage, such as UV-induced damage, the rate of loss of 5-metC and oxidative damage has been established in various studies[25,27].

Genomic instability and altered gene expression is the hallmark of cancer cells. There are no cancer cells or immortalised cells which are genetically normal. Although subtle alterations at the level of single gene or other specific regions of the genome may occur in ageing cells, normal cells have a remarkable capacity to maintain their gross genomic stability[6,25]. However, it is the failure of maintenance at the level of the genome which is the fundamental basis of cell transformation, immortalisation and carcinogenesis.

4. MOLECULAR MARKERS OF CELLULAR AGEING

Since cell division is regulated by both positive and negative signals, it is possible that cellular senescence results from changes in the balance of expression of positive and negative regulatory genes. However, it remains unclear whether the overexpression of a DNA inhibitor is primary or secondary event, since these events frequently appear when most of the age-related changes have already occurred. Moreover, it is unlikely that the expression of a single inhibitor is sufficient to trigger senescence. However, a large proportion of the experimental effort has been directed towards identification and characterisation of inhibitors of cellular proliferation. Molecular experiments in senescent cells and cell fusion experiments between senescent and young cells, as well as between normal and transformed cells, brought up many such candidates[6].

Several chromosomes have been highlighted as directly implicated in the multi-step processes of senescence and immortalisation in hybridization experiments. Experiments on hybrids between normal and transformed cells of human and rodent origin, DNA transfections, microcell fusions and karyotypic analyses, revealed the crucial role the human chromosomes 1, 3, 4, 6 and X play in the processes conferring senescence[1, 6]. It has been suggested that these genetic markers of mortality or immortality are tumour suppressor genes or oncogenes, respectively, and that a number of genes may act individually or in group for determining a particular cellular phenotype. Several genes have also been associated with ageing as they were detected overexpressed both in normal senescent cells and in cells derived from premature ageing syndromes, and which could block DNA synthesis when introduced in proliferating and immortal cells[28]. These genes include prohibitin, interleukin-1α, vimentin, p21, SAG, mot-1 and the gas genes[6]. In addition, by using differential screening and subtractive hybridisation techniques we have identified nine genes encoding for fibronectin, osteonectin, αI-procollagen, laminin receptor, SM22, sulphated glycoprotein-2 (apolipoprotein J), cytochrome-C oxidase, GTP binding protein-α and a novel gene, which are overexpressed in senescent rat embryo fibroblasts and in ageing cultures of human osteoblasts (Gonos et al., Kveiborg et al., manuscripts in preparation). While all these genes identified by different research groups using variable cloning methods associate with either cellular ageing or longevity, it remains to be demonstrated for nearly all of them whether they can induce or block senescence in primary cultures or immortalised cell lines, respectively.

Another approach in the search for the ageing regulating genes has revolved around attempts of identification of longevity-related genetic markers. Genes like the apolipoprotein B and E (Apo B and Apo E), angiotensin converting enzyme (ACE), renin and thyroid peroxidase (TPO) genes are expected to be of great importance as some of their allelic variants have been associated with the extended lifespan of a group of centenarians[29]. In other experiments, the introduction of genes, such as the genes for superoxide dismutase and catalase have resulted in a longer lifespan in transgenic fruitflies[30].

5. SOME THEORETICAL ASPECTS OF MORTALITY AND IMMORTALITY

The expression of ageing markers in cultures of normal cells brings up the question of the genetic nature of senescence. The lifespan of an individual, whether an organism or a cell, is terminated within a so-called species-specific limited period and therefore genes are likely to be involved. Another indication that, at least, cellular ageing may be genetically determined comes from cell fusion experiments, where hybrids between normal and immortalised cells exhibited a limited lifespan. Since the rate and the manifestations of the mortal cellular phenotype and cell death are subject to alterations due to the effects of environmental factors, such as the external stimuli, agents of transformation and physiological conditions, these rule out the exclusive operation of a strictly genetic programme as the only cause of ageing. In contrast, it is appealing to suggest that the causes of ageing lie with genes involved in the regulation of homeostasis; such genes may be activated either by an "internal biological clock"[31] or by external stimuli. It is in this context that the term gerontogenes has been suggested to describe such genetic elements involved in the regulation of lifespan[32]. Furthermore, it has been suggested that genes implicated in the optimal maintenance of homeostasis are the most likely candidates qualifying for being virtual gerontogenes[32].

The demonstration of the dominant nature of the senescent phenotype led to a proposal of a two-stage model for immortalisation of human fibroblasts[33, 34]. Mortality stage-1 (M1) is the first stage of cellular growth arrest at the G_1/S boundary and parallel loss of mitogen responsiveness. Mortality stage-2 (M2) is the stage when crisis, manifested with failure of cell replication, is induced unless a second event occurs which takes the cell into a path of immortalisation. Genes such as SV40 large T antigen can override the M1 stage and thus, are capable of stimulating cellular DNA synthesis, but the cells still undergo crisis because M2 is still active. Inactivation of M2 is very rare because it requires a rare independent event. According to this hypothesis, SV40 large T antigen immortalised cells would therefore have bypassed the M1 mechanism, since the large T antigen itself can stimulate cellular DNA synthesis, and also have an inactivated M2 mechanism. This model also proposes that binding of the cellular tumour suppressor proteins RB and wild type p53 is involved in the bypass of the M1 mechanism by SV40 large T antigen[35].

An important timing mechanism regulating M1 may be the telomere-shortening. According to the telomere hypothesis, which has been based on several observations that normal senescent cells carry shorter telomeres and lack telomerase activity, as compared to immortal cells and sperm cells, there is a certain telomere length, a "checkpoint" that evokes the proliferative decline[36,37]. In addition, the stabilisation of the telomere length in immortal cells due to the action of telomerase has been suggested to be implicated in immortalisation as a required but not a sufficient factor[38].

The transformation of a normal cell into a malignant is, however, a multi-step process that arises as result of the cumulative effects of multiple genetic insults. These come as a consequence of the alterations in cellular proliferation control, as well as due to disorders in the interaction of cells with their local environment. As oncogenes are defined as the genes whose manipulation engenders a transforming allele, the transformation by an oncogene is considered a gain of function, alteration and a dominant genetic damage. Genetic damage of recessive nature also occurs with the loss of functions of tumour suppressor genes. The development of a malignant phenotype is considered to be the end-result of the collaborative effects of both dominant and recessive lesions. Aged cells have to carry such a genetic information for being capable and self-sufficient to revert the transformed phenotype. However, how a mortal cell is dominant over a transformed cell, remains to be elucidated.

Senescence as a process is likely to be the outcome of some kind of a genetic regulation (programme?) ensuring the normality of a given biological system against immortalisation. Moreover the regulatory mechanisms implicated in ageing are the ones employed to maintain the fundamental homeostasis of a biological entity and it is the failure of efficiency and coordination of this programme that brings about the gradual homeostatic collapse. This failure of homeostasis initiates a struggle of tendencies in the cell either to become immortal or to undergo ageing and die out. Whatever the outcome for the individual cell, death is the ultimate consequence for the organism.

REFERENCES

1. Derventzi, A., Rattan, S.I.S. and Gonos, E.S. (1996) Molecular links between mortality and immortality. Anticancer Res (in press).
2. Gonos, E.S., Powell, A.J., Ikram, Z., Cline, M. and Jat, P.S. (1992) A molecular genetic approach towards senescence and immortalisation. In: Current Perspectives on Molecular and Cellular Oncology. JAI Press Ltd., pp 163–195.
3. Holliday, R (1995) Understanding Ageing. Cambridge University Press, UK.
4. Grana, X. and Reddy, E.P. (1995) Cell cycle control in mammalian cells: role of cyclins CDKs, growth suppressor genes and CKIs. Oncogene, 11, 211–219.
5. Vaux, D.L. and Strasser, A. (1996) The molecular biology of apoptosis. Proc. Natl. Acad. Sci. USA 93, 2239–2244.
6. Rattan, S.I.S. (1995) Ageing – a biological perspective. Molec. Aspects Med. 16, 439–508.
7. Kaufman, W.K. and Kaufman, D.G. (1994) Cell cycle control, DNA repair and initiation of carcinogenesis. FASEB J. 7, 1188–1191.
8. Derventzi, A. and Rattan, S.I.S. (1991) Homeostatic imbalance during cellular ageing: altered responsiveness. Mutat. Res. 256, 191–202.
9. Derventzi, A., Rattan, S.I.S. and Clark, B.F.C. (1992) Senescent human fibroblasts are more sensitive to the effects of a phorbol ester on macromolecular synthesis and growth characteristics. Biochem. Inter. 27, 903–911.
10. Derventzi, A., Rattan, S.I.S. and Clark, B.F.C. (1993) Phorbol ester PMA stimulates protein synthesis and increases the levels of active elongation factors EF-1a and EF-2 in ageing human fibroblasts. Mech. Ageing Dev. 69, 193–205.
11. De Tata, V., Ptasznik, A. and Cristofalo, V.J. (1993) Effect of the tumour promoter phorbol-12-myristate-13-acetate (PMA) on proliferation of young and senescent WI-38 human diploid fibroblasts. Exp. Cell. Res. 205, 261–269.
12. Gonos, E.S., Powell, A.P. and Jat, P.S. (1992) Human and rodent fibroblasts: model systems for studying senescence and immortalisation. Int. J. Oncol. 1, 209–213.
13. Atadja, P.W., Stringer, K.F. and Riabowol, K.T. (1994) Loss of serum response element binding activity and hyperphosphorylation of serum response factor during cellular ageing. Mol. Cell Biol. 14, 4991–4999.
14. Dimri, G.P. and Campisi, J. (1994) Altered profile of transcription factor-binding activities in senescent human fibroblasts. Exp. Cell Res. 212, 132–140.
15. Bowlus, C.L., McQuillan, J.J. and Dean, D.C. (1991) Characterisation of three different elements in the 5'flanking region of the fibronectin gene which mediate a transcriptional response to cAMP. J. Biol. Chem. 266, 1122–1127.
16. Marshall, C.J. (1991): Tumour suppressor genes. Cell 64, 303–312.
17. Gonos, E.S. and Spandidos, D.A. (1994) The retinoblastoma and p53 onco-suppressor genes and their role in human cancer. Cancer Mol. Biol. 1, 27–34.
18. El Deiry, W.S., Tokino, T., Velculescu, V.E., Levy, D.B., Parsons, R., Trent, J.M., Lin, D., Mercer, E., Kinzler, K.W. and Vogelstein, B. (1993) WAF1, a potential mediator of p53 tumour suppression. Cell 75, 817–825.
19. Wong, H. and Riabowol, K. (1996) Differential CDK-inhibitor gene expression in aging human diploid fibroblasts. Exp. Cell Res. 31, 311–326.

20. Stein, G.H., Drullinger, L.F., Robetorye, R.S., Pereira-Smith, O.M. and Smith, J.R. (1991) Senescent cells fail to express cdc2, cyc A, and cyc B in response to mitogen stimulation. Proc. Natl. Acad. Sci. USA 88, 11012–11016.

21. Gonos, E.S. and Spandidos, D.A. (1996) The role of cdc2, cdc25 and cyclin A genes in the maintenance of immortalisation and growth arrest in a rat embryonic fibroblast conditional cell line. Cell Biol. Int. in press.

22. Weinberg, R.A. (1992) The retinoblastoma gene and gene product. Cancer Surveys 12, 43–57.

23. Stein, G.H., Beeson, M. and Gordon, L. (1990) Failure to phosphorylate the RB gene product in senescent human fibroblasts. Science 249, 666–669.

24. La Thangue, N.B. (1994) DRTF1/E2F: an expanding family of heterodimeric transcription factors implicated in cell cycle control. TIBS 19: 108–114.

25. Rattan, S.I.S. (1989) DNA damage and repair during cellular ageing. Int. Rev. Cytol. 116, 47–88.

26. Weirich-Schwaiger, H., Weirich, H.G., Gruber, B., Schweiger, M. and Hirsch-Kauffmann, M. (1994) Correlation between senescence and DNA repair in the cells from young and old individuals and in premature ageing syndromes. Mutat. Res. 316, 37–48.

27. Catania, J. and Fairweather, D.S. (1991) DNA methylation and cellular ageing. Mutat. Res. 256, 283–293.

28. Lecka-Czernik, B., Moerman, E.J., Jones, R.A. and Goldstein, S. (1996) Identification of gene sequences overexpressed in senescent and Werner syndrome human fibroblasts. Exp. Gerontol. 31, 159–174.

29. Schachter, F.L., Faure-Delanef, L., Guenot, F., Rouger, H., Frouguel, P., Lesueur-Ginot, L. and Cohen, D. (1994) Genetic association with human longevity at the APO E and ACE loci. Nature Genet. 6, 29–35.

30. Orr, W.C. and Sohal, R.S. (1994) Extension of lifespan by overexpression of superoxide dismutase and catalase in Drosophila melanogaster. Science 263, 1128–1130.

31. Murray, A.W. and Kirschner, M.W. (1989) Dominoes and clocks: the union of two views of the cell cycle. Science 246, 614–621.

32. Rattan, S.I.S. (1995) Gerontogenes: real or virtual? FASEB J. 9, 284–286.

33. Wright, W.E., Pereira-Smith, O.M. and Shay, J (1989) Reversible cellular senescence: implications for immortalisation of normal human diploid fibroblasts. Mol. Cell. Biol. 9, 3088–3092.

34. Wright, W.E. and Shay, J.W. (1992) The two-stage mechanism controlling cellular senescence and immortalisation. Exp. Gerontol. 27, 383–389.

35. Shay, J.W., Pereira-Smith, O.M. and Wright, W.E. (1991) A role for both RB and p53 in the regulation of human cellular senescence. Exp. Cell Res. 196, 33–39.

36. Levy, M.Z., Allsopp, R.C., Futcher, A.B., Greider, C.W. and Harley, C.B. (1992) Telomere end replication problem and cell ageing. J. Mol. Biol. 225, 951–960.

37. Allsopp, R.C. (1996) Models of initiation of replicative senescence by loss of telomeric DNA. Exp. Gerontol. 31, 235–244.

38. Sharma, H.W., Sokoloski, J.A., Perez, J.R., Maltese, J.Y., Sartorelli, A.C., Stein, C.A., Nichols, G., Khaled, Z., Telang, N.T. and Narayanan, R. (1994) Differentiation of immortal cells inhibits telomerase activity. Proc. Natl. Acad. Sci. USA 92: 12343–12346.

3

MAINTAINING THE STABILITY OF THE GENOME

Alexander Bürkle

Deutsches Krebsforschungszentrum
Abt. 0610, D-69120 Heidelberg, Germany

1. WHAT IS AGEING?

Ageing may be defined as the time-dependent general decline of physiological functions of an organism, which is associated with a progressively rising risk of morbidity and mortality. For many different types of cancer, for instance, old age is a very important, if not the single most important risk factor[1]. The same holds true for a large number of degenerative diseases, whether they affect the cardiovascular system, the brain, or bone, joints and connective tissue, to mention but a few. So it is obvious that detailed knowledge of the basic biological mechanisms of ageing is indispensible for the full understanding of the pathogenesis and for rational and effective prophylaxis and therapy of all of these chronic conditions, which together place an already enormous, yet rapidly growing (and potentially overwhelming) burden on the social and economical systems in developed countries.

While there will be only little dispute about the definition of ageing, fundamentally different views exist in the explanation of its causes. I am not going to dwell on the vast number of different ageing theories that have been forwarded so far and which fall into two major classes, the so-called error theories and the programme theories of ageing. Instead, I should like to make the point that the nuclear genome of cells has some unique properties which suggest that it may serve both as a prominent target and, at the same time, as a mediator for ageing-associated changes: Unlike most other biological macromolecules (e.g., RNA, proteins, lipids), nuclear genes as the individual genome components are mostly present in only two copies per diploid cell, i.e., they are informationally very little redundant, thus entailing a high risk of permanent loss of genetic information in case of damage. Moreover, in nondividing cells DNA is not turned over in a scheduled fashion, thus allowing for accumulation of damage with time, unless repair activities come into operation (see below). (N.B. The mitochondrial genome represents a special case and is the topic of the chapter by H.D. Osiewacz, this volume.)

In the following paragraphs, I am going to list a few arguments taken from or compatible with various ageing theories as well as supportive experimental evidences, which will help to explain our research strategy.

1.1. Association of Genomic Instability with Ageing

Genomic instability is a term collectively describing alterations in the genome, such as point mutations in DNA, microsatellite expansions or contractions, amplifications and deletions of DNA sequences, gene rearrangements, and finally structural or numerical chromosomal aberrations. A lot of data have accumulated which show an age-dependent increase of chromosomal damages and instabilities in a broad sense[2,3]: For instance, it has long been known that in mouse liver the number of hepatocytes with chromosomal abnormalities increased with age[4]. Likewise, an increased frequency of structural and numerical chromosomal abnormalities was recorded in lymphocytes from elderly humans[5,6]. Besides, complete or partial loss of specific chromosomes is a characteristic feature of malignant tumor cells and plays an important role in tumor formation and progression, through the loss of growth control genes ("tumor suppressor genes"[7,8]). Very interestingly, cells from patients suffering from Werner syndrome (a rare, autosomal recessive disease defined by premature ageing of several organ systems and tissues) displayed a substantially elevated level of genomic instability. This was evident from their high frequency of structural chromosomal abnormalities and high mutation rate due to extensive DNA deletions[9]. The latter may be linked mechanistically with a disturbance of DNA ligation processes which renders them error-prone[10].

Also at the level of specific genes or DNA sequences very intriguing observations concerning age-related changes have been made: Already in the seventies, the selective loss of tandemly repeated genes for ribosomal RNA (r-DNA) in terminally differentiated human tissues (brain, heart) has been described. This phenomenon was supposed to contribute directly to age-dependent functional deficiency of cells[11,12]. Likewise, the passaging of human fibroblasts in tissue culture was associated with a continuous loss of certain repetitive DNA sequences, termed alphoid family/EcoRI repeat[13], and of telomeric sequences. Telomeres are stretches of highly reiterated characteristic short sequences units (*e.g.*, TTAGGG in human telomeres) which may be several kilobasepairs long and are located at the very ends of chromosomes of eucaryotic cells[14]. On the one hand, the presence of telomeres (plus telomerase activity, see below) allows for complete replication of the 5′-end of chromosomal DNA, on the other hand telomeres seem to protect chromosome ends from being involved in (undesirable) recombination events. In normal human fibroblasts, there was a progressive shortening of telomeres with increasing donor age or with increasing number of cell doublings *in vitro*[15]. On the average, 65 basepairs of telomeric DNA were lost per cell division[16]. By contrast, telomere length did not decrease in the germline nor in most immortalised cells lines studied. Seemingly, the latter phenomenon is due to the presence of telomerase, an enzyme which adds new telomere units to the 5′-end of telomeric DNA. Very recently, telomerase activity has been detected also in some normal, non-immortalised human cell types[17,18]. These findings may complicate the picture somewhat, but anyhow, it is tempting to speculate that telomeres function as a molecular counting device that determines the number of doublings a cell may still undertake before reaching replicative crisis (the "Hayflick limit").

With increasing age, there is not only *loss* of DNA sequences, but also the opposite, *i.e.* the enrichment of certain DNAs: Lymphocytes from old rats displayed increased amounts of extrachromosomal circular DNAs of variable length, compared with cells from young rats. The same phenomenon was detected in cultured human lung fibroblasts at late passages compared with early passages[19]. In skin fibroblasts taken from old people or from Werner syndrome patients, extrachromosomal circular DNAs of a particular size class (about 1.5 kilobasepairs) have been found predominantly.

1.2. Possible Induction of Ageing-Associated Genomic Instability by DNA Damages

Some of the manifestations of genomic instability mentioned above are obviously programmed changes, and these may be related with the turning on/off of specific enzyme activities, as is most evidently the case with telomere shortening. Other phenomena, such as the whole spectrum of ageing-associated DNA and chromosome instabilities, may be triggered by stochastic events rather.

From experimental cancer research we know that many manifestations of genomic instability can easily be *induced* by treatment of cells with a wide variety of DNA-damaging agents (carcinogens), and that this may be viewed as a cellular response to cope with unrepaired damages (*e.g.*, recombination events) and/or to adapt to an unfavourable microenvironment (*e.g.*, gene amplification and drug resistance).

So it was obvious to search for endogenous, "physiological" agents/events that could provoke genomic instability in association with ageing. In fact, plenty of agents have already been described that continuously attack the genomes of living organisms, such as reactive oxygen intermediates (ROI), nitric oxide, reducing sugars and other metabolites, all of which may have a potential to trigger genomic instability. For instance, the endogenous formation of mutagenic and carcinogenic N-nitroso compounds has been demonstrated to occur in humans and has been correlated with the incidence of specific cancer types[20]. N-nitroso compound formation is in fact one of the pathways by which nitric oxide, an extremely important physiological regulator of cell and tissue function, can damage cells. Furthermore, the endogenous formation of highly mutagenic etheno-derivatised DNA bases has recently been shown to occur in normal liver[21]. In addition, it is worth mentioning that the "endogenous" dysfunction of DNA polymerases can lead to the insertion of wrong deoxynucleotides (particularly in the context of "high-risk" sequences like simple repeats), leading to mismatches in DNA[22]. The spontaneous deamination of 5-methyl-cytosine in DNA leads to the inappropriate formation of thymine bases opposite guanine and represents yet another mechanism for mismatch formation[23].

However, most studies on the causes for ageing-associated DNA damage have focussed on ROI. A whole range of physiological and pathological sources of ROI have been identified, and the following enumeration is likely to be incomplete: ROI are produced (i) in mitochondria as by-products of the electron transport in the respiratory chain; (ii) by activated macrophages; (iii) in cytochrome P450-dependent and in other enzymatic reactions; (iv) in the auto-oxidation of hemoglobin and lipids; (v) in the radiolysis of water induced by ionising radiation; and (vi) by so-called advanced glycation end products (AGE). The latter are stable end products of the non-enzymatic, covalent coupling of sugar moieties with proteins. For instance, glycated tau-protein, a component of neurofibrillary tangles found in brains of patients with Alzheimer's disease, is able to generate ROI[24]. Interestingly, fragments of Alzheimer βA4 amyloid peptide also induce ROI formation[25]. ROI are chemically very reactive compounds that can damage all kinds of biological macromolecules to a substantial extent: In normal rat liver, about 1 in 10^5 bases of nuclear DNA and about 1 in 10^4 bases of mitochondrial DNA were found to be oxidised[26]. Oxidative DNA base lesions proved to be highly mutagenic[27].

Based on the above reasoning, the following may be deduced:

i. A low rate of endogenous ROI formation will keep the risk of DNA damage low. In fact, the rate of ROI formation, which has been correlated with the mitochondrial oxygen consumption under resting conditions (stage 4 of mitochondrial function), seems to be much lower in long-lived *versus* short-lived vertebrate species[28,29].

ii). Efficient anti-oxidative defences will prevent at least some ROI from damaging DNA. To some extent, cells are protected against the damaging effects of ROI by means of nonenzymatic and enzymatic antioxidant activities through which oxidants are detoxified before they can damage cellular macromolecules. Among the enzymatic antioxidants are superoxide dismutases (SOD), catalase, and gluthathione peroxidases. Experimentally, in at least one system, the critical impact of these enzymes for the ageing rate has already been demonstrated: Orr & Sohal[30] reported on transgenic *Drosophila melanogaster* strains carrying an additional gene copy for the *Drosophila* cytosolic copper/zinc SOD and catalase, respectively. Simultaneous overexpression of both enzymes led to the following phenotype: an increase of maximal life span by up to 34%; retardation of the ageing rate; reduction of the level of oxidative protein damage; increased total body oxygen uptake; and increased physical fitness.

iii. DNA repair mechanisms will antagonise the accumulation of DNA damages. Once DNA damages have already been inflicted, at least some of them may be removed by DNA repair activities. The pivotal importance of efficient DNA repair systems in the protection against physical or chemical carcinogenesis has been demonstrated in a number of experimental settings. By contrast, only correlative studies exist on the relationship between DNA repair and ageing: Hart & Setlow[31] were the first to describe a positive correlation between unscheduled DNA synthesis (repair-type DNA synthesis) after UV-irradiation of mammalian fibroblasts with the species-specific life span of the donor species. In the meantime, this "classical" finding was reproduced in a number of different cell systems[32], although opposing views have been published[33].

1.3. Loss of Cellular Functions, Programmed Cell Death, and Ageing of Tissues

It is likely that cells in a living organism may accumulate some genome damages or instabilities slowly over time without any significant phenotypic effect. Only when a critical threshold of cumulative damage/instability is reached will the cells cease to function properly. Nonfunctional/dysfunctional cells are typical targets for elimination by apoptosis (programmed cell death). At first, the consequences will be different among different cell types: For postmitotic cells (*e.g.*, neurons, muscle cells), replacement of the dead cell by another cell of the same lineage is not possible. Instead, neighbouring cells will have to take over the function of the lost one as much as they can, while the filling of the "empty space" will rather occur by proliferating connective tissue, equivalent to scar formation. For mitotic cell types (*e.g.*, stem cell systems; fibroblasts), the lost cell may initially be perfectly replaced via cell proliferation, thus preventing any immediate structural or functional impairment of the tissue. However, when the proliferative self-renewal capacity of the respective lineage declines and eventually becomes exhausted, be it through reaching the Hayflick limit ("replicative ageing") or, in an accelerated fashion, through the effect of ROI[34], the same situation arises as outlined for postmitotic cells, and the loss of parenchymal cell mass as well as functional impairment will become manifest. The latter are typical features of ageing in many tissues or organs.

2. EXPERIMENTAL APPROACH

Out of the many and extremely interesting questions that come up from the above points, the Heidelberg group has chosen to focus on the relationship between DNA repair and ageing/longevity, an important question which, nevertheless, has been much neglected in recent years. What makes this topic particularly attractive is the potential to reconcile "error theories" of ageing with "programme theories": The life span of an organism may turn out to be dependent on the life-long interplay or balance between "stochastic" damaging events on the one hand, and the particular genetic "programme" of DNA repair activities (and/or other protective functions, such as antioxidants) of a given animal species on the other hand.

Our starting point was a particular repair-related biochemical reaction which is directly triggered by the presence of DNA strand breaks, namely poly(ADP-ribosyl)ation.

3. PRESENT STATE OF KNOWLEDGE IN THE AREA

Poly(ADP-ribosyl)ation is one of the immediate responses of eukaryotic cells to oxidative and other types of DNA damage. Chemically speaking, it is a posttranslational protein modification (Fig. 1) catalysed by the nuclear enzyme poly(ADP-ribose) polymerase (PARP) which uses NAD^+ as substrate[35,36]. A few key features of this unique enzyme are listed in Table 1. Its DNA-binding domain, located at the aminoterminus, specifically binds to DNA single- or double-strand breaks. This binding is mediated through two zinc fingers and causes a drastic activation of the catalytic center in the carboxyterminal NAD^+-binding domain. A number of "acceptor" proteins covalently modified with poly(ADP-ribose) have been identified in vivo and in vitro, including PARP itself. PARP automodification is thought to preferentially occur on a separate protein domain, located between those for DNA-binding and enzyme catalysis. Treatment of cells with certain chemical or physical DNA-damaging agents, including ROI, alkylating agents, cisplatin[37], or γ-radiation induces a dose-dependent stimulation of poly(ADP-ribose) synthesis. High ADP-ribose polymer concentrations, in turn, stimulate enzymatic polymer degradation. As a consequence, polymer turnover may be accelerated drastically under conditions of massive DNA breakage. The life cycle of poly(ADP-ribose) is depicted schematically in Fig. 2.

To understand the biological function(s) of poly(ADP-ribosyl)ation, NAD^+ analogs have often been employed to inhibit PARP activity of intact cells in a competitive manner. However, such inhibitors cannot be considered specific for PARP. Therefore, it is hard to draw any definitive conclusions on the basis of inhibitor experiments only.

One molecular genetic strategy to interfere selectively with the expression of a particular gene is the expression of antisense RNA in transfected cells. Depletion of cellular PARP protein by this approach was reported to result in a disturbance of DNA repair[38,39]. Furthermore, mice with a homozygously disrupted PARP gene (PARP knock-out mice) have recently been described. Perhaps surprisingly, these mice developed normally and were fertile. However, initial results indicate that cells derived from such animals may be hypersensitive to environmental stress[40]. Taken together with data from many other laboratories, an involvement of poly(ADP-ribosyl)ation in DNA repair[41], specifically in DNA base-excision repair[42-44], was clearly documented, but the relevant molecular mechanisms are a matter of debate[42,43,45]. In addition to its role in DNA repair, poly(ADP-ribosyl)ation has been proposed to be involved in cell differentiation, integration of transfected foreign DNA into the cell genome[46], intrachromosomal homologous recombination, and finally apoptotic[47-49] or necrotic[50,51] cell death.

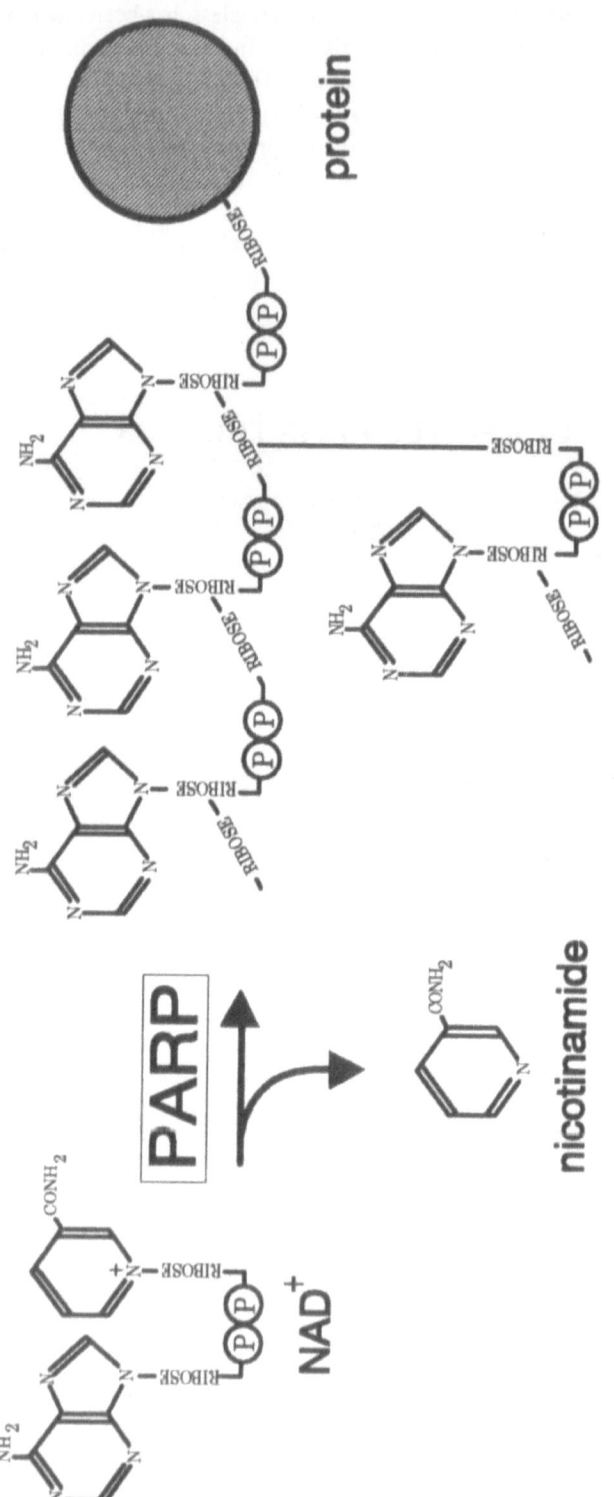

Figure 1. Chemical structure of poly(ADP-ribose).

Table 1. Some key features of poly(ADP-ribose) polymerase

- Chromatin-associated enzyme; M_r about 113 K; catalytically active as dimer
- Encoded by a single-copy gene (chromosomal location: 1q41-42 in humans)
- Present in many eukaryotic organisms (vertebrates, arthropods, mollusks, plants, and some lower eukaryotes like *Crypthecodinium cohnii, Dictyostelium discoideum* and *Physarum polycephalum*, but not in *Saccharomyces cerevisiae, Schizosaccharomyces pombe*)
- Not found in procaryotes
- Primary structure highly conserved during evolution
- Highly abundant (about 1 enzyme molecule per 80 kilobasepairs of nuclear DNA)
- catalyses
  - Transfer of ADP-ribosyl moieties from NAD^+ to acceptor proteins
  - Elongation to poly(ADP-ribose) chains (n<190)
  - Branching
- Catalysis drastically stimulated by DNA single- or double-strand breaks
- Domain structure:
  - DNA-binding domain (two zinc fingers, nuclear location signal)
  - Automodification domain
  - Catalytic domain (NAD^+ binding)
- Acceptor proteins:
  - Poly(ADP-ribose) polymerase itself
  - Histones and other chromosomal proteins
  - Many other nuclear enzymes (*in vitro*)

---

4. PAST, CURRENT, AND FUTURE RESEARCH ACTIVITIES OF THE RESEARCH GROUP

4.1. Elucidation of the Relationship between Poly(ADP-Ribosyl)ation and Ageing / Longevity of Mammalian Species

DNA base excision repair, a repair pathway in which poly(ADP-ribosyl)ation has been show to be involved, plays a pivotal role in removing "small" DNA lesions, such as

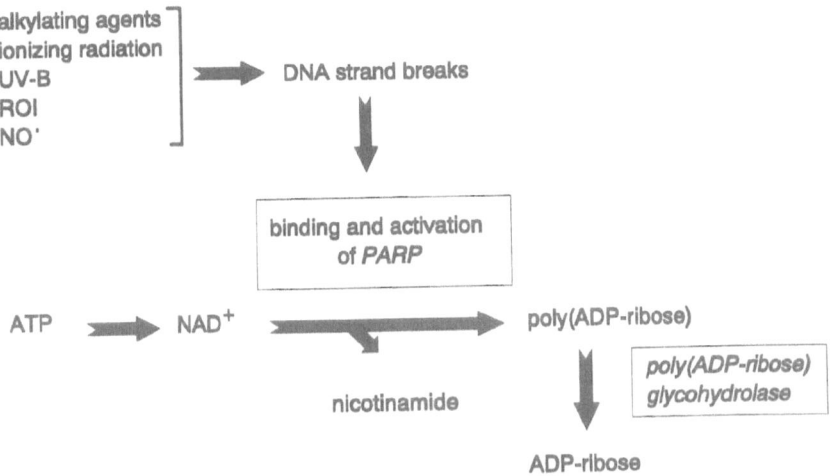

Figure 2. Poly(ADP-ribose) life cycle.

damage from alkylation or oxidation. On the other hand, oxidative DNA damage is particularly prominent in and relevant for the ageing process[52]. We have measured maximal oligonucleotide-stimulatable PARP activity[53] in permeabilized mononuclear leukocytes of 13 mammalian species of different life span. The data showed a strong, positive correlation of enzyme activity with the species-specific life span[54]. Interestingly, maximal PARP activity in different species was *not* correlated with the level of immunoreactive PARP protein as detected on Western blots. More recently, by using an improved immunofluorescence technique for assaying the poly(ADP-ribose) synthesis of living cells[37], we performed a comparison between γ-irradiated, living mononuclear leukocytes of two species with vastly different life span (rat and man). The percentage of polymer-positive cells was significantly higher in the human samples[55], supporting a *biological* relevance of our data on permeabilised cells. We are now going to extend this interspecies comparison by using an HPLC-based method[56] to precisely quantitate poly(ADP-ribose) synthesised in living cells. Furthermore, we are interested in the molecular mechanism(s) underlying the species-dependent and longevity-related differences in PARP activity. One possibility may be that differences in the primary structure of this (rather well conserved) enzyme are responsible. Thus we are going to compare recombinant expressed human enzyme with that of rat, given the vast difference in longevity between these species. Should we then see the same differences in enzyme activity as we did in permeabilised cells, it will be interesting to define the relevant aminoacid exchanges that occured during evolution.

4.2. Molecular Genetic Systems to Study the Biological Functions of Poly(ADP-Ribosyl)ation

We could show that overexpression of the PARP DNA-binding domain (DBD) in transfected cell cultures leads to a *trans*-dominant inhibition of poly(ADP-ribosyl)ation by competition for DNA breaks[57]. On the one hand, this feature of the DBD was employed to characterise the DNA break-binding ability of the enzyme in the context of living cells[44]. On the other hand, it was interesting to perform a detailed analysis of biological consequences of inhibiting poly(ADP-ribosyl)ation. In order to do this, we have established stably transfected cell lines which overexpress the PARP DNA-binding domain (DBD) under the control of the glucocorticoid-inducible mouse mammary tumour virus (MMTV) promoter. Induction of the DBD led to a 90-percent reduction of steady-state polymer levels after γ-irradiation. This inhibition is associated with a clear-cut sensitization of the cells against γ-radiation and alkylating agents, while there is no significant influence on normal cell growth nor on the episomal replication of a supertransfected shuttle plasmid[58]. Very recently, we could show that *trans*-dominant inhibition of poly(ADP-ribosyl)ation also potentiates carcinogen-induced amplification of chromosomally integrated SV40 DNA sequences in these cells[59], in agreement with our earlier data obtained with PARP-inhibitory drugs[60,61]. These results provide evidence for a role of cellular poly(ADP-ribosyl)ation as a mechanism that limits the extent of DNA damage-induced genomic instability.

As a complementary approach, we also generated stably transfected hamster cell clones that overexpress full-length human PARP. These clones show higher-than-normal poly(ADP-ribose) production even at relatively low doses of γ-radiation and, surprisingly, are also sensitized to the cytotoxic effects of γ-radiation or alkylating agents (L. van Gool *et al.*, submitted).

In our current and future work, the characterisation of our transfected cell lines will be completed. Attention will be given to several cellular responses to DNA damage, such as DNA repair, mutagenesis, gene amplification, and apoptosis.

In analogy with the cell transfection experiments but at a higher level of complexity, we will attempt to modulate PARP activity in transgenic (Tg) mice by overexpressing either the PARP DBD or the full-length enzyme in single or multiple tissues. The parameters to be studied here are susceptibility to carcinogens and ageing of the respective tissues.

5. WORKING HYPOTHESIS ON THE ROLE OF POLY(ADP-RIBOSYL)ATION IN AGEING / LONGEVITY

Ageing is associated with genetic instability in a variety of organisms studied, as outlined above. It appears very likely that DNA damages generated by endogenous agents that continuously attack the genomes of living organisms, (e.g., ROI, nitric oxide, reducing sugars and other physiological cell metabolites) play an important role in the induction of genetic instability. This view is fully consistent with the correlation between mammalian life span and DNA repair[32]: DNA repair would antagonise the accumulation of damages more efficiently in longer-lived species, and thus genome integrity and stability could be maintained over a longer time scale. Very likely, this is also one of the reasons for the obvious delay in tumor formation in long-lived species as compared with short-lived ones. Since poly(ADP-ribosyl)ation is involved in (at least) one segment of DNA repair, namely base-excision repair, we speculate that the species-specific, longevity-related variation in poly(ADP-ribosyl)ation capacity may represent one mechanism through which the longevity-related variation in global DNA repair capacity is mediated.

6. POTENTIAL FOR MEDICAL APPLICATION

It is commonplace that solid knowledge on basic mechanisms is a prerequisite for rational intervention. At the (still rather low) levels of mechanistic knowledge on ageing and longevity on the one hand, and on DNA repair and poly(ADP-ribosyl)ation on the other, it is too early to design prophylactic / therapeutic options based on poly(ADP-ribose) metabolism. Only the results of a lot more work in fundamental research will tell us about the real potential for practical medical application, be it at the level of pharmacology or of gene therapy. However, the magnitude and severity of the problems our "greying society" is facing with the rapidly growing incidence of age-related health disorders is an excellent motivation to go to work.

ACKNOWLEDGMENTS

I thank Professor Harald zur Hausen for continuous encouragement and support as well as for critical reading of the manuscript. Our work cited in this chapter was supported by grants from the Deutsche Forschungsgemeinschaft (Bu 698/2–1, 2–2, and 2–3) and from the Deutsches Krebsforschungszentrum Heidelberg.

7. REFERENCES

1. Riggs, J.E. (1994) Carcinogenesis, genetic instability and genomic entropy: insight derived from malignant brain tumor age specific mortality rate dynamics. J. Theor. Biol., 170, 331–338.
2. Slagboom, P.E., and Vijg, J. (1989) Genetic instability and aging: theories, facts, and future perspectives. Genome, 31, 373–385.

3. Osiewacz, H.D. (1990) Molecular analysis of aging processes in fungi. Mutat. Res., 237, 1–8.

4. Crowley, C., and Curtis, H. (1963) The development of somatic mutations in mice with age. Proc. Natl. Acad. Sci. USA, 49, 626–628.

5. Jarvik, L.F., Yen, T.-K., Fu, F.-S., and Matsuyama, S.S. (1976) Chromosomes in old age: a six year longitudinal study. Hum. Genet., 33, 17–22.

6. Dutkowski, R.T., Lesh, R., Staiano-Coico, L., Thaler, H., Darlington, G.J., and Weksler, M.E. (1985) Increased chromosomal instability in lymphocytes from elderly humans. Mutat. Res., 149, 505–512.

7. Nigro, J.M., Baker, S.J., Preisinger, A.C., Jessup, J.M., Hostetter, R., Cleary, K., Bigner, S.H., Davidson, N., Baylin, S., Devilee, P., et al. (1989) Mutations in the p53 gene occur in diverse human tumour types. Nature, 342, 705–708.

8. zur Hausen, H. (1991) Viruses in human cancers. Science, 254, 1167–1173.

9. Fukuchi, K., Martin, G.M., and Monnat, R.J. (1989) Mutator phenotype of Werner syndrome is characterized by extensive deletions. Proc. Natl. Acad. Sci. USA, 86, 5893–5897.

10. Runger, T.M., Bauer, C., Dekant, B., Moller, K., Sobotta, P., Czerny, C., Poot, M., and Martin, G.M. (1994) Hypermutable ligation of plasmid DNA ends in cells from patients with Werner syndrome. J. Invest. Dermatol., 102, 45–48.

11. Strehler, B.L., and Chang, M.-P. (1979) Loss of hybridizable ribosomal DNA from human post-mitotic tissues during aging: II. Age-dependent loss in human cerebral cortex - hippocampal and somatosensory comparison. Mech. Ageing Dev., 11, 379–382.

12. Strehler, B.L., Chang, M.P., and Johnson, L.K. (1979) Loss of hybridizable ribosomal DNA from human post-mitotic tissues during aging: I. Age-dependent loss in human myocardium. Mech. Ageing Dev., 11, 371–378.

13. Shmookler Reis, R.J., and Goldstein, S. (1980) Loss of reiterated DNA sequences during serial passage of human diploid fibroblasts. Cell, 21, 739–749.

14. Blackburn, E.H. (1994) Telomeres: no end in sight. Cell, 77, 621–623.

15. Harley, C.B., Futcher, A.B., and Greider, C.W. (1990) Telomeres shorten during ageing of human fibroblasts. Nature, 345, 458–460.

16. Counter, C.M., Avilion, A.A., LeFeuvre, C.E., Stewart, N.G., Greider, C.W., Harley, C.B., and Bacchetti, S. (1992) Telomere shortening associated with chromosome instability is arrested in immortal cells which express telomerase activity. EMBO J., 11, 1921–1929.

17. Counter, C.M., Gupta, J., Harley, C.B., Leber, B., and Bacchetti, S. (1995) Telomerase activity in normal leukocytes and in hematologic malignancies. Blood, 85, 2315–2320.

18. Härle-Bachor, C., and Boukamp, P. (1996) Telomerase activity in the regenerative basal layer of the epidermis in human skin and carcinoma-derived skin keratinocytes. Proc. Natl. Acad. Sci. USA, in press.

19. Kunisada, T., Yamagishi, H., Ogita, Z.-I., Kirakawa, T., and Mitsui, Y. (1985) Appearance of extrachromosomal circular DNAs during in vivo and in vitro ageing of mammalian cells. Mech. Ageing Dev., 29, 89–99.

20. Bartsch, H., Ohshima, H., Shuker, D.E., Pignatelli, B., and Calmels, S. (1990) Exposure of humans to endogenous N-nitroso compounds: implications in cancer etiology. Mutat. Res., 238, 255–267.

21. Nair, J., Barbin, A., Guichard, Y., and Bartsch, H. (1995) 1,N6-ethenodeoxyadenosine and 3,N4-ethenodeoxycytine in liver DNA from humans and untreated rodents detected by immunoaffinity/^{32}P-postlabeling. Carcinogenesis, 16, 613–617.

22. Modrich, P. (1995) Mismatch repair, genetic stability, and cancer. Science, 266, 1959–1960.

23. Schmutte, C., Yang, A.S., Beart, R.W., and Jones, P.A. (1995) Base excision repair of U:G mismatches at a mutational hotspot in the p53 gene is more efficient than base excision repair of T:G mismatches in extracts of human colon tumors. Cancer Res., 55, 3742–3746.

24. Yan, S.D., Chen, X., Schmidt, A.M., Brett, J., Godman, G., Zou, Y.S., Scott, C.W., Caputo, C., Frappier, T., Smith, M.A., et al. (1994) Glycated tau protein in Alzheimer disease: a mechanism for induction of oxidant stress. Proc. Natl. Acad. Sci. USA, 91, 7787–7791.

25. Hensley, K., Carney, J.M., Mattson, M.P., Aksenova, M., Harris, M., Wu, J.F., Floyd, R.A., and Butterfield, D.A. (1994) A model for beta-amyloid aggregation and neurotoxicity based on free radical generation by the peptide: relevance to Alzheimer disease. Proc. Natl. Acad. Sci. USA, 91, 3270–3274.

26. Richter, C., Park, J.-W., and Ames, B.N. (1988) Normal oxidative damage to mitochondrial and nuclear DNA is extensive. Proc. Natl. Acad. Sci. USA, 85, 6465–6467.

27. Feig, D.I., Sowers, L.C., and Loeb, L.A. (1994) Reverse chemical mutagenesis: identification of the mutagenic lesions resulting from reactive oxygen species-mediated damage to DNA. Proc. Natl. Acad. Sci. USA, 91, 6609–6613,

28. Ku, H.H., Brunk, U.T., and Sohal, R.S. (1993) Relationship between mitochondrial superoxide and hydrogen peroxide production and longevity of mammalian species. Free Radic. Biol. Med., 15, 621–627.

29. Barja, G., Cadenas, S., Rojas, C., Perez-Campo, R., and Lopez-Torres, M. (1994) Low mitochondrial free radical production per unit O_2 consumption can explain the simultaneous presence of high longevity and high aerobic metabolic rate in birds. Free Radic. Res., 21, 317–327.

30. Orr, W.C., and Sohal, R.S. (1994) Extension of life-span by overexpression of superoxide dismutase and catalase in Drosophila melanogaster. Science, 263, 1128–1130.

31. Hart, R.W., and Setlow, R.B. (1974) Correlation between deoxyribonucleic acid excision-repair and life-span in a number of mammalian species. Proc. Natl. Acad. Sci. USA, 71, 2169–2173.

32. Bernstein, C., and Bernstein, H. (1991) Aging, Sex, and DNA Repair. Academic Press, New York.

33. Promislow, D.E. (1994) DNA repair and the evolution of longevity: a critical analysis. J. Theor. Biol., 170, 291–300.

34. Chen, Q., and Ames, B.N. (1994) Senescence-like growth arrest induced by hydrogen peroxide in human diploid fibroblast F65 cells. Proc. Natl. Acad. Sci. USA, 91, 4130–4134.

35. de Murcia, G., and Ménissier de Murcia, J. (1994) Poly (ADP-Ribose) polymerase: a molecular nick sensor. Trends in Biochem. Sci., 19, 172–176.

36. Lindahl, T., Satoh, M.S., Poirier, G.G., and Klungland, A. (1995) Post-translational modification of poly(ADP-ribose) polymerase induced by DNA strand breaks. Trends in Biochem. Sci., 20, 405–411.

37. Bürkle, A., Chen ,G., Küpper, J.-H., Grube, K., and Zeller, W.J. (1993) Increased poly(ADP-ribosyl)ation in intact cells by cisplatin treatment. Carcinogenesis, 14, 559–561.

38. Ding, R., Pommier, Y., Kang, V.H., and Smulson, M. (1992) Depletion of poly(ADP-ribose) polymerase by antisense RNA expression results in a delay in DNA strand break rejoining. J. Biol. Chem., 267, 12804–12811.

39. Stevnsner, T., Ding, R., Smulson, M., and Bohr, V.A. (1994) Inhibition of gene-specific repair of alkylation damage in cells depleted of poly(ADP-ribose) polymerase. Nucl. Acids Res., 22, 4620–4624.

40. Wang, Z.Q., Auer, B., Stingl, L., Berghammer, H., Haidacher, D., Schweiger, M., and Wagner, E.F. (1995) Mice lacking ADPRT and poly(ADP-ribosyl)ation develop normally but are susceptible to skin disease. Genes. Dev., 9, 509–520.

41. Durkacz, B.W., Omidiji, O., Gray, D.A., and Shall, S. (1980) (ADP-ribose)$_n$ participates in DNA excision repair. Nature, 283, 593–596.

42. Satoh, M.S., and Lindahl, T. (1992) Role of poly(ADP-ribose) formation in DNA repair. Nature, 356, 356–358.

43. Satoh, M.S., Poirier, G.G., and Lindahl, T. (1993) NAD^+-dependent repair of damaged DNA by human cell extracts. J. Biol. Chem., 268, 5480–5487.

44. Molinete, M., Vermeulen, W., Bürkle, A., Ménissier-de Murcia, J., Küpper, J.-H., Hoeijmakers, J.H.J., and de Murcia, G. (1993) The poly(ADP-ribose) polymerase DNA-binding domain blocks alkylation-induced DNA repair synthesis in living cells. EMBO J., 12, 2109–2117.

45. Althaus, F.R. (1992) Poly ADP-ribosylation: a histone shuttle mechanism in DNA excision repair. J. Cell Sci., 102, 663–670.

46. Farzaneh, F., Panayotou, G.N., Bowler, L.D., Hardas, B.D., Broom, T., Walther, C., and Shall, S. (1988) ADP-ribosylation is involved in the integration of foreign DNA into the mammalian cell genome. Nucleic. Acids Res., 16, 11319–11326.

47. Kaufmann, S.H., Desnoyers, S., Ottaviano, Y., Davidson, N.E., and Poirier, G.G. (1993) Specific proteolytic cleavage of poly(ADP-ribose) polymerase: an early marker of chemotherapy-induced apoptosis. Cancer Res., 53, 3976–3985.

48. Nicholson, D.W., Ali, A., Thornberry, N.A., Vaillancourt, J.P., Ding, C.K., Gallant, M., Gareau, Y., Griffin, P.R., Labelle, M., Lazebnik, Y.A., Munday, N.A. Raju ,S.M., Smulson, M.E., Yamin, T-T., Yu, V.L., and Miller, D.K. (1995) Identification and inhibition of the ICE/CED-3 protease necessary for mammalian apoptosis. Nature, 376, 37–43.

49. Tewari, M., Quan, L.T., O'rourke, K., Desnoyers, S., Zeng ,Z., Beidler, D.R., Poirier, G.G., Salvesen, G.S., and Dixit, V.M. (1995) Yama/CPP32 beta, a mammalian homolog of CED-3, is a CrmA-inhibitable protease that cleaves the death substrate poly(ADP-ribose) polymerase. Cell, 81, 801–809.

50. Berger, N.A. (1985) Poly(ADP-ribose) in the cellular response to DNA damage. Radiat. Res., 101, 4–15.

51. Heller, B., Wang, Z.-Q., Wagner, E.F., Radons, J., Bürkle, A., Fehsel, K., Burkart, V., and Kolb, H. (1995) Inactivation of the poly(ADP-ribose) polymerase gene affects oxygen radical and nitric oxide toxicity in islet cells. J. Biol. Chem., 270, 11176–11180.

52. Shigenaga, M.K., Hagen, T.M., and Ames, B.N. (1994) Oxidative damage and mitochondrial decay in aging. Proc. Natl. Acad. Sci. USA, 91, 10771–10778.

53. Grube, K., Küpper, J.H., and Bürkle, A. (1991) Direct stimulation of poly(ADP ribose) polymerase in permeabilized cells by double-stranded DNA oligomers Anal Biochem., 193, 236–239.

54. Grube, K., and Bürkle, A. (1992) Poly(ADP-ribose) polymerase activity in mononuclear leukocytes of 13 mammalian species correlates with species-specific life span. Proc. Natl. Acad. Sci. USA, 89, 11759–11763.

55. Bürkle, A., Müller, M., Wolf, I., and Küpper, J.-H. (1994) Poly(ADP-ribose) polymerase activity in intact or permeabilized leukocytes from mammalian species of different longevity. Mol. Cell. Biochem., 138, 85–90.

56. Jacobson, M.K., Payne, D.M., Alvarez-Gonzalez, R., Juarez-Salinas, H., Sims, J.L., and Jacobson, E.L. (1984) Determination of in vivo levels of polymeric and monomeric ADP-ribose by fluorescence methods. Methods Enzymol., 106, 483–494.

57. Küpper, J.H., de Murcia, G., and Bürkle, A. (1990) Inhibition of poly(ADP-ribosyl)ation by overexpressing the poly(ADP-ribose) polymerase DNA-binding domain in mammalian cells. J. Biol. Chem., 265, 18721–18724.

58. Küpper, J.-H., Müller, M., Jacobson, M.K., Tatsumi-Miyajima, J., Coyle, D., Jacobson, E.L., and Bürkle, A. (1995) Trans-dominant inhibition of poly(ADP-ribosyl)ation sensitizes cells against γ-irradiation and N-methyl-N′-nitro-N-nitrosoguanidine but does not limit DNA replication of a polyoma virus replicon. Mol. Cell. Biol., 15, 3154–3163.

59. Küpper, J.-H., Müller, M., and Bürkle, A. (1996) Trans-dominant inhibition of poly(ADP-ribosyl)ation potentiates carcinogen-induced gene amplification in SV40-transformed Chinese hamster cells. Cancer Res., 56, 2715–2717.

60. Bürkle, A., Meyer, T., Hilz, H., and zur Hausen, H. (1987) Enhancement of N-methyl-N′-nitro-N-nitrosoguanidine-induced DNA amplification in a Simian Virus 40-transformed Chinese hamster cell line by 3-aminobenzamide. Cancer Res., 47, 3632–3636.

61. Bürkle, A., Heilbronn, R., and zur Hausen, H. (1990) Potentiation of carcinogen-induced methotrexate resistance and dihydrofolate reductase gene amplification by inhibitors of poly(adenosine diphosphate-ribose) polymerase. Cancer Res., 50, 5756–5760.

4

NUCLEAR–MITOCHONDRIAL INTERACTIONS INVOLVED IN BIOLOGICAL AGEING

Heinz D. Osiewacz

Johann Wolfgang Goethe-Universität
Botanisches Institut
Marie-Curie-Str.9, D-60439 Frankfurt am Main, Germany.

1. INTRODUCTION: BIOLOGICAL AGEING HAS A GENETIC BASIS

Biological ageing, as the time-dependent physiological changes of a biological system, is a complex process which is controlled by both, environmental as well as genetic factors. The genetic basis of ageing becomes evident from the observation that all living beings are characterized by a typical mean and maximum life-span. Moreover, in the different systems, individuals can be selected with a life-span which significantly differs from the species-specific limits. This phenotype is inherited to the progeny.

What is the basis of these observations? And why are the later periods in life often characterized by various degenerations and severe health problems? These are only two of many questions of interest related to a complex process which is of high significance for both, the individual as well as for society. Knowing the answers certainly will provide clues for the development of efficient interventions into those processes leading to degeneration and age-related diseases. But what is currently known about the mechanisms of ageing and, in particular, about the genetic basis of this process?

2. THE GENETIC INFORMATION OF EUKARYOTES

From genetic investigations with plants early in this century it became clear that genetic traits do not always follow the Mendelian rules[1,2]. Later, the work on the yeast *Saccharomyces cerevisiae* led conclude that the genetic information of this eukaryotic microorganism is found in two different cellular compartments: the nucleus and in the cytoplasm. In the meantime it is clear that this holds true for all eukaryotes. In both, autotrophic as well as heterotrophic eukaryotes the wast majority of the genetic information is found in the nucleus and a much smaller complement in mitochondria. In plants, the plastids, organelles which are involved in CO_2 assimilation and the conservation of solar energy, also contain genetic information.

Molecular Gerontology, edited by Rattan and Toussaint
Plenum Press, New York, 1996

Table 1. Comparison of the approximate size of the nuclear and the mitochondrial genome of selected species from different taxa. The size is indicated in kilo base pairs (kbp). Genome size of different strains of a given species may differ significantly. Only one example is presented in this table. The indicated data are derived from [3-6]

Group of organism	Species	Size of the genome (kbp)	
		Nuclear	Mitochondrial
Bacteria	*Escherichia coli*	4,700	—
Yeasts	*Saccharomyces cerevisiae*	15,000	75
Filamentous Fungi	*Podospora anserina*	33,400	94
Mammals	*Homo sapiens*	3,000,000	15

As may be seen from Table 1, significant differences in the size of the nuclear genome are found between lower and higher eukaryotes (e.g., yeast and humans). From investigations of the last years it became clear that these differences cannot be explained by the higher organization and the requirement for extra-genes. The observed size differences were found to be due to sequences which can be found as repetitive units. The number of these sequences may differ from species to species or even from strain to strain. Interestingly, although repetitive sequences account for a significant portion of the genome of most eukaryotes, it is amazing how little is clear about the function of this type of sequences. In many cases almost nothing is known and the corresponding sequences are thought to represent "selfish" DNA, or "molecular parasites," sequences which appear to be exclusively concerned with their own propagation. However, although no specific function can be attributed to these sequences, their multicopy nature appears to be of significance for the flexibility of the genome since repetitive sequences constitute nucleotide stretches of homology which are the targets of the cellular general recombination system.

The genetic information of mitochondria, the chondriome (or mitochondrial genome), is several factors of magnitudes smaller than the nuclear genome and consequently its coding capacity is rather limited. However, the size of the genome of different species or even of different strains of the same species, may differ significantly. From the nucleotide sequences of several completely sequenced chondriomes it is clear that these differences, like in the case of the nuclear genome, are generally not due to the presence or absence of extra-genes (Table 2). Basically, all mitochondrial genomes code for a specific gene set. One group of genes encodes components of the mitochondrial protein synthesis apparatus (rRNAs, tRNAs) whereas only a few genes code for proteins (Table 2, Figure 1). Most of these proteins are part of the energy-generating machinery found in mitochondria. In organisms with larger mtDNAs, additional sequences are found. In particular, intervening sequences (introns) may constitute a significant portion of the mtDNA. In some cases, these introns code for proteins which are required for the maturation of their own pre-messenger RNA (RNA maturases). In addition, in the last decade it was shown that some introns code for proteins which allow them to move from one location in the genome to another one. As discussed below, this intron transposition can be of high significance for the ageing process.

From the organization of the mitochondrial genome it is apparent that about 90–95% of the genetic information coding for components of this cellular compartment must be located in the nucleus and the corresponding products need to be transported into mitochondria. The elucidation of the involved apparatus is currently a subject of intensive investigations.

Table 2. Comparison of the coding capacity of the mitochondrial DNA (mtDNA) of the filamentous fungus *Podospora anserina* 3 and of *Homo sapiens* (cited from: 4)

	Podospora anserina	*Homo sapiens*
Cytochrome oxidase		
COI	+	+
COII	+	+
COIII	+	+
Apocytochrome b		
Cytb	+	+
ATP Synthase		
ATPase6	+	+
ATPase8	+	+
ATPase9	–	+
NADH Dehydrogenase		
ND1	+	+
ND2	+	+
ND3	+	+
ND4	+	+
ND4L	+	+
ND5	+	+
ND6	+	+
Ribosomal RNAs		
LSU	+	+
SSU	+	+
Number of tRNAs	25	22
ORFs*	32*	–
Introns	31	–
Size (kbp)	94	15

*Among the different ORFs some are characterized by significant homology of the deduced polypeptides to endonucleases, maturases, and reverse transcriptases [7, 8, 19]

3. MITOCHONDRIA: A CELLULAR ACHILLES HEEL

The mitochondria are the cellular compartments (organelles) of eukaryotic cells which are essential for energy production in aerobic organisms (e.g., humans). The matrix space of mitochondria, in which different metabolic processes occur (e.g., citrate cycle), is surrounded by two biomembranes. The inner membrane contains the respiration chain, a highly organized system of four protein complexes which are involved in electron transport. During this process, protons derived from energy-rich biomolecules which are degraded in the cytoplasm and the mitochondrial matrix, are transported out of the mitochondrion leading to an electro/chemical gradient. This gradient is the driving force by which adenosine triphosphate (ATP), the cellular "currency," is synthesized at another membrane-bound protein complex, the ATP-synthase (Fig.2).

The generation of ATP is the essential process from which all energy-consuming cellular processes are dependent. The efficient function of this complex machinery is dependent on the coordinated generation of the involved components and their subsequent assembly. Since most of the relevant components are encoded by nuclear genes, another important process in the biogenesis of mitochondria has to be considered: the controlled transport of proteins from the cytoplasm into these organelles.

As a whole, the various processes involved in the biogenesis of mitochondria are prone to errors. In the last two decades, it became clear that impairment of mitochondrial

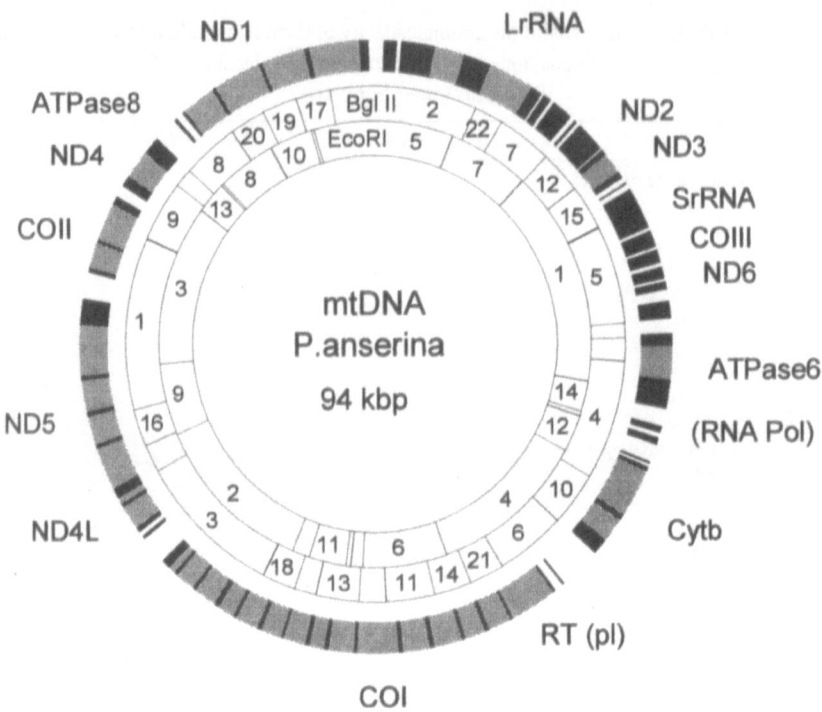

Figure 1. Physical and genetic map of the mtDNA of *Podospora anserina*. The inner circle represents the physical map with the recognition sites for endonucleases *Bgl*II and *Eco*RI. The location of the individual genes is indicated on the outer circle. Coding sequences for proteins, rRNAs and tRNAs, are indicated in black, intron sequences in grey. For simplicity, the location of the individual tRNA genes, which appear in clusters, is not indicated individually. The position of the first intron ("mobile intron") of the cytochrome oxidase subunit I gene (COI) is indicated (pl). This intron codes for a protein with a reverse transcriptase activity (RT) 19. (RNAPol) indicates three open reading frames which represent remnants of a viral-type RNA polymerase gene 9. The figure was prepared according to the published complete nucleotide sequence of the *P. anserina* mtDNA [3].

functions is of major significance for degenerative processes. Today, an steadily growing body of evidence indicates that mitochondria constitute a type of a cellular "achilles heel." In the next parts of this review, I shall give a brief overview of certain aspects of our current knowledge about the role of mitochondria in biological ageing. In particular, special emphasizes is put on the semiautonomous nature of these organelles. Some major nuclear-mitochondrial interactions will be discussed and perspectives to elucidate this complex molecular network will be drawn.

4. AGE-RELATED INSTABILITIES OF THE MITOCHONDRIAL GENOME

4.1. Mobile Elements in *Podospora* and *Neurospora*

The first clear evidence for a major role of the mitochondrial genome in the control of ageing is derived from genetic and molecular investigations of the ageing process in the fila-

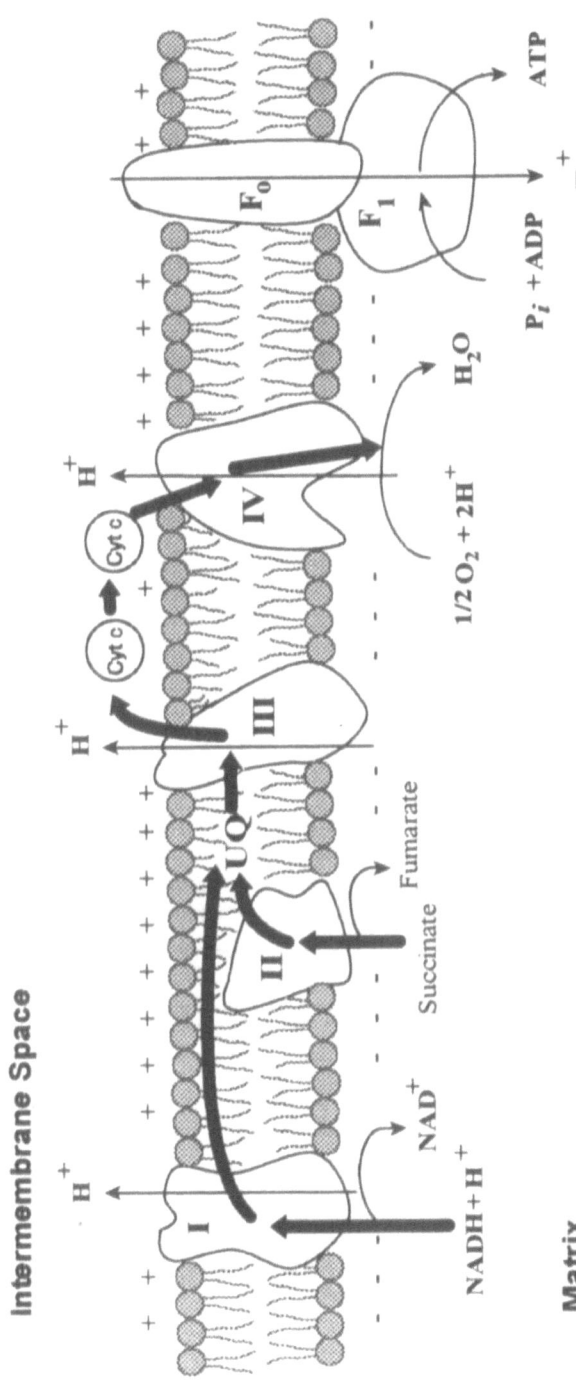

Figure 2. The respiration chain and oxidative phosphorylation is located in the inner membrane of the mitochondrion. Four protein complexes (I–IV) are involved in electron acceptance and transport (thick black arrows). As a consequence of a translocation of protons across the membrane (thin arrows) an electro-chemical gradient is generated which is the driving fource to synthesize ATP at another membrane-bound protein complex, the ATP-synthase. The different complexes are encoded in part by the mtDNA and by the nuclear genome. Impaired electron transport leads to the production of reactive oxygen species (ROS) which lead to cellular damage.

mentous fungus *Podospora anserina*. In contrast to most related fungi, this simple eukaryotic microorganism is not able to grow indefinitely. After a short period of growth, which is strain-specific and lies in the order of several weeks to a few month, cultures of *P. anserina* show various symptoms of senescence. Finally, they stop growing and die[10, 11]. A comparative analyses of the mtDNA revealed, that in senescent cultures a specific DNA species accumulates as an autonomous element of circular structure. In juvenile cultures this element, which was termed plDNA (for plasmid-DNA), is derived from a structural gene of the high molecular weight mtDNA[12–14]. It corresponds to the first intron of the gene coding for subunit I of the cytochrome oxidase (*COI*), a component of the mitochondrial respiration chain. During ageing of *Podospora* cultures the intron becomes liberated and amplified. Due to this unusual behavior, the pl-intron was termed a "mobile intron"[7,15]. Subsequently, the ability of the pl-intron to transpose from its resident position in the genome to another locus was demonstrated[16]. The consequence of this process is similar to what is known from "classical" transposons which move by replicative transposition, a process leading to the duplication of the element. Recombination processes between the resulting homologous sequences which are controlled by the general recombination apparatus may now account for age-related reorganizations of the mtDNA. As demonstrated in other systems, these processes may lead to the formation of DNA subcircles. Depending on whether these subcircles contain an origin of replication or not, they may either be retained or may be lost. In fact, it was demonstrated that in senescent *Podospora* cultures large parts of the mtDNA are deleted and as a consequence senescent cultures are respiratory deficient and die[14, 17, 18].

Apart from the ability of the pl-intron to transpose, some clues about this element are derived from its coding potential. Like various mitochondrial introns, the first *COI* intron contains an long open reading frame (ORF)[7]. The deduced amino acid sequence of a putative protein was shown to display significant homology to reverse transcriptases of different mobile genetic elements[8]. A reverse transcriptase activity of a polypeptide derived from the intron ORF was demonstrated after overexpression of the ORF in yeast[19]. In addition, an intron-encoded polypeptide was identified immunologically in senescent mycelia of *P. anserina*[20]. These data further support the idea that the pl-intron is related to a group of mobile genetic elements which, like retrotransposons, transposes via an RNA intermediate. Whether the transposition of this element occurs as a reversion of the normal intron splicing process on the RNA level or whether it takes place after a DNA strand break was generated remains to be shown. The latter situation was reported for a mobile mitochondrial intron of *S. cerevisiae*[21]. Finally, the generation of autonomous plDNA molecules which are regularly found in senescent *Podospora* cultures may be explained by the formation of tandem copies of plDNA in the mtDNA and subsequent homologous recombination processes[18].

Apart from this extensively investigated example, several other cases of mtDNA reorganizations which are related to circular DNA species were reported in *Neurospora crassa* and *N. intermedia*, two fungi which are closely related to *P. anserina* (for a review see: 22, 23). Also in these cases, the integration of circular plasmids leads to the formation of defective mitochondrial genomes and consequently to mitochondrial dysfunction. Interestingly, the plasmids Mauriceville and Varkud appear to integrate via an RNA intermediate and code, like the pl-intron, for a reverse transcriptase[24]. However, although distinctive sequence stretches with homology to intron consensus sequences were identified, these elements were not found as regular introns in the mtDNA of *Neurospora*.

Finally, some other less intensively studied degenerative processes were reported in a number of other filamentous fungi (e.g., *Aspergillus amstelodami*, *Cochliobolus heterostrophus*, *Podospora curvicolla*), processes which are also correlated with mtDNA reorganizations. In these cases, additional circular molecules which are derived from the mtDNA were reported to accumulate in impaired strains (for review see: 25).

Mobile introns and circular plasmids are not the only genetic trait which were demonstrated to modulate the life-span in filamentous fungi. Another group of elements which share structural characteristics with certain types of DNA viruses were found to lead to mitochondrial dysfunction. The first elements for which this was clearly demonstrated were the two linear plasmids kalilo and maranhar of *Neurospora crassa* and *Neurospora intermedia*[26–28]. In juvenile strains, these plasmids represent autonomous, self-replicating elements. During ageing they become integrated into the mtDNA. Plasmid insertion leads to the duplication of mtDNA sequences at the integration site and may lead to the inactivation of essential mitochondrial genes. Following integration, the defective mtDNA molecules carrying the integrated plasmid become suppressive to wild-type mtDNA. The mechanism which leads to suppressiveness is still unclear but appears to play a key role in ageing of senescence-prone *Neurospora* strains since it leads to the accumulation of defective mtDNA molecules and consequently to mitochondrial dysfunction and death of senescent cultures.

At this point it needs to be stressed that, despite the clear role certain extrachromosomal elements do play in molecular mechanisms leading to senescence, the majority of all elements of this type which are currently known appear to be neutral. For example, a survey of 171 natural isolates of *N. crassa* and *N. intermedia* revealed that only 21 strains became senescent although also in nonsenescing strains one or more linear or circular plasmids were identified[29]. Even more interestingly, investigations in our laboratory revealed that one long-lived strain of *P. anserina* in which longevity is maternally inherited contains a linear plasmid[30, 31]. Life-span modulation was observed although this element, as the two senescing-inducing linear plasmids kalilo and maranhar of *Neurospora*, is able to integrate into the mtDNA. Specifically, we identified integrated pAL2-1-sequences in the apocytochrome b gene (*Cytb*) which codes for an essential component of the mitochondrial respiration chain. However, in contrast to the situation in kalilo and maranhar strains of *Neurospora*, defective mtDNA molecules with integrated plasmid copy do not become suppressive[31–33]. The long-lived mutant remains heteroplasmic: it contains both, mtDNA molecules with a functional and a nonfunctional *Cytb* gene. Recently, we demonstrated that it is indeed pAL2-1 which is able to modulate the life-span of *Podospora* strains[32, 33]. In a series of experiments, transfer of pAL2-1 into a short-lived strain which did not contain a plasmid led to longevity. Vice versa, elimination of the element from a long-lived strain led to a decreased life-span. Interestingly, transfer of the autonomous element into mitochondria of strains without the linear plasmid was always found to be correlated with the formation of heteroplasmons with mtDNAs containing integrated plasmid copies and other molecules without this element. Finally, recently we analyzed a *Podospora* strain isolated from nature which contains a whole family of linear plasmids. These plasmids are very similar to pAL2-1 but do not integrate into the mtDNA of the corresponding strain, a strain which displays a short-lived phenotype[34]. Thus, life-span modulation appears to be linked to the ability of a linear plasmid to integrate into the mitochondrial genome. This ability is dependent on a certain nuclear genetic background of the host strain and, most likely, on the structure of the plasmid itself implying that nuclear-mitochondrial interactions are involved in this type of mtDNA reorganization.

5. MITOCHONDRIAL DNA INSTABILITIES IN MAMMALS

5.1. Neuromuscular Diseases

In comparison to the mtDNA of fungi, the mitochondrial genome of mammals appears to be stable. Extensive and almost quantitative reorganizations of the mitochondrial

Table 3. Selection of human mitochondrial disorders. The type of mutation identified in the mtDNA of selected individuals is indicated

Disease	Type of mutation in the mtDNA
Chronic progressive external ophthalmogplegia (CPEO)	Deletions
	Point mutations (tRNA)
Diabetes melitus and deafness	Deletions
	Point mutations (tRNA)
Kearn-Sayre syndrome (KSS	Deletions, Duplications
	Point mutations
Leber's hereditary optic neuropathy (LHON)	Point mutations (protein encoding genes)
Mitochondrial encephalopathy lactic acidosis and stroke like episodes (MELAS)	Point mutations (tRNA)
Myoclonic epilepsy associated with ragged fibers (MERRF)	Point mutations (tRNA)
Pearson syndrome	Deletions
Parkinson syndrome	Deletions

genome, as they are found during ageing of different fungi, cannot be observed. From what we know today about the organization of the genetic information in mitochondria and about the different types of genetic elements found in these organelles in different organisms, this observation is not surprising. In fact, until now no mobile genetic elements were reported as part of the genetic complement found in mammalian mitochondria. However, in the last several years, various alterations of the mitochondrial genome were reported to accumulate in particular in a number of neuromuscular disorders and since the first reports, the number of publications dealing with alterations of the mammalian mtDNA is increasing enormously[35, 36].

As may be seen from Table 3, various human disorders were related to mtDNA alterations. Both, gross mtDNA reorganizations as well as single point mutations were reported. In some cases, the mutations were shown to be maternally inherited[37, 38], in others, autosomal inheritance was found[39, 40]. Yet other mutations seem to have arisen sporadically during development of the individual[41, 42].

Gross mtDNA rearrangements, in particular deletions and duplications, lead to heteroplasmons, cells containing both normal and rearranged mtDNA molecules. The ratio of wild-type to defective mtDNA molecules varies between different patients and between different tissues of the same species. As an example, mtDNA deletions were identified in about 65% of all mtDNA molecules of the skeletal muscle of one patient, only 16%, 15% and 5% deleted molecules were found in the heard muscle, kidneys and liver, respectively, of the same individual[43]. Furthermore, in contrast to what might be expected, the ratio of affected molecules does not necessarily relate to the severity of symptoms of the disorder.

Apart from the identification of mtDNA rearrangements in postmitotic tissues (e.g., muscles, tissues of the nerval system) a few publications reported heteroplasmic mtDNA reorganizations in the blood of certain patients with Pearson's syndrome and Kearns-Sayre syndrome (KSS)[39, 44, 45].

Although a rather important issue, currently only little is known about the molecular mechanisms leading to mtDNA deletions and most explanations remain still speculative. However, one interesting observation is that many deletion borders are located at short direct repeats in the mtDNA (Table 4). This finding led to the suggestion that these short sequences somehow may relate to the corresponding deletion mechanism. In particular, the so-called "common deletion" of 4.98 kbp, which is located between the two origins of rep-

Table 4. Selection of human mtDNA deletions located between short direct repeats (underlined) according to [47-49]. Deletion borders are indicated by brackets (underlined)

Size of deletion	Organ	Sequence surrounding the site of deletion
3610 bp	Muscle	CAAGGACTAACCCC(TAT...TTCCTCCCC)ACACTCAT
4977 bp	Muscle	TACCTCCCTCACCA(AAG...AACCTCCCTCACCA)TTG
	Liver	
5756 bp	Muscle	AATAGAAACCG(TCTG...CATCGAAACCG)CAAACATA
5827 bp	Muscle	ACATACTTCCCCC(ATTA...ACAATCCCCC)TCTACCT
7635 bp	Muscle	CCTCATCACCCA(ACTA...GACTCACCCA)TCAACAAC
8041 bp	Brain	GAAGCCCCCAT(CCAT...CTCACCCAT)TCAACAACCG

lication of the human mtDNA, may be explained by a slip-replication mechanism. During replication, the H-strand of the mtDNA first represents a single-stranded DNA until the light strand origin of replication on the mtDNA molecule is reached. Shoffner et al.[46] suggested this elegant model which explains recombination of the two direct repeats surrounding the deletion site as the result of an illegitimate pairing of two direct repeats during the single-stranded stage of the replication intermediate. A subsequent introduction of a strand-break in one molecule and a partial degradation of the corresponding broken DNA strand may now produce the observed deletion. Finally, the ligation of the ends of the molecule lead to one shortened strand of the replicating molecule. After replication is complete, two mtDNAs arise, one carrying the deletion, the other one represents a wild-type molecule. It is evident at this point, that mtDNA deletions resulting from a mechanism of this kind are not only depending on the structure of the mtDNA alone (e.g., the direct repeats at the recombination boundaries) but also on factors encoded by the nuclear genome (e.g., components of the replication machinery, endonucleases, ligases).

In addition to gross mtDNA rearrangements, single point mutations of the mtDNA were correlated with various mitochondrial disorders, including LHON, MERRF, MELAS, Parkinson disease, and late onset of Alzheimers's disease (e.g.[38, 50, 51]). Mutations may lead to severe, moderate, and mild disorders. Mutations can either be heteroplasmic or homoplasmic and can effect different mitochondrial genes, including rRNA and tRNA genes, but also protein encoding genes (e.g., genes for *Cvtb*, *NADH* dehydrogenes). Mutations were identified predominantly in muscle tissue and in blood. Little is known about the definitive mechanisms by which mitochondrial point-mutations are generated. It is reasonable to assume that they are due to the action of reactive oxygen species (ROS) which are produced in mitochondria in higher amounts than in other cellular compartments.

5.2. Ageing

At the same time when the first correlations between human neuromuscular diseases and mtDNA alterations were identified, first data about an age-related increase of mtDNA rearrangements were reported by Piko[52]. These data were derived from heteroduplex analyses of mice, rats and human cells utilizing heteroduplex analyses. With the introduction of polymerase chain reaction technology (PCR) many correlative data were reported indicating that mtDNA reorganizations, although in only very low amounts, were reported. MtDNA deletions accumulating during ageing were reported in the range of 0.002% and 2% [53]. These numbers make it rather difficult to understand how they can significantly contribute to the ageing process. However, the low numbers of alterations may for several reasons be underestimations of the real alterations found in mtDNA of individuals of older age. First, the identification of one particular deletion by PCR is biased, since it is depending on the combination of primers used. Second, severely impaired cells may be eliminated from an tissue via apoptosis as soon as a particular threshold of altered mtDNA molecules is reached. Anyhow, the identification of a progressive accumulation of mtDNA mutations during ageing of mammals is of particular interest, but needs further data in order to firmly elucidate the causal role of these instabilities of the genetic information in higher biological systems. Furthermore, the observation that mtDNA mutations accumulate exponentially during ageing leads to another important issue since this observation may suggest that a preferential replication of defective mtDNA molecules may take place during later times in life of humans. This relates to the above mentioned, unexplained phenomenon of suppressiveness found during degeneration of fungal cultures, and, in addition, to a nuclear genetic control of mtDNA instabilities in different biological systems.

Taken together, despite the rather low abundance of altered mtDNA in individuals of older age, these rearrangements seem to be an important contributor to the ageing process. It was suggested that segregation processes lead to certain tissue mosaics, consisting of nonaffected, affected, and grossly defective capacity[54, 55]. These changes are believed to result in a progressive drop in the energy generating capability of older tissues and, after reaching a critical threshold, to death of the organism.

6. NUCLEAR CONTROL OF MITOCHONDRIAL DNA INTEGRITY

The integrity of the mtDNA is dependent on various factors. Among these, factors involved in mtDNA replication are obviously important. In particular, the fidelity of the involved machinery is of significance. An age-related decrease in the performance of this nuclear encoded apparatus must necessarily lead to mtDNA alterations. Other factors may more directly lead to age-related mtDNA reorganizations and to the progressive accumulation of these alterations. These nuclear encoded factors may control transposition of mobile genetic elements (e.g., mobile introns), the integration of extrachromosomal elements (e.g., circular and linear plasmids), or to a mechanism involved in outcompeting normal wild-type mtDNA molecules. Moreover, the general recombination apparatus leading to homologous recombination between repeated DNA-sequences is another component of the machinery. Conversely, the mitochondrial repair system, which is much less developed in mitochondria than in the nucleus is importantly involved in repairing DNA alterations. Since all these factors, but also others, are encoded by nuclear genes, they need to be transported into the mitochondrion. Consequently, the corresponding molecular apparatus represents another level of a molecular network which may be involved in the genetic control of biological ageing.

6.1. mtDNA Damage by Reactive Oxygen Species and Defense Systems

One ageing theory which has been proposed more than forty years ago, the "free radical theory"[56], explains ageing as the result of a progressive damage of different cellular components. A modification of this original concept is the "mitochondrial theory of aging" suggesting that an age-related decline of the respiration chain leads to cellular degenerations. An important contributor leading to an impairment of this apparatus is oxidative stress. During normal metabolism, in particular in mitochondria, reactive oxygen species (ROS) are produced. These substances contain a highly reactive, unpaired electron which is able to remove a partner electron from other molecules. The resulting reaction products themselves are highly reactive and as a consequence lead to different reactive components which are able to damage nucleic acids, proteins, lipids and other molecules. As a consequence, these molecules are affected in their biological function and cellular processes are impaired. Most importantly, damage of mitochondrial genes may lead to an impaired respiration chain which can increase ROS production[57] and consequently increases cellular damage. This type of a self-perpetuating decline was suggested to lead to an error-catastrophe and to a failure of the cellular system[36]. The idea is further supported by the fact that mtDNA, unlike nuclear DNA, is not protected by histones. Both characteristics of the mitochondrial system, together with those of the mitochondrial repair system (see below), may be the reasons why mtDNA alterations are found to be much more abundant than comparable changes in the nuclear DNA[58, 59]. However, a cellular defense system has to be incorporated into this framework. This system, at least within certain limits, is able to cope with the hazardous effect of ROS.

Today a huge body of experimental data is available pointing to a significant role of cellular ROS in a molecular framework leading to biological ageing. Out of the many data only a few key observations shall be stressed here:

Many species which consume oxygen faster than others and therefore produce larger amounts of reactive oxygen species, are characterized by a shorter life-span (e.g., mouse - humans)[60].

At least in rodents, a reduction of cellular metabolism as a result of calorie reduction leads to an increase in life-span[61].

Species with increased levels of enzymes which react with ROS (e.g. superoxide dismutase) have an increased life span[62].

Long-lived strains of *Drosophila* or *Caenorhabditis elegans*, selected by breeding or mutagenesis, are characterized by increased levels of superoxide dismutase and catalase[63, 64].

The introduction of additional copies of the two genes coding for superoxide dismutase (SOD) and catalase led to transgenic *Drosophila* strains with an increased life span[65].

6.2. mtDNA Repair

Although controversially discussed for a long time, it is clear today that mitochondria contain enzymes required for different forms of DNA repair. Moreover, evidence accumulates that these enzymes indeed are involved in mtDNA repair although they may also play a role in the degradation of damaged DNA. It appears that the enzymatic apparatus for recombination repair, mismatch repair, and base excision repair exists in mitochondria. However, in comparison to the repair of nuclear DNA, mtDNA repair is rather week. Moreover, certain types of repair which are efficiently used in the nucleus are totally absent in mitochondria. All the repair enzymes used in mitochondria are encoded by the nuclear DNA. Interestingly, mitochondrial enzymes (e.g., yeast uracil DNA glycosylase, mammalian AP endonuclease) appear to be encoded by different genes than their nuclear counterparts[66, 67]. However, in contrast to what is known for the nuclear compartment (see the chapter of A. Bürkle, this volume) nothing is known about the definitive contribution of this molecular system in respect to ageing. In fact, two questions which need to be addressed are whether or not the efficiency of the mitochondrial repair systems decreases during ageing of a given species and whether differences in the efficiency can be observed in short-lived and long-lived species.

6.3. mtDNA Replication

It is trivial to mention that the fidelity of the machinery involved in mtDNA is of high significance in respect to cellular degeneration. A reduced fidelity must necessarily lead to a progressive increase of DNA alterations. Moreover, the mode of replication as it is found in mammals appears to be prone to errors. As suggested by a number of mtDNA deletions located between the two origins of replication of mammalian mtDNA, replication intermediates which partially consist of single stranded DNA may be highly susceptible to damage and to erroneous replication processes (e. g., slip-replication).

6.4. The Protein Transport Machinery of Mitochondria

Finally, the last network of molecular processes which should be noted in this overview is the molecular machinery involved in the transport of nuclear-encoded mitochondrial proteins into the organelle. In this process, well-ordered membrane-bound protein complexes are required[68–70]. In addition, certain proteins located both in the cytoplasma and in the mitochondrial matrix assist in the corresponding pathway. These proteins, so-called chaperones, are involved in maintaining the protein which needs to be transported across the membranes in a transport-competent conformation[70, 71]. All components of this complex molecular machinery are encoded by nuclear genes and thus are part of the nuclear-mitochondrial interactions which may be considered to be of significance for biological ageing. Among the various components of such a system, chaperones belonging to the so-called heat stress proteins (HSP) appear to be of particular interest. These proteins which, in addition to elevated temperatures, are induced by various stress conditions (e.g., against toxic substances, oxidative stress), play a key role in protecting biological systems against stress. Some HSPs are part of the mitochondrial protein transport machinery. Moreover, HSP expression appears to decline during ageing[72]. And finally, among the various long-lived strains from different species, most appear to be more resistant against different stress conditions. Currently, the relevance of these observations is far from being understood in detail but the elucidation of the corresponding network appears to be a key to unravel the molecular basis of biological ageing in general.

7. CURRENT INTERESTS AND PERSPECTIVES

7.1. Identification and Characterization of Nuclear Factors Involved in the Biogenesis of Mitochondria and in mtDNA Maintenance

As mentioned above, it is known for a long time that both nuclear as well as mitochondrial genetic traits are involved in the control of ageing in the different biological systems. However, although various components and various mechanisms may be involved in these mitochondrial-nuclear interactions today only very limited data are available shedding some light onto the corresponding processes. The reason for this is obvious. Only a few nuclear genes selected as life-span affecting genes were cloned and characterized in some detail[73–75].

The current main interest of the group in Frankfurt is the elucidation of the genetic basis of biological ageing in *Podospora anserina*. We have chosen this simple eukaryotic model systems because it allows us to generate and investigate longevity mutants, both with an decreased and an increased life-span. The mode of inheritance, either Mendelian or maternal, can be determined easily and points to the site of mutation: the nucleus or the mitochondrion. In the past (see above), the organism has demonstrated to have many advantages as a model in experimental gerontology. Among these are: a short life-span (a few weeks to several months), the existence of longevity mutants, the accessibility to formal and molecular genetic experiments, the possibility to transform protoplasts with exogenous DNA (see ref. 6). In the last few years, we put special emphasis on the cloning of nuclear gerontogenes, genes which, after mutation, lead to an increased life-span of the corresponding individual. This was possible since various relevant mutants of this type were isolated in the past[76]. As the first gene of this type, we recently cloned gerontogene *grisea*. Mutation of *grisea* gives rise to an increased life-span (about 56%)[77]. *Grisea* turned out to code for a putative copper-activated transcription activator which appears to be involved in the control of copper homeostasis. In the mutant decreased levels of cellular copper appear to lead to decreased levels of cellular ROS[78] and consequently to reduced oxidative stress. Cloning of the corresponding mutant gene copy of *grisea* revealed, that the long-lived mutant is a loss-of-function mutant in which the gerontogene is not expressed. Currently we investigate the role of the *grisea* transcription factor in respect to the control of the mtDNA stability. In addition, we are utilizing different approaches to identify the different target genes of this transcription factor. Among other candidate genes, gerontogene *vivax* is one of the putative targets. This is suggested by the observation that the combination of the two single mutations leads to a double mutant, *grisea vivax*, which appears to be immortal[79]. Currently we are in the process of cloning gerontogene *vivax*. From this approach we expect to unravel an unknown part of the nuclear-mitochondrial network involved in the control of ageing in *P. anserina*.

Recapitulating the above mentioned nuclear-mitochondrial interactions which constitute a complex cellular network contributing to the complex biological problem of ageing, it should be stressed, that impairment of mitochondrial functions first analyzed in filamentous fungi in detail, appears to be also of importance in degenerative processes in other biological systems, although the specific molecular mechanisms leading to the impairments may differ in detail. One of the most important challenges of the next years is the cloning and characterization of further age-related genes from the diverse biological systems, in particular of genes which are part of the mitochondrial-nuclear interactions described above. Specifically, those genes are of particular interest, which are known to be linked to mitochondrial biogenesis and molecular processes located in these organelles. From what we know today, one group of candidate genes appears to be involved in the control of the stability/instability of the mitochondrial genome in the broadest sense (e.g., replication, recombination, repair). It is advisable that, although the cloning of the corresponding genes involved in the control of mtDNA instabilities in humans is obviously of special interest and promising approaches toward this goal exist[40, 80], the analyses of various biological systems should be followed. This strategy will finally settle a controversially discussed, unsolved question, the question of whether or not a universal molecular mechanisms of ageing exists (or a few), or whether every species or group of organisms has developed a different mechanism.

ACKNOWLEDGMENTS

I wish to thank the different collaborators of my group for excellent work. The experimental work of the group was supported by a grant of the Deutsche Forschungsgemeinschaft, Bonn (Os75/2-1, -2, -3, 4), and from the Deutsches Krebsforschungszentrum, Heidelberg.

REFERENCES

1. Baur E. (1909) Das Wesen und die Erblichkeitsverhälnisse der 'Varietates albo-marginatae hort' von Pelargonium zonale. Z. Vererbungsl. 1: 330.
2. Correns C. (1909) Vererbungsversuche mit blaß (gelben), grünen und buntblättrigen Sippen bei Mirabilis, Urtica und Lunaria. Z Vererbungsl 1: 291.
3. Cummings D.J., McNally K.L., Domenico J.M, Matsuura E.T. (1990) The complete DNA sequence of the mitochondrial genome of Podospora anserina. Curr. Genet. 17: 375–402.
4. Bonen L. (1991) The mitochondrial genome: so simple yet so complex. Curr. Opin. Genet. Devel. 4: 515–522.
5. Domdey H. (1991) Biochemische Aspekte der Genomanalyse In: Genomanalyse; (Ellermann R., Opolka U, Eds.), pp 13–33, Campus Verlag: Frankfurt.
6. Osiewacz H.D., Clairmont A., Huth, M. (1990) Electrophoretic karyotype of the ascomycete Podospora anserina. Curr. Genet. 18: 481–483.
7. Osiewacz H.D., K. Esser K. (1984) The mitochondrial plasmid of Podospora anserina: a mobile intron of a mitochondrial gene. Curr. Genet. 8: 299–305.
8. Michel F., Lang B. (1985) Mitochondrial class II introns encode proteins related to reverse transcriptases of retroviruses. Nature 316: 641–643.
9. Hermanns J., Osiewacz H.D. (1994) Three mitochondrial unassigned open reading frames of Podospora anserina represent remnants of a viral-type RNA polymerase. Curr. Genet. 25: 150–157.
10. Rizet G. (1953) Sur l'impossibilité d'obtenir la multiplication végétative ininterrompue et illimité de l'ascomycete Podospora anserina. C.R.Acad.Sci. (Paris) 237, 838–855.
11. Marcou D. (1961) Notion de longevite et nature cytoplasmatique du determinant de la senescence chez quelque chapignons. Ann. Sci. Natur. Bot. Ser. 12,2: 653–764.
12. Stahl U., Lemke P.A., Tudzynski P., Kück U., Esser K. (1978) Evidence for plasmid like DNA of the ascomycete Podospora anserina. Mol. Gen. Genet. 178, 639–646.
13. Cummings D.J., Belcour L., Grandchamps C. (1979) Mitochondrial DNA from Podospora anserina. II. Properties of mutant DNA and multimeric circular DNA from senescent cultures. Mol. Gen. Genet. 171:239–250.
14. Kück U., Stahl U., Esser K. (1981) Plasmid-like DNA is part of the mitochondrial DNA in Podospora anserina. Curr. Genet. 3: 151–156.
15. Kück U., Osiewacz H.D., Schmidt U., Kappelhoff B., Schulte E., Stahl U., Esser K. (1985) The onset of senescence is affected by DNA rearrangements of a discontinuous mitochondrial gene in Podospora anserina. Curr. Genet. 9: 373–382.
16. Sellem C.H., Lecellier G, Belcour L. (1993) Transposition of a group II intron. Nature 366: 176–178.
17. Belcour L, Begel O., Keller A.M., Vierny C. (1982) Does senescence in Podospora anserina result from instability of the mitochondrial genome In Mitochondrial Genes (Slonimski P.P., Borst P., Attardi G., Eds.); pp 415–422, Cold Spring Harbor Laboratory Press: Cold Spring Harbor.
18. Osiewacz H.D. (1992) The genetic control of aging in the ascomycete Podospora anserina. In: Biology of Aging. (Zwilling R., Balduini C., Eds.) pp 153. Springer-Verlag, Heidelberg.
19. Fassbender S., Br,hl K.-H., Ciriacy M, Kück U. (1994) Reverse transcriptase activity of an intron encoded polypeptide. EMBO J. 13: 2075–2083
20. Sellem C.H., Sainsard-Chanet A., Belcour L. (1990) Detection of a protein encoded by a class II mitochondrial intron of Podospora anserina. Mol. Gen. Genet. 224: 232–240.
21. Zimmerly S., Guo H., Perlman P.S., Lambowitz A.M. (1995) Group II intron mobility occurs by target DNA-primed reverse transcription. Cell 82: 545–554.
22. Osiewacz H.D. (1990) Fungal model systems for the analysis of aging processes. Mutat. Res. 237: 1–8.
23. Griffiths A.J.F. (1992) Fungal senescence. Annu. Rev. Genet. 26: 351–357.
24. Akins R.A., Kelley R.L., Lambowitz A.M. (1986) Mitochondrial plasmids of Neurospora: integration into mitochondrial DNA and evidence for reverse transcription in mitochondria. Cell 47: 505–516.
25. Esser K., Kück U., Lang-Hinrichs C., Lemke P., Osiewacz H.D., Stahl U., Tudzynski P. (1986) Plasmids of Eukaryotes. Springer-Verlag, Heidelberg.
26. Bertrand H., Chan B.S.S., Court D.A., Griffiths A.J.F. (1985) Insertion of a foreign nucleotide sequence into mitochondrial DNA causes senescence in Neurospora intermedia. Cell 41: 877–884.
27. Chan B.S.-S., Court D.A., Vierula P.J., Bertrand H. (1991) The kalilo linear senescence-inducing plasmid of Neurospora is an invertron and encodes DNA and RNA polymerases. Curr. Genet. 20:225–237.
28. Court D.A., Griffiths A.J.K., Krauss S.R., Russel P.J., Bertrand H. (1991) A new senescence-inducing mitochondrial linear plasmid in field-isolated Neurospora crassa strains from India. Curr. Genet. 19:129–137.

29. Yang X, Griffiths A (1993) Plasmid diversity in senescent and nonsenescent strains of Neurospora. Mol. Gen. Genet. 237: 177–186.
30. Osiewacz H.D., Hermanns J., Marcou D., Triffi M., Esser K. (1989) Mitochondrial DNA rearrangements are correlated with a delayed amplification of the mobile intron (plDNA) in a long-lived mutant of Podospora anserina. Mutat. Res. 219: 1–7.
31. Hermanns J., Osiewacz H.D. (1992) The linear mitochondrial plasmid pAL2–1 of a long-lived Podospora anserina mutant is an invertron encoding a DNA and RNA polymerase. Curr. Genet. 22: 491–500.
32. Hermanns J., Asseburg A., Osiewacz H.D. (1994) Evidence for a life span prolonging effect of a linear plasmid in longevity mutant Podospora anserina. Mol. Gen. Genet. 243, 297–307.
33. Hermanns J., Osiewacz H.D. (1996) Induction of longevity by cytoplasmic transfer of a linear plasmid in Podospora anserina. Curr. Genet., 29: 250–256.
34. Hermanns J., Debets F., Hoekstra R., Osiewacz H.D. (1995) A novel family of linear plasmids with homology to plasmid pAL2–1 of Podospora anserina. CMol. Gen. Genet. 246: 638–647.
35. Osiewacz H.D., Hermanns J. (1992) The role of mitochondrial DNA rearrangements in human disease and aging. Aging Clin. Exp. Res. 4: 273–286.
36. Wallace D.C. (1995) Mitochondrial DNA mutations in human disease and aging. In Molecular aspects of aging (Esser K, Martin G.M., Eds.), pp 163–177, Wiley & Sons: Chichester.
37. Ozawa T., Yoneda M., Tanaka M., Ohno K., Sato W., Suzuki H., Nishikimi M., Yamamoto M., Nonaka I., Horai S. (1988) Maternal inheritance of deleted mitochondrial DNA in a family with mitochondrial myopathy. Biochem. Biophys. Res. Commun. 154: 1240–1247.
38. Shoffner J.M., Lott M.T., Lezza A.M.S., Seibel P., Ballinger S.W., Wallace D.C. (1990) Myoclonic epilepsy and ragged-red fiber disease (MERRF) is associated with a mitochondrial DNA tRNA Lys mutation. Cell 61: 931–937.
39. Poulton J., Deadman M.E., Gardiner R.M. (1989) Duplications of mitochondrial DNA in mitochondrial myopathy. Lancet I: 236–239.
40. Zeviani M., Servidei S., Gellera C., Bertini E., DiMauro S. (1989) An autosomal dominant disorder with multiple deletions of mitochondrial DNA starting at the D-loop region. Nature 339: 309–311.
41. Mita S., Rizzuta R., Moares C., Shanske S., Arnaudo E., Fabrizi G.M., Koga Y., DiMauro S., Schon E.A. (1990) Recombination via flanking direct repeats is a major cause of large-scale deletions of human mitochondrial DNA. Nucleic Acids Res. 18: 561–567.
42. Degoul F., Nelson I., Amselem S., Romero N., Obermaier-Kusser B., Ponsot G., Marsac C., Lestienne P. (1991) Different mechanism inferred from sequences of human mitochondrial DNA deletions in ocular myopathies. Nucleic Acids Res. 19: 493–496.
43. Obermaier-Kusser B., Höcker-Müller J., Nelson I., Lestienne P., Enter C.H., Riedele T., Gerbitz K.D. (1990) Different copy numbers of apparently identically deleted mitochondrial DNA in tissues from a patient with Kearns-Sayre syndrome detected by PCR. Biochem. Biophys. Res. Commun. 169: 1007–1015.
44. Moares C.T., DiMauro S., Zeviani M., Lombes A., Shanske S., Miranda F., Nakase H., Bonilla E., Werneck L.C., Servidei S., Nonaka I., Koga Y., Spiro A., Brownell K.W., Schmidt K.W., Schotland D.L., Zupanc M., deVito D.C., Schon E.A., Rowland L.P. (1989) Mitochondrial deletions in progressive external ophthalmoplegia and Kearns-Sayre-Syndrom. N. Engl. J. Med. 320: 1293–1299.
45. Fishel-Ghodsian N., Bohlman M.C., Prezant T.R., Graham J.M.Jr, Cederbaum S.D., Edwards M.J. (1992) Deletions of blood mitochondrial DNA in the Kearns-Sayre Syndrome. Pediatr. Res. 31: 557–560.
46. Shoffner J.M., Lott M.T., Voljavec A.S., Soueidan S.A., Costigan D.A., Wallace D.C. (1989) Spontaneous Kearns-Sayre / chronic external ophthalmoplegia plus syndrome associated with mitochondrial DNA deletion: a slip-replication model and metabolic therapy. Proc. Natl. Acad. Sci. USA 86: 7952–7956.
47. Cortopassi G.A., Arnheim N. (1990) Detection of a specific mitochondrial DNA deletion in tissues of older humans. Nucleic Acids Res. 18: 6927–6933.
48. Katayama M., Tanaka M., Yamamoto H., Ohbayashi T., Nimura Y., Ozawa T. (1991) Deleted mitochondrial DNA in the skeletal muscle of aged individuals. Biochem. Int. 25: 47–56.
49. Zhang C., Baumer A., Maxwell R.J., Linnane A.W., Nagley P. (1992) Multiple mitochondrial DNA deletions in an elderly human individual. FEBS Lett. 297: 34–38.
50. Goto Y., Nonaka I., Horai S. (1990) A mutation in the tRNA Leu(UUR) gene associated with the MELAS subgroup of mitochondrial encephalomyopathies. Nature 348: 651–653.
51. Wallace D.C., Singh G., Lott M.T., Hodge J.A., Schurr T.G., Lezza A.M.S., Elas L.J., Nikoskelainen E.K. (1988) Mitochondrial DNA mutation associated with Leber's hereditary optic neuropathy. Science 242: 1427–1430.
52. Piko L., Hougham A.J., Bulpitt K.J. (1988) Studies of sequence heterogeneity of mitochondrial DNA from rat and mouse tissues: evidence for an increased frequency of deletions/additions with aging. Mech. Ageing Dev. 43: 279–293.

53. Wallace D.C., Bohr V.A., Cortopassi G., Kadenbach B., Linn S., Linnane A.W., Richter C., Shay J.W. (1995) The role of bioenergetics and mitochondrial DNA mutations in aging and age-related diseases. In: Molecular Aspects of Aging; (Esser K, Martin G.M., Eds.), pp 199–225, Whiley & Sons: Chichester.

54. Linnane A.W., Marzuki S., Ozawa T., Tanka M. (1989) Mitochondrial DNA mutations as an important contribution to ageing and degenerative diseases. Lancet I: 642–645.

55. Kadenbach B., Höcker-Müller J. (1990) Mutations of mitochondrial DNA and human death. Naturwissenschaften 77: 221–225.

56. Harman D. (1956) Aging: A theory based on free radical and radiation chemistry. J. Gerontol. 11: 298–300.

57. Bandy B, Davison A.J. (1990) Mitochondrial mutations may increase oxidative stress: implications for carcinogenesis and aging? Free Rad Biol. & Medicine 8: 523–539.

58. Richter C, Park J.W., Ames B.N. (1988) Normal oxidative damage to mitochondrial and nuclear DNA is extensive. Proc. Natl. Acad. Sci USA 85: 6465–6467.

59. Ames B.N., Shigenaga M.K., Hagen T.M. (1993) Oxidants, antioxidants, and the degenerative diseases of aging. Proc. Natl. Acad. Sci. USA 90: 7915–7922.

60. Adelman, R., Saul R.L., Ames B.N. (1988) Oxidative damage to DNA: Relation to species metabolic rate and life span. Proc. Natl. Acad. Sci. USA 85: 2706–2708.

61. Weindruch R., Walford R.L., Fligiel S., Guthrie D. (1986) The retartation of aging in mice by dietary restriction: Longevity, cancer, immunity and lifetime energy intake. J. Nutr. 116: 641–654.

62. Tolmasoff J.M., Ono T., Cutler R.G. (1980) Superoxide dismutase: Correlation with life span and specific metabolic rate in primate species. Proc. Natl. Acad. Sci. USA 77: 2777–2781.

63. Rose M.R., Vu L.N., Park S.U., Graves J.L (1992) Selection of stress resistance increases longevity in Drosophila melanogaster. Exp. Gerontol. 27: 241–250.

64. VanFleteren J.R., De Vreese A. (1995) The gerontogenes age-1 and daf-2 determine metabolic rate potential in aging Caenorhabditis elegans. FASEB J. 9: 1355–1361.

65. Orr W.C., Sohal R.S. (1994) Extension of life-span by overexpression of superoxide dismutase and catalase in Drosophila melanogaster. Science 263: 1128–1130.

66. Burgers P.M.J., Klein M.D. (1986) Selection by genetic transformation of a Saccharomyces cerevisiae mutant defective for the nuclear uracil-DNA-glycosylase. J. Bacteriol. 166: 905–913.

67. Tomkinson A.E., Bonk T., Linn S. (1988) Mitochondrial endonuclease activities for apurinic/apyrimidinic sites in DNA from mouse cells. J. Biol. Chem. 263: 12532–12537.

68. Glick B.G., Schatz G. (1991) Import of proteins into mitochondria. Annu. Rev. Genet. 25: 21–44.

69. Pfanner N, Rassow J, van der Klei I.J., Neupert W. (1992) A dynamic model of the mitochondrial protein import machinery. Cell 68: 999–1002.

70. Külbrich M., Dietmeier K., Pfanner N. (1995) Genetic and biochemical dissection of the mitochondrial protein-import machinery. Curr. Genet. 27: 393–403.

71. Stuart R.A., Cyr D.M., Craig E.A., Neupert W. (1994) Mitochondrial molecular chaperones: their role in protein translocation. Trends Biochem. Sci. 19: 87–92.

72. Fargnoli J. Kunisada T., Fornace A.J., Schneider E.L., Hohlbrook N.J. (1990) Decreased expression of heat shock protein 70 mRNA and protein after heat treatment in cells of aged rats. Proc. Natl. Acad. Sci. USA 87: 846–850.

73. D'Mello N., Childress A.M., Franklin D.S., Kale S.P., Pinswasdi C., Jazwinski S.M. (1994) Cloning and characterization of LAG1, a longevity assurance gene in yeast. J. Biol. Chem. 269: 1545–15459.

74. Sun J, Kale S.P., Childress A.M., Pinswasdi C., Jazwinski S.M. (1994) Divergent roles of RAS1 and RAS2 in yeast longevity. J. Biol. Chem. 28: 18638–18645.

75. Kennedy B.K., Austriaco N.R., Zhasng J., Guarente L. (1995) Mutation in the silencing gene SIR4 can delay aging in S. cerevisiae. Cell 80: 485–496.

76. Tudzynski P., Esser K. (1979) Chromosomal and extrachromosomal control of senescence in the ascomycete Podospora anserina. Mol. Gen. Genet. 173: 71–84.

77. Prillinger H., Esser K. (1977) The phenoloxidases of the ascomycete Podospora anserina. XIII. Action and interaction of genes controlling the formation of laccase. Mol. Gen. Genet. 156: 333–345.

78. Halliwell B., Gutteridge M.C. (1984) Oxygen toxicity, oxygen radicals, transition metals and disease. Biochem. J. 219: 1–14.

79. Tudzynski P., Stahl U., Esser, K. (1982) Development of a eukaryotic cloning system in Podospora anserina. I. Long-lived mutants as potential recipients. Curr. Genet. 6: 219–222.

80. Suomalainen, A., Kaukonen, J., Amati, P., Timonen, R., Haltia, M., Weissenbach, J., Zeviani, M., Somer, H., Peltonen, L. (1995) An autosomal locus predisposing to deletions of mitochondrial DNA. Nature Genetics 9, 146–151.

5

FROM GENES TO FUNCTIONAL GENE PRODUCTS DURING AGEING

Lise Brock Andersen, Ann Lund, Marie Kveiborg, Brian F. C. Clark, and Suresh I. S. Rattan

Laboratory of Cellular Ageing
Department of Molecular and Structural Biology
University of Aarhus, DK-8000 Aarhus-C, Denmark

INTRODUCTION

For the survial of any living system it is crucial that its homeostatic processes function accurately and efficiently. Homeostatic balance in the cell is attained through a molecular network of repair and maintenance processes. Although the genomic information for these processes is encoded in the DNA, this information becomes functionally meaningful only when it is accurately transcribed and translated into gene products. Whereas two types of RNA, transfer (t) RNA and ribosomal (r) RNA, are themselves functional molecules, the genetic information transcribed into the third RNA, messenger (m) RNA, has to be generally translated from a language of nucleic acids into a language of amino acids in order to produce proteins which are the functional gene products. Since we consider ageing as the progressive accumulation of damage due to a failure of maintenance mechanisms[1], we think that it is important that the formation, functioning and turnover of the primary molecules constituting the maintenance network are studied with respect to ageing.

It has been estimated that in a human cell there are about 80,000 genes per haploid genome, of which about 22,000 are housekeeping genes and the rest are tissue-specific[2]. Furthermore, in order to become a functional protein, a newly synthesised polypeptide chain has to undergo a wide variety of post-translational modifications that determine its activity, stability, specificity and transportability. Misregulation of genetic information transfer at any of these steps can be critical for the failure of homeostasis, and can be the basis of ageing and death.

1. TRANSCRIPTION

The eukaryotic genome consists largely of the chromatin, which is a complex of DNA, histones and non-histone proteins that are arranged in a series of repeating units,

Molecular Gerontology, edited by Rattan and Toussaint
Plenum Press, New York, 1996

termed nucleosomes. The stability of the structure and organization of chromatin is crucial both for the maintenance of the state of differentiation of a cell and for its function as an active source of genetic information[3]. There are three types of RNA in a cell, of which about 70–80% is rRNA, 10–15% is tRNA and 5–7% is mRNA. The transcription of RNA from DNA involves RNA polymerases I, II and III in complex with different factors in holoenzymes which synthesise tRNA, mRNA and rRNA, respectively.

Initiation of transcription is a very important regulatory step involving many factors. A large number of transcription factors regulate transcription by binding to DNA sequences distal to the 5′end of transcribed genes prior to transcription initiation. Some of these factors, such as CAAT-binding transcription factor (CTF), promotor-specific transcription factor-1 (SP-1), transcription factor-IID, activator protein factor-1 (AP-1), cAMP- and glucocorticoid-response element-binding proteins (CREBP and GREBP) and octamer binding protein, have been investigated in relation to cellular ageing, but no clear pattern of change emerges from such studies[4]. Similarly, an age-related decline in the expression of transferrin gene in the livers of transgenic mice carrying chimeric human transferrin transgenes has been related to both an increase in the binding activities of YY1-a and YY1-b proteins and a decrease in SP-1-like binding activity[5]. There are several other factors, such as TATA binding protein (TBP), TATA associated factors (TAF) and supressors of RNA polymerase II proteins (SRB) which participate in the transcription initiation reactions. However, the sequence of assembly of RNA polymerase II and transcription factors in the transcription of mRNA is not completely elucidated.

Studies on the synthesis of RNA and on its processing during ageing have been few. Although the level of total transcription is generally reduced during ageing, the proportion of different RNAs does not change significantly. Furthermore, the endogenous nucleotide pool and the activities of the enzymes involved in RNA synthesis are also reduced during ageing[6].

For each type of RNA some age-related changes have been observed in various ageing systems[7] (Tables 1 and 2). Levels of tRNAs and aminoacyl-tRNA synthetases (aaRS) have been considered to be rate limiting for protein synthesis[8]. According to one of the molecular theories of ageing, called the codon restriction theory[9], a random loss of various isoaccepting tRNAs will progressively restrict the readability of codons resulting in the inefficiency and inaccuracy of protein synthesis. There is some evidence that a shift in the pattern of isoaccepting tRNAs occurs during development and ageing in some plants, nematodes, insects and rat liver and skeletal muscle[10], but its significance in ageing is not well understood. Similarly, a 30- to 60-fold increase in the amount of UAG suppressor tRNA has been reported in the brain, spleen and liver of old mice, and has been related to increased expression of Moloney murine leukemia virus (MO-MuLV) in fibroblasts[11].

Other characteristics of tRNAs that have been studied during ageing include the rate of synthesis, total levels, aminoacylation capacity and nucleoside composition (Table 1). There is no generalised pattern that emerges from these studies, and the reported changes vary significantly among different species. The aminoacylation capacity of different tRNAs varies to different extents during ageing, and the reasons for such variability are not known[12]. However, the fidelity of aminoacylation did not differ significantly in cell-free extracts prepared from young and old rat livers[13]. Therefore, more studies are required to establish the changes in the structural and functional aspects of individual tRNAs, including their stability, accuracy and turnover, in order to elucidate their role in the regulation of protein synthesis during ageing.

In the case of aaRS, an increase or decrease in the specific activities of almost all of them has been reported in various organs of ageing mice without any apparent correlation

Table 1. Alterations in tRNA and mRNA characteristics during ageing

Characteristic	Change	Ageing systems[*]
tRNA		
Total levels	decrease	mouse liver, kidney, heart, muscle
Rate of synthesis	decrease	mouse liver, kidney, heart, muscle
Capacity to accept amino acids	variable	rat liver
Methylation	decrease	nematodes; rat and mouse liver, kidney; human fibroblasts
Pattern of isoacceptors	unstable	soybean cotyledon; nematodes; Drosophila; rat liver
Nucleoside composition	no change	mosquitoes; mouse liver
mRNA		
Total poly(A)$^+$ RNA	decrease	rat brain, liver; rabbit liver
Length of poly(A)	decrease	Quail oviduct, heart, hepatocytes; rat liver
Cap structure	no change	rat liver parenchymal cells
Overall half life	decrease	Drosophila
In vitro translatability	no change	rat liver

[*]For complete references to each system, see ref. 31.

with tissue/cell type and its protein synthetic activity. A significant decline in the specific activities of 17 aaRS has been reported in the liver, lung, heart, spleen, kidney, small intestine and skeletal muscle of ageing female mice[14] and during development and ageing of *C. elegans*[15]. Similarly, an increase in the proportions of the heat-labile fraction of several of these enzymes has been reported in the liver, kidney and brain of old rats[12]. However, no universal pattern can be seen for the changes in the activities of various synthetases in different organs and in different animals. Although an age-related decrease in the efficiency of aaRS can be crucial in determining the rate and accuracy of protein synthesis, direct evidence in this respect is lacking at present.

Another step in the transfer of genetic information that can be rate-limiting is the availability of mRNAs for translation. Post-transcriptional processing of eukaryotic mRNA is a kind of "fine control" of the information to be expressed and only a small fraction of the originally synthesised sequences reaches the cytoplasm[6]. These post-transcriptional fine control mechanisms include the formation of the cap, addition of the poly(A) tail, splicing and removal of the introns, and the transport of mature mRNA to the cytoplasm. Only a few reports are available on changes in various aspects of mRNA-processing during ageing (Table 1). Although it appears that at a gross level there are no major alterations in mRNA characteristics, it is possible that individual mRNA species do undergo changes, including splicing, transport from nucleus to cytoplasm, binding to ribosomes, stability and turnover during ageing. An indication that differential regulation of mRNA availability occurs during ageing is clear from numerous studies reporting an increase or decrease in the levels of various mRNA transcripts hybridizing to their cDNA probes.

In the case of rRNA, a significant decline in the content and synthesis has been reported in ageing beagles, rodent organs and cultured cells[7]. This decline in rRNAs was previously thought to be associated with the loss of ribosomal genes. However, no age-related decline in the gene copy number of rRNA has been observed in human fibroblasts or in mouse myocytes. Furthermore, the number of rRNA genes in a cell is already in great excess (between 200 and 1000 copies), and a small loss with age may not have any serious consequences for cell function and survival. However, whether there is a differential loss of various rRNA species during ageing, and what effects such a loss might have, is not known.

2. TRANSLATION

Protein synthesis is one of the most complex processes in the cell, and requires about 200 components divided in three groups: ribosomes, translational factors and the tRNA-charging system. The process uses large quantities of cellular energy (GTP and ATP), and can be divided into three steps (i) initiation; (ii) elongation; and (iii) termination followed by post-translational modifications and folding. A decline in the rate of total protein synthesis (both cytoplasmic and mitochondrial) is one of the most common age-associated biochemical changes that has been observed in a wide variety of cells, tissues, organs and organisms, including human beings[16, 17].

The implications and consequences of slower rates of protein synthesis are manifold in the context of ageing and age-related pathology. These consequences include decreased availability of enzymes, inefficient removal of intracellular damaged products, inefficient intra- and inter-cellular communication, decreased production of hormones and growth factors, decreased production of antibodies, and altered nature of the extracellular matrix. Although there is a considerable variability among different tissues and cell types in the extent of decline (varying from 20% to 80%), the fact remains that the bulk protein synthesis slows down during ageing Furthermore, it has been shown that the conditions, such as calorie-restriction, that increase the lifespan and retard the ageing process in many organisms, also slow down the age-related decline in protein synthesis[17]. These observations reinforce the view that slowing down of protein synthesis is an integral part of the ageing process and may be crucial for the failure of homeostasis.

However, it should be pointed out that age-related slowing down of bulk protein synthesis does not mean that the synthesis of each and every protein becomes slower uniformly during aging. Although no senescence-specific unique proteins have been detected during ageing, a significant increase in the heterogeneity of protein synthesis during ageing has been observed[17]. Furthermore, even though bulk protein synthesis slows down with age, total protein content of the cell generally increases because of an accumulation of undegraded abnormal proteins during ageing.

2.1. Initiation of Translation

Association of 40S and 60S ribosomal subunits, an initiator tRNA called methionyl (Met)-tRNA$_i$ and the initiation factors (eIFs) results in the formation of an active 80S initiation complex placed at the start codon of the mRNA. The process takes about 2 to 3 seconds cell-free assays and is probably much faster in vivo[18]. Each mRNA can participate in multiple rounds of initiation, thus giving rise to a string of ribosomes called polysomes, engaged at different stages of translation. It is estimated that an efficiently translated mRNA at 37°C initiates protein synthesis once every 5 to 6 seconds[19]. How many times an mRNA can be translated depends on several aspects of its structure including the context surrounding the AUG codon and its lifespan expressed in terms of the rate of degradation[18].

What is important in a biological context is that the initiation step is considered to be a major target for the regulation of protein synthesis during cell cycle, growth, development, hormonal response and under stress conditions including heat shock, irradiation and starvation[20, 21]. With respect to ageing however the rate of translation initiation appears to remain unaltered[22, 23]. On the other hand, since polysomal fraction of the ribosomes decreases during ageing, it implies that the activity of an anti-ribosomal-association factor eIF-3 may increase during ageing[22, 23]. Similarly, the activity of eIF-2, which is required for the formation of the ternary complex of Met-tRNA$_i$, GTP and eIF-2, has been reported to decrease in rat tissues during development and ageing[24, 25]. A decline in the amount and activity of GDP/GTP exchange factor eIF-2β has been reported in the brains and livers of 10 months old Sprague-Dawley rats as compared with 1 and 4 months old animals[26].

Recent developments in our understanding of the functioning and of post-translational modifications of various eIFs during cell growth, proliferation, tumorigenesis, stress and pathological conditions have made it necessary that detailed studies on eIFs are also undertaken in the context of ageing and the question of the regulation of protein synthesis at the level of initiation is reinvestigated.

2.1.1. Ribosomes. The eukaryotic ribosome is one of the most complex components of the protein synthetic apparatus. It consists of two subunits: a small 40S subunit and a large 60S subunit, which combine to form the 80S ribosome. The ribosome from the three taxonomic kingdoms share several common features, including physical shape, rRNA structure, ribosomal proteins and rRNA-protein interaction. The small subunit (40S) contains one rRNA (18S) and about 30 ribosomal proteins whereas the large subunit (60S) contains 3 rRNAs (5S, 5.8S, and 28S) and about 45 ribosomal proteins. The exact role of various rRNAs and proteins and their interactions in determining the activity, efficiency and accuracy of the ribosomes are poorly understood at present[27].

There is a slight decrease in the number of active ribosomes during ageing, although this may not be rate limiting for the total protein synthesis because of ribosomal abundance in the cell (Table 2). On the other hand, the translational capacity of ribosomes does show an age-associated decline which, however, is highly variable in various parts of the body. Reasons for such variability are not clear at present but may be related to variable protein synthetic activity of different organs.

Compared with rRNAs, there is relatively more information available regarding ribosomal proteins in connection with hormonal-, growth-factor-related and developmental regulation. For example, during the development of *Xenopus laevis* there is a co-ordinated expression of ribosomal proteins regulated both by transcriptional feed-back which controls the processing and stability of their mRNAs, and by translational utilization of the mRNAs for ribosomal proteins[27]. With respect to ageing, although the amount of L7

Table 2. Alterations in ribosomes and rRNAs during ageing

Characteristic	Change	Ageing systems*
rRNA synthesis	decrease	mouse liver; rat heart
rRNA content	increase	human fibroblasts
Number of ribosomes	decrease	rat brain, liver; Drosophila; nematodes
Ribosomal protein pattern	no change	mouse liver; Drosophila
Thermal stability	decrease	nematodes; Drosophila
Binding to aminoacyl-tRNA	decrease	rat liver, kidney; nematodes; Drosophila
Translational capacity	decrease	nematodes; rat liver
Fidelity of poly(U) translation	no change	rat liver, brain, kidney; mouse liver
Sensitivity to aminoglycosides	increase	human fibroblasts; rat liver

*For complete references to each system, see ref. 31.

mRNA increases in human fibroblasts and in rat preadipocytes, no differences between the two dimensional gel electrophoretic patterns of the ribosomal proteins from young and old *Drosophila* and mouse liver have been observed[22, 23]. However, these observations require careful reinvestigations using better developed and more sensitive methods of protein pattern analysis.

2.2 Elongation and Termination of Translation

A repetitive cyclic event of peptide chain elongation, which is a series of reactions catalysed by elongation factors (EFs; also abbreviated as eEFs for the eukaryotic factors) follows the formation of the 80S initiation complex. In terms of energy consumption, addition of each new amino acid to a growing polypeptide chain costs 4 high-energy phosphates from 2 molecules of ATP during aminoacylation and 2 molecules of GTP during elongation. Various estimates of the elongation rates in eukaryotic cells give a value in the range of 3–6 amino acids incorporated per ribosome per second, which is several times slower than the prokaryotic elongation rate of 15–18 amino acids incorporated per second[28, 29].

The regulation of protein synthesis can also occur totally and differentially at the level of polypeptide chain elongation. Examples of differential regulation include the rapid translation of heat shock-induced mRNAs in *Drosophila* and chick reticulocytes, translation of viral S1 mRNA in reovirus-infected cells, synthesis of vitellogenin in cockerel liver after estradiol injection and the synthesis of tyrosine aminotransferase in cultured hepatoma cells treated with cAMP[30]. The regulation of bulk protein synthesis at the level of elongation has been reported for normal and transformed cells during cell cycle transition, amino acid starvation, serum stimulation and phorbol ester treatment[30]. Similarly, alterations in the rates of elongation have also been reported in full-term human placenta from diabetic mothers, and in rat livers during fasting and refeeding[31].

During ageing, a slowing-down of the elongation phase of protein synthesis has been suggested to be crucial in bringing about the age-related decline in total protein synthesis. This is because a decline of up to 80% in the rate of protein elongation has been reported by estimating the rate of phenylalanyl-tRNA binding to ribosomes in poly(U)-translating cell-free extracts from old *Drosophila*, nematodes and rodent organs[22, 23, 32]. In vivo, a two-fold decrease in the rate of polypeptide chain elongation in old WAG albino rat liver and brain cortex has been reported[33]. Similarly, a decline of 31% in the rate of protein elongation in the livers of male Sprague-Dawley rats has been reported by measuring the rate of polypeptide chain assembly which was 5.7 amino acids per second in young animals and was 4.5 amino acids per second in 2 year old animals[34]. However, these estimates of protein elongation rates have been made for "average" size proteins. It will be important to see if there is differential regulation of protein elongation rates for different proteins during ageing.

In principle, elongation rates can be regulated through changes in the concentration of aa-tRNAs, modifications of ribosomes and changes in the amounts and activities of elongation factors. The peptide chain elongation cycle continues until one of the three stop codons is reached. There is no aa-tRNA to read these codons and instead a release factor (eRF1) binds to the ribosome and GTP and releases the peptide chain. The ribosome-mRNA complex thus dissociates and the components can start another round of synthesis. Regulation of the rates of protein synthesis at the level of termination is considered to be the least likely target. With respect to ageing the termination process in cell free-extract from *Drosophila* and rat liver showed no change[35, 36].

2.2.1 Elongation Factors. Elongation factors are considered to be the prototypes of the superfamily of G proteins and much information is available on the structural and functional aspects of bacterial factors, EF-Tu, EF-Ts and EF-G as compared with their eukaryotic counterparts. In eukaryotes, the addition of an amino acid to the growing polypeptide chain is facilitated by at least two elongation factors, EF-1 and EF-2[37]. A third factor, EF-3, is reported only in some fungi and yeast. EF-1 is composed of two distinct parts: a G-binding protein, EF-1α and a nucleotide exchange protein complex, EF-1βγδ. EF-1 usually occurs in multiple molecular forms, composed of varying amounts of EF-1α and EF-1βγδ.

EF-1α is a ubiquitous, highly expressed and very conserved translational factor. Sequences of EF-1α genes, cDNAs and protein from more than 15 different species have been determined and found to be highly conserved during evolution[37]. The human EF-1α gene family consists of two actively transcribed isoform genes EF-1α1 and EF-1α2 and more than eighteen pseudogenes. EF-1α2 is expressed in a tissue specific manner whereas EF-1α1 is expressed ubiquitously, and both of them can function in translation[38, 39]. EF-1α1 and EF-1α2 are very similar with respect to amino acid sequence (96% similarity) but differ in their 3'-UTR region and the expression pattern. The tissue specificity has been characterized in both humans and rodents[38, 39]. Whereas EF-1α1 is expressed to varying extents in every tissue examined, the EF-1α2 mRNA is detected only in the brain, muscle and heart tissues and furthermore, only in the terminally differentiated cells of these tissues. Although the EF-1α1 mRNA is found in the muscle, only EF-1α2 protein can be purified from muscle.

Some other interesting features of EF-1α include its high abundance (between 3 and 10% of the soluble protein), and having several other biological activities in addition to its requirement in protein synthesis (Table 3). Although EF-1α is among the most abundant proteins in the cell, much of EF-1α mRNA is stored in mRNP and is translated only under

Table 3. Multiple biological activities of elongation factor-1[*]

Protein synthesis
Binding of aminoacylated-tRNA to the ribosome.
Maintaining the accuracy of decoding.
Part of the valyl-tRNA synthetase complex.
Part of messenger ribonucleoprotein (mRNP) particle.
Protein degradation
Ubiquitin-dependent degradation.
Intracellular chanelling
Associated with endoplasmic reticulum.
Binding to cytoskeletal actin elements and regulating polymerization.
Rearrangement and severing of microtubules.
Binding to intermediate filaments.
Signal transduction
Calcium-dependent calmodulin binding.
Binding to cell membrane.
Phosphoinositol kinase activation.
Mitosis
Binding to the mitotic apparatus.
Other biological effects
Increasing cellular susceptibility to transformation.
Modulating ageing and lifespan.

[*]For complete references to each function, see refs. 22, 28.

stress conditions such as starvation or exposure of cells to cycloheximide[40, 41]. Recently, EF-1α was found to be necessary for ubiquitin dependent degradation of certain N-acetylated proteins. A homodimer of EF-1α (also called FH) was shown to be required along with the 26S protease complex for degradation of ubi-conjugates of histone H2A and probably also for α-crystallin and actin[42]. These diverse roles qualify it to be considered as a general "sensor" molecule for keeping track of the overall health status of the cell.

With regard to ageing, the activity of EF-1 declines with age in rat livers and *Drosophila*, and the drop parallels the decrease in protein synthesis[35, 36]. This decline in the activity of EF-1 has been correlated only to EF-1α as no changes were observed in the EF-1βγδ-mediated activity. We have also observed a 35–45% decrease in the activity and amounts of active EF-1α in serially passaged senescent human fibroblasts, old mouse and rat livers and brains, but not in senescent human keratinocytes[22, 23]. Experiments on the germ line insertion of an extra copy of EF-1α gene under the regulation of a heat shock promoter resulted in a better survival of transgenic *Drosophila* at high temperature as compared with their controls[43]. However, this relative increase in the lifespan of transgenic insects at high temperature was not accompanied by any increase in the levels of mRNA, amount and activity of EF-1α[44, 45]. Furthermore, various other transgenic combinations of Drosophila having altered levels of EF-1α expression did not show any relationship with lifespan although large changes in other fitness components, including fecundity could be observed[46, 47]. Similarly, no increased expression of EF-1α genes was observed in *Drosophila* with extended longevity phenotype in a long-lived strain[48]. Although increased longevity of EF-1α high-fidelity mutants of a fungus *Podospora anserina* has been reported which suggests that the life prolonging effects of EF-1α may be due to its role in maintaining the fidelity of protein synthesis[49], future studies on other ageing systems, particularly human cells and rodents will clarify the role of EF-1α in the regulation of both protein synthesis and longevity.

The other elongation factors, EF-1βγδ and EF-2, are involved, respectively, in the post-hydrolytic exchange of GDP with GTP and in the translocation of peptidyl-tRNA on the ribosome. Of these, EF-2 has a unique characteristic in the form of a histidine residue at position 715 modified into diphthamide, as a result of which it can be ADP-ribosylated either endogenously or by bacterial toxins such as diphtheria toxin[37]. Conflicting data are available regarding the changes in EF-2 during ageing. For example, a lack of difference in the rate of translocation has been observed during the translation of poly(U) by cell-free extracts prepared from young and old *Drosophila* and from rodent organs[35, 36]. Similarly, although the proportion of heat-labile EF-2 increases during ageing, the specific activity of EF-2 purified from old rat and mouse liver remains unchanged[50]. However, we have observed a decline of more than 60% in the amount of active EF-2 during ageing of human fibroblasts in culture, but not in rat livers, measured by determining the content of diphtheria toxin-mediated ADP-ribosylatable EF-2 in cell lysates[51, 52].

ADP-ribosylation of diphthamide residue results in the abolition of the translocation activity of EF-2. Phosphorylation of EF-2 by a calcium/calmodulin-dependent protein kinase III (CaM PK III), also known as the EF-2 kinase[53] is considered to be another mode of regulation of EF-2 activity. Changes in the amounts of phosphorylated EF-2 during the mammalian cell cycle have been correlated with the changes in protein-synthetic rates[30, 54]. We have reported an age-related increase in the activity of CaM PKIII in the livers of calorie-restricted and freely-fed rats, which may account for the loss of activity of EF-2 during ageing[55]. Furthermore, this change appears to be irreversible which cannot be compensated by dephosphorylation of EF-2 by the protein phosphatase PP2A, whose activity remains unaltered during ageing[56].

Recently, significant advances have been made in our understanding of some of the structural and functional aspects of prokaryotic elongation factors, which are the prototypes of the superfamily of G proteins[57]. Future research on the genetic regulation of the structure and function of eukaryotic elongation factor genes and proteins will unravel their pluripotent roles in various biological processes, including ageing.

3. ACCURACY OF PROTEIN SYNTHESIS

The rate and accuracy of protein synthesis (as also of DNA and RNA synthesis) have been presumably gone through natural selection and evolved to optimal levels according to the overall life history of an organism. Since, the error frequency of amino acid misincorporation is generally considered to be quite high (10^{-3} to 10^{-4}) as compared with nucleotide misincorporation, the role of protein error feedback in ageing has been a widely discussed issue.

At present, no direct estimates of protein error levels in any ageing system have been made primarily due to the lack of appropriate methods to determine spontaneous levels of errors in a normal situation. Several indirect attempts however have been made to determine the accuracy of translation in cell-free extracts, using synthetic templates or natural mRNAs. Studies on the accuracy of protein synthesis during ageing that have been performed on animal tissues, such as chick brain, mouse liver, and rat brain, liver and kidney, did not reveal any major age-related differences in the capacity and accuracy of ribosomes to translate poly(U) in cell-free extracts (for cross references see[23]). However, these attempts to estimate the error frequencies during translation in vitro of poly(U) template were inconclusive because the error frequencies encountered in the assays were several times greater than the estimates of natural error frequencies (for a detailed discussion of this, see refs.[58, 59]).

The accuracy of mouse liver ribosomes did not change with age in cell-free assays measuring the incorporation of radioactive lysine during the translation of trout protamine mRNA which does not have codons for lysine[60]. In contrast to this, using mRNA of CcTMV coat protein for translation by cell extracts prepared from young and old human fibroblasts, a seven-fold increase in cysteine misincorporation during cellular ageing has been observed[61, 62]. These studies also showed that an aminoglycoside antibiotic paromomycin (Pm), which is known to reduce ribosomal accuracy during translation in vivo and in vitro induces more errors in the translation of CcTMV coat protein mRNA by cell extracts prepared from senescent human fibroblasts than those from young cells. Further indirect evidence that indicates the role of protein errors in cellular ageing can be drawn from studies on the increase in the sensitivity of human fibroblasts to the life-shortening and ageing-inducing effects of Pm and another aminoglycoside antibiotic G418[63, 64]. Similarly, increased longevity of high-fidelity mutants in *Podospora anserina* indicate the role of protein errors in lifespan[49].

Another indirect method that has been used to detect misincorporation of amino acids during ageing is the method of 2D-gel electrophoresis of proteins, by which at least one kind of error, that is the misincorporation of a charged amino acid for an uncharged one (or vice versa) can be demonstrated because of "stuttering" of the protein spot on 2D gels. Using this method, no age-related increase in amino acid misincorporation affecting the net charge on proteins was observed in histidine-starved human fibroblasts and in *Caenorhabditis elegans*[65, 66].

Although a global "error catastrophe" as a cause of ageing due to errors in each and every macromolecule is considered unlikely, it is not ruled out that some kind of errors in various components of protein synthetic machinery including tRNA charging may have long-term effects on cellular stability and survival[67]. Better methods are still required for measuring the basal levels of translational errors in young and old cells, tissues and organisms.

4. POST-TRANSLATIONAL MODIFICATIONS

Almost all proteins undergo post-translational modifications, which is the final step in the transfer of genetic information from a gene into a functional gene product. Faithful translation of the genetic information encoded in mRNA into a polypeptide chain is not enough to guarantee efficient functioning of the protein. More than 200 types of post-translational modifications of proteins have been described that determine the activity, stability, specificity, transportability and lifespan of a protein. Numerous studies on the biochemical basis of protein functioning and turnover during various biological processes indicate the crucial role of post-translational modifications.

The term "post-translational modification" includes: (i) covalent modifications that yield derivatives of individual amino acid residues, for example, phosphorylation, methylation, ADP-ribosylation, oxidation and glycation; (ii) proteolytic processing through reactions involving the polypeptide backbone; and (iii) nonenzymic modifications, for example, deamidation, racemization and spontaneous changes in protein conformation.

4.1 Phosphorylation

Phosphorylation of serine, threonine and tyrosine residues is one of the best studied modifications of proteins. The coordinated activities of protein kinases which catalyse phosphorylation, and protein phosphatases which catalyse dephosphorylation, regulate several biological processes, including protein synthesis, cell division, signal transduction, cell growth, development and ageing (Table 4).

During cellular ageing several putative inhibitors of DNA synthesis have been identified in senescent cells[68-70]. It is possible that the activity of several of these inhibitors is

Table 4. Some proteins whose activities are modulated by phosphorylation

Biological process	Phosphorylated proteins	
	Increased activity	Decreased activity
DNA synthesis, repair cell proliferation	DNA polymerase α, histones	products of *cdc*-series genes, retinoblastoma gene product, some DNA repair proteins
Protein synthesis	S6 ribosomal protein, initiation factors, eIF-3, 4B, 4F, aminoacyl-tRNA-synthetases	initiation factor eIF-2, elongation factors EF-1α, β, EF-2
Structural organization and metabolism	vimentin, lamin, neurofilament proteins, microtubules, glycogen phosphorylase, tyrosine hydroxylase	synapsin
Signal transduction	growth factor receptors, G protein subunits α_{i-2}, αz	adrenergic and muscarinic receptors, cyclin-dependent kinases

regulated by phosphorylation. For example, age-related alterations in cell-cycle-regulated gene expression of various genes such as *c-fos, c-jun, JunB, c-myc, c*-Ha-*ras, p53, cdc2, cycA, cycB, cycD* and retinoblastoma gene *RB1*, may be due to alterations in their phosphorylation status. A decrease in phosphorylated cyclin E and Cdk2, and failure to phosphorylate *RB1* gene product p110^{Rb}, and *cdc2* product p34^{cdc2} during cellular ageing have been reported at present[71–73]. It will be important to find out if there are age-related alterations in the phosphorylation state of other cell cycle related gene products including proteins involved in DNA and RNA synthesis, and various transcription factors.

Various components of the protein synthetic apparatus undergo phosphorylation and dephosphorylation and thus regulate the rates of protein synthesis[28]. For example, phosphorylation of eIF-2 correlates with inhibition of initiation reactions and consequently the inhibition of protein synthesis. Conditions like starvation, heat shock and viral infection, which inhibit the initiation of protein synthesis, induce the phosphorylation of eIF-2 in various cells. Stimuli such as insulin and phorbol esters modulate the phosphorylation of eIF-3, eIF-4B and eIF-4F by activating various protein kinases.

At the level of protein elongation, the phosphorylation of elongation factors EF-1α and EF-2 appears to be involved in regulating their activities[29]. Since it has been reported that the activity and amounts of active EF-1α and EF-2 decrease significantly during ageing, it will be interesting to see whether this decline is accompanied by a parallel change in the extent of phosphorylation of these enzymes. There is indirect evidence that alterations in the phosphorylation and dephosphorylation of EF-2 due to changes in the activities of EF-2-specific protein kinase III[55], and PP2A phosphatase[56] may affect the rates of protein synthesis during ageing in rat livers.

Phosphorylation also occurs in other proteins that participate in the translational process. For example, the regulatory role of phosphorylation of aa-tRNA synthetase in protein synthesis has been suggested[74]. However, to what extent the decline in the activity and the accumulation of heat-labile aa-tRNA synthetases reported in studies performed on various organs of ageing mice and rats is related to their phosphorylation is not known. Furthermore, since the phosphorylation of the S6 ribosomal protein correlates with the activation of protein synthesis, failure to phosphorylate S6 protein in senescent human fibroblasts in response to serum[75] can be one of the reasons for the decline in the rate of protein synthesis observed during ageing.

Pathways of intracellular signal transduction depend on sequential phosphorylation and dephosphorylation of a wide variety of proteins. Studies performed on ageing cells have not shown any deficiency in the amount, activity or ability of PKC to elicit signalling pathway[76]. Furthermore, there is no age-related decline in the autophosphorylation activity of various growth factor receptors [77, 78]. Similarly, most of the PKC-mediated pathways of intracellular signal transduction in response to various mitogens including phorbol esters appear to remain unaltered in senescent fibroblasts [77, 79]. However, a decline in both serine/threonine- and tyrosine-specific protein kinase signals after activation has been observed in the case of T lymphocytes in ageing mice [80].

Thus, phosphorylation of a wide variety of proteins has significant influence in biological processes and it will be extremely useful to undertake detailed studies on this posttranslational modification of various proteins in relation to the process of ageing.

4.2. ADP-Ribosylation

The structure and function of many proteins such as nuclear proteins topoisomerase I, DNA ligase II, endonuclease, histones H1, H2B and H4, DNA polymerases α and β, and cytoplasmic proteins adenyl cyclase and elongation factor EF-2 is modulated by ADP-ribosylation[37, 81, 82]. Indirect evidence suggests that poly-ADP-ribosylation of proteins may decrease during ageing because the activity of poly(ADP)ribose polymerase (PARP) decreases in ageing human fibroblasts both as a function of donor age and during serial passaging in vitro[83]. Similarly, the direct relationship observed between maximum lifespan of a species and the activity of PARP in mononuclear leukocytes of 13 mammalian species indicates its important role in ageing and longevity[84, 85].

One cytoplasmic protein that can be specifically ribosylated by diphtheria toxin and exotoxin A is the protein elongation factor EF-2. ADP-ribosylation of the diphthamide (modified histidine 715) residue of EF-2 results in the complete abolition of its catalytic activity[37]. The amount of EF-2 that can be ADP-ribosylated in the presence of diphtheria toxin in cell-free extracts decreases significantly during ageing of human fibroblasts in culture[51]. Another protein which appears to be mono-ADP-ribosylated is the antiproliferative protein prohibitin located primarily in the mitochondria[86]. However, further studies are required to establish the role of ADP-ribosylation on the activity of various proteins during ageing.

4.3. Methylation

Methylation of nitrogens of arginine, lysine and histidine, and carboxyls of glutamate and aspartate residues is a widely observed post-translational modification that is involved in many cellular functions. Proteins whose activities are increased by methylation include alcohol dehydrogenase, histones, ribosomal proteins, cytochrome C, elongation factor EF-1α, myosin, myelin and rhodopsin. Of these, decreased methylation of histones in livers and brains of ageing rats has been reported. On the other hand, there is no difference in the extent of methylation of newly synthesised histones during cellular ageing of human fibroblasts in culture. Studies on the levels of methylated histidine, arginine and lysine of myosin isolated from the leg muscles of ageing rats, mice and hamsters showed unchanged levels of histidine, decreased levels of arginine and trimethyllysine, and increased levels of monomethyllysine (for details, see ref. [87]).

During the ageing of erythrocytes, there is an increase in the number of methyl groups per molecule of band 2.1 (ankyrin) and band 3 protein, which correlates with increased membrane rigidity of erythrocytes during ageing[88]. There is a several fold increase in the number of methyl acceptor proteins in the eye lenses from aged humans and persons suffering from cataract[89]. The number of carboxylmethylatable sites of cerebral membrane-bound proteins also increases in rat brain during ageing[90]. At present, age-related changes in the methylation of other proteins such as ribosomal proteins, calmodulin, cytochrome C and myosin have not been studied. It is clear that protein methylation is involved in diverse functions including protein synthesis and turnover, and that it should be studied thoroughly in relation to the process of ageing.

4.4. Oxidation

One of the main reasons for the inactivation of enzymes during ageing can be their oxidative modification by oxygen free radicals and by mixed-function oxidation (MFO) systems or metal catalyzed oxidation (MCO) systems. Since some amino acid residues, particularly proline, arginine and lysine, are oxidized to carbonyl derivatives, the amount of carbonyl content of proteins has been used as an estimate of protein oxidation during ageing[91].

Table 5. Some proteins which are oxidatively inactived or denatured during ageing

Collagen
Ceruloplasmin
Lens crystallin
Liver malic enzyme
Lactate dehydrogenase
Superoxide dismutase
Glutamine synthetase
Ornithine decarboxylase
Fructose-1,6-diphosphatase
Glucose-6-phosphate dehydrogenase
6-Phosphogluconate dehydrogenase
Glyceraldehyde-3-phosphate dehydrogenase

Increased levels of oxidatively modified proteins (Table 5) have been reported in old human erythrocytes of higher density, and in cultured human fibroblasts from normal old donors and from individuals suffering from progeria and Werner's syndrome[91]. An age-related increase in the carbonyl content has also been reported for houseflies[92], mouse organs[93], and *Drosophila*[94]. It has also been reported that the concentration of the oxidation products of human lens proteins and skin collagen increases along with the accumulation of oxidative forms of α-crystallin in patients with age-related cataract[95].

Structural alterations introduced into proteins by oxidation can lead to the aggregation, fragmentation, denaturation, distortion of secondary and tertiary structure, increasing thereby the proteolytic susceptibility of oxidized proteins. Thus, the accumulation of abnormal proteins during ageing may be due to an impairment of the protein degradation processes and/or defective protection from oxidative damage.

4.5. Glycation

Glycation is one of the most prevalent covalent modifications in which the free amino groups of proteins react with glucose forming a ketoamine called Amadori product. This is followed by a sequence of further reactions and rearrangements producing the so-called AGE or advanced glycosylation end products[96]. Most commonly, it is the long-lived structural proteins such as lens crystallins, collagen and basement membrane proteins which are more susceptible to glycation. The glycated proteins are then more prone to form crosslinks with other proteins, leading to structural and functional alterations.

An increase in the levels of glycated proteins during ageing has been observed in a wide variety of systems. There is an increase in the level of glycated lysine residues of rat sciatic nerve, aorta and skin collagen during ageing[97]. Similarly, there is an increase in the glycation of human collagen and osteocalcin during ageing[98]. It has been observed that pentosidine (cross-linked glycated lysine and arginine) and carboxylmethyllysine increase with age in humans[99, 100]. Since the mechanistic formation of these products involves both glycation and oxidation, the term "glycoxidation product" has been suggested[101]. An age-related increase in collagen pentosidine, has been reported in eight mammalian species, and the rate of increase was inversely related with the maximum lifespan of the species[101]. Pyrroline, another AGE protein, has been shown to increase in diabetics[100]. By using AGE-specific antibodies, an AGE-modified form of human hemoglobin has been identified whose levels increase during ageing and in patients with diabetes-induced hypergly-cemia[102]. More studies are required in order to understand differences in the rates of for-mation and removal of glycated proteins in different species with different lifespans and rates of ageing.

4.6 Deamidation, Racemization and Isomerization

Age-related changes in the catalytic activity, heat stability, affinity for substrate and other physical characteristics, such as the conformation of proteins may also be due to the charge change introduced by conversion of a neutral amide group to an acidic group by deamidation. Spontaneous deamidation of asparaginyl and glutaminyl residues of several proteins has been related with the observed accumulation of their inactive and heat labile isoforms during ageing[103]. For example, the sequential deamidation of two asparagine residues of triphosphate isomerase is responsible for the differences of the isoenzymes present in ageing cells and tissues, such as bovine eye lens, and human skin fibroblasts from old donors and patients with progeria and Werner's syndrome[104]. Deamidation of glucose-6-phosphate isomerase produces the variant of the enzyme that accumulates in ageing bovine lenses[104]. There is a constant isopartate formation of tubulin in vivo and its levels may accumulate during ageing[105].

The interconversion of optical isoforms of amino acids, called racemization, has been reported to increase during ageing. The concentration of D-aspartate in protein hy-drolysates from human teeth, erythrocytes and eye lens increases with age [106]. Racemiza-tion of tyrosine has been reported to occur in the ageing brunescent human cataract lenses[107]. The spontaneous prolyl *cis-trans* isomerization in proteins that may cause some of the so-called spontaneous conformational changes has been implicated in the age-re-lated decline in the activity of certain enzymes.

4.7. Proteolytic Processing

Many newly synthesised proteins undergo post-translational proteolytic processing by which certain conformational restraint on the inactive precursor is released and a bio-logically active protein is generated. Several inactive precursors of enzymes called zymo-gens, precursors of growth factors, peptide and protein hormones such as insulin, precursors of extracellular matrix and many other secretory proteins including various proteases such as collagenase undergo proteolytic processing.

There are no systematic studies performed on age-related changes in post-transla-tional proteolytic processing of any proteins. However, there is some evidence that altera-tions in proteolytic processing may be one of the reasons for the appearance or disappearance of certain proteins during ageing. For example, the appearance of the "se-nescent cell antigen" on the surface of a wide variety of aging cells is considered to be de-rived from the proteolysis of band 3 protein[108]. The exposure of senescent cell-specific epitopes on fibronectin[109], may also be due to altered proteolytic processing. Progressive proteolysis of a 90 kDa protein, Tp-90 terminin, into Tp-60 and Tp-30 terminin in senes-cent cells and in cells committed to apoptosis has been reported[110]. Proteolytic cleavage of the β-amyloid precursor protein is well known to play an important role in the pathogene-sis of Alzheimer's disease[111]. Increased proteolysis of a conformationally more labile sin-gle-chain form of the lysosomal protease cathepsin B has been suggested as a reason for the age-related decline in its activity during ageing of human fibroblasts[112]. Similarly, alterations in the activity of collagenase during ageing of human fibroblasts has been suggested to be due to structural and catalytic changes[113, 114].

4.8. Other Modifications

In addition to the types of post-translational modifications mentioned above, there are some other modifications that determine the structure and function of various proteins and may have a role to play during ageing. For example, the incorporation of ethanolamine into protein elongation factor EF-1α may be involved in determining its stability and interaction with intracellular membranes[37]. Similarly, the protein initiation factor eIF-5A contains an unusual amino acid, hypusine, which is synthesised post-translationally as a result of a series of enzymatically catalysed alterations of a lysine residue[115]. Since the absence of hypusine in eIF-5A blocks the initiation of protein synthesis. it will be interesting to investigate changes in this modification during ageing when total protein synthesis slows down.

Protein tyrosine sulfation is another post-translational modification that may have significance in protein alteration during ageing because it is involved in determining the biological activity of neuropeptides and the intracellular transportation of a secretory protein[116]. Similarly, prenylation, the covalent attachment of isoprenoid lipids on cysteine-rich proteins, is involved in the regulation of the activity of Ras family of GTP-binding proteins designated Rab, and the nuclear lamins A and B[117]. These studies have indicated a critical role for prenylation in the regulation of oncogenesis, nuclear structure, signal transduction and cell cycle progression, functions very much related with the causative aspects of ageing. Recent evidence shows an age-dependent decrease in the activity of prenyltransferases in the rat liver, which may account for the changes in the synthesis and turnover of mevalonate pathways lipids including cholestrol, ubiquinone and dolichol[118].

Detyrosination of microtubules affecting the cytoskeletal organization and many other cellular functions, may also be important during ageing. Furthermore, the roles of chaperones in protein folding and conformational organization are yet to be studied in relation to the ageing process. According to the crosslinking theory of ageing, the progressive linking together of large vital molecules, especially the proteins, results in the loss of cellular functions[119]. There is some evidence that both the pentose-mediated protein crosslinking[99] and transglutaminase-mediated crosslinking[120] of proteins is involved in ageing. For example, there is a high correlation between pentosidine protein crosslinks and pigmentation in senescent and cataract affected human lens[121]. Similarly, an increase in transglutaminase activity during cellular apoptosis, differentiation and ageing of human keratinocytes[122, 123] indicates an important role of this modification in the process of ageing.

5. Protein Turnover

Efficient macromolecular turnover is integral to the normal functioning and survival of a biological system. Protein degradation during ageing is a relatively little researched topic as compared with RNA and protein synthesis. Although there are several hundred-fold variations in the rates of degradation of individual proteins, it is generally believed that protein turnover slows down during ageing (for a recent comprehensive review, see [17]). The physiological consequences of decreased protein turnover include the accumulation of altered and abnormal proteins in the cell, an altered pattern of post-translational modifications due to increased dwell time, and a disruption of the organisation of the cytoskeleton and extracellular matrix.

Age-related decline in protein turnover is generally due to a decrease in the proteolytic activity of various lysosomal and cytoplasmic proteases. It is only recently that the molecular details of various pathways of protein degradation, such as the proteasome-mediated, ubiquitin-mediated and the lysosome-mediated pathways, have begun to be studied in relation to ageing. For example, ubiquitin marking of proteins for degradation and ubiquitin-mediated proteolysis did not decline in ageing human fibroblasts and no change in the levels of ubiquitin mRNA and ubiquitin pools was detected[124]. In contrast, the amount of ubiquitin-protein conjugates increased in all regions of the brain from senescent mice, but not much in other tissues[125].

Other reasons for age-related changes in the activities of various proteases leading to a decrease in the rate of protein turnover include slower transcription, reduced rates of synthesis and altered pattern of post-synthetic modifications, as discussed above. Furthermore, there is evidence that certain inhibitors of various proteases, such as tissue inhibitor of metalloproteinases (TIMP), plasminogen activator inhibitor and trypsin inhibitor had increased levels of expression and activities during ageing of human fibroblasts[126, 127]. This will also lead to a decrease in the activities of proteases leading to decreased protein degradation during ageing.

Finally, the synthesis, modifications and turnover of proteins are interdependent processes that practically set a limit on the efficiency of genetic information transfer from coded molecules to functional molecules. Ageing as the failure of maintenance can only be fully understood by studying various enzymic components of the networks of repair and maintenance, which progressively become inefficient.

REFERENCES

1. Rattan, S.I.S. (1995) Ageing – a biological perspective. Molec. Aspects Med., 16, 439–508.
2. Antequera, F. and Bird, A. (1993) Number of CpG islands and genes in human and mouse. Proc. Natl. Acad. Sci. USA, 90, 11995–11999.
3. Macieira-Coelho, A. (1995) Reorganization of the genome during aging of proliferative cell compartments. In: Molecular Basis of Aging. (Macieira-Coelho, A., Ed.), pp. 21–69, CRC Press, Boca Raton.
4. Dimri, G. and Campisi, J. (1994) Altered profile of transcription factor-binding activities in senescent human fibroblasts. Exp. Cell Res., 212, 132–140.
5. Adrian, G.S., Seto, E., Fischbach, K.S., Rivera, E.V., Adrian, E.K., Herbert, D.C., Walter, C.A., Weaker, F.J. and Bowman, B.H. (1996) YY1 and Sp1 transcription factors bind the human transferrin gene in an age-related manner. J. Gerontol. Biol. Sci., 51A, B66-B75.
6. Müller, W.E.G., Agutter, P.S. and Schröder, H.C. (1995) Transport of mRNA into the cytoplasm. In: Molecular Basis of Aging. (Macieira-Coelho, A., Ed.), pp. 353–388, CRC Press, Boca Raton.
7. Medvedev, Z.A. (1986) Age-related changes of transcription and RNA processing. In: Drugs and Aging. (Platt, D., Ed.), pp. 1–19, Springer-Verlag, Berlin.
8. Lapointe, J. and Giegé, R. (1991) Transfer RNAs and aminoacyl-tRNA synthetases. In: Translation in Eukaryotes. (Trachsel, H., Ed.), pp. 35–69, CRC Press, Boca Raton.
9. Strehler, B.L., Hirsch, G., Gusseck, D., Johnson, R. and Bick, M. (1971) Codon restriction theory of ageing and development. J. Theor. Biol., 33, 429–474.
10. Vinayak, M. (1987) A comparison of tRNA populations of rat liver and skeletal muscle during aging. Biochem. Int., 15, 279–285.
11. Schröder, H.C., Ugarkovic, D., Müller, W.E.G., Mizushima, H., Nemoto, F. and Kuchino, Y. (1992) Increased expression of UAG suppressor tRNA in aged mice: consequences for retroviral gene expression. Eur. J. Gerontol., 1, 452–457.
12. Takahashi, R., Mori, N. and Goto, S. (1985) Alteration of aminoacyl tRNA synthetases with age: accumulation of heat-labile moleculaes in rat liver, kidney and brain. Mech. Ageing Dev., 33, 67–75.
13. Takahashi, R. and Goto, S. (1988) Fidelity of aminoacylation by rat-liver tyrosyl-tRNA synthetase. Effect of age. Eur. J. Biochem., 178, 381–386.

14. Gabius, H.J., Goldbach, S., Graupner, G., Rehm, S. and Cramer, F. (1982) Organ pattern of age-related changes in the aminoacylation synthetase activities of the mouse. Mech. Ageing Dev., 20, 305–313.

15. Gabius, H.-J., Graupner, G. and Cramer, F. (1983) Activity patterns of aminoacyl-tRNA synthetases, tRNA methylases, arginyltransferase and tubulin:tyrosine ligase during development and ageing of Caenorhabditis elegans. Eur. J. Biochem., 131, 231–234.

16. Ward, W. and Richardson, A. (1991) Effect of age on liver protein synthesis and degradation. Hepatol., 14, 935–948.

17. Van Remmen, H., Ward, W.F., Sabia, R.V. and Richardson, A. (1995) Gene expression and protein degradation. In: Handbook of Physiology: Aging (Masoro, E., Ed.), pp. 171–234. Oxford University Press.

18. Kozak, M. (1994) Determinants of translational fidelity and efficiency in vertebrate mRNAs. Biochim., 76, 815–821.

19. Hershey, J.W.B. (1991) Translational control in mammalian cells. Annu. Rev. Biochem., 60, 717–755.

20. Hershey, J.W.B. (1994) Expression of initation factor genes in mammalian cells. Biochim., 76, 847–852.

21. Pain, V.M. (1996) Initiation of protein synthesis in eukaryotic cells. Eur. J. Biochem., 236, 747–771.

22. Rattan, S.I.S. (1995) Translation and post-translational modifications during aging. In: Molecular Basis of Aging. (Macieira-Coelho, A., Ed.), pp. 389–420. CRC Press, Boca Raton, Florida.

23. Rattan, S.I.S. (1996) Synthesis, modifications and turnover of proteins during aging. Exp. Gerontol., 31, 33–47.

24. Calés, C., Fando, J.L., Azura, C. and Salinas, M. (1986) Developmental studies of the first step of the initiation of brain protein synthesis, role for initiation factor 2. Mech. Ageing Dev., 33, 147–156.

25. Casteñeda, M., Vargas, R. and Galván, S.C. (1986) Stagewise decline in the activity of brain protein synthesis factors and relationship between this decline and longevity in two rodent species. Mech. Ageing Dev., 36, 197–210.

26. Kimball, S.R., Vary, T.C. and Jefferson, L.S. (1992) Age-dependent decrease in the amount of eukaryotic initiation factor 2 in various rat tissues. Biochem. J., 286, 263–268.

27. Wool, I.G. (1991) Eukaryotic ribosomes: structure, function, biogenesis, and evolution. In: Translation in eukaryotes. (Trachsel, H., Ed.), pp. 3–33. CRC Press, Boca Raton.

28. Merrick, W.C. (1992) Mechanism and regulation of eukaryotic protein synthesis. Microbiol. Rev., 56, 291–315.

29. Ryazanov, A.G., Rudkin, B.B. and Spirin, A.S. (1991) Regulation of protein synthesis at the elongation stage. New insights into the control of gene expression in eukaryotes. FEBS Lett., 285, 170–175.

30. Spirin, A.S. and Ryazanov, A.G. (1991) Regulation of elongation rate. In: Translation in Eukaryotes. (Trachsel, H., Ed.), pp. 325–350. CRC Press, Boca Raton.

31. Rattan, S.I.S. (1992) Regulation of protein synthesis during ageing. Eur. J. Gerontol., 1, 128–136.

32. Richardson, A. and Semsei, I. (1987) Effect of aging on translation and transcription. Rev. Biol. Res. Aging., 3, 467–483.

33. Khasigov, P.Z. and Nikolaev, A.Y. (1987) Age-related changes in the rates of polypeptide chain elongation. Biochem. Int., 15, 1171–1178.

34. Merry, B.J. and Holehan, A.M. (1991) Effect of age and restricted feeding on polypeptide chain assembly kinetics in liver protein synthesis in vivo. Mech. Ageing Dev., 58, 139–150.

35. Webster, G.C. (1985) Protein synthesis in aging organisms. In: Molecular Biology of Aging: Gene Stability and Gene Expression. (Sohal, R.S., Birnbaum, L.S. and Cutler, R.G., Eds.), pp. 263–289, Raven Press, New York.

36. Webster, G.C. (1986) Effect of aging on the components of the protein synthesis system. In: Insect Aging. (Collatz, K.G. and Sohal, R.S., Eds.), pp. 207–216, Springer-Verlag, Berlin.

37. Riis, B., Rattan, S.I.S., Clark, B.F.C. and Merrick, W.C. (1990) Eukaryotic protein elongation factors. Trends Biochem. Sci., 15, 420–424.

38. Ann, D.K., Lin, H.H., Lee, S., Tu, Z.J. and Wang, E. (1992) Characterization of the statin-like S1 and rat elongation factor 1a as two distinctly expressed messages in rat. J. Biol. Chem., 267, 699–702.

39. Knudsen, S.M., Frydenberg, J., Clark, B.F.C. and Leffers, H. (1993) Tissue-dependent variation in the expression of elongation factor-1alpha isoforms: isolation and characterisation of a cDNA encoding a novel variant of human elongation factor 1alpha. Eur. J. Biochem., 215, 549–554.

40. Slobin, L.I. and Jordan, P. (1984) Translational repression of mRNA or eukaryotic elongation factors in Friend erythroleukemia cells. Eur. J. Biochem., 145, 143–150.

41. Slobin, L.I. and Rao, M.N. (1993) Translational repression of EF-1α mRNA in vitro. Eur. J. Biochem., 213, 919–926.

42. Gonen, H., Smith, C.E., Siegel, N.R., Kahana, C., Merrick, W.C., Chakraburtty, K., Schwartz, A.L. and Ciechanover, A. (1994) Protein synthesis elongation factor EF-1α is essential for ubiquitin-dependent deg-

radation of certain Nα-acetylated proteins and may be substituted for by the bacterial elongation factor Tu. Proc. Natl. Acad. Sci. USA, 91, 7648–7652.

43. Shepherd, J.C.W., Walldorf, U., Hug, P. and Gehring, W.J. (1989) Fruitflies with additional expression of the elongation factor EF-1α live longer. Proc. Natl. Acad. Sci. USA, 86, 7520–7521.

44. Shikama, N., Ackermann, R. and Brack, C. (1994) Protein synthesis elongation factor EF-1α expression and longevity in Drosophila melanogaster. Proc. Natl. Acad. Sci. USA, 91, 4199–4203.

45. Shikama, N. and Brack, C. (1996) Changes in the expression of genes involved in protein synthesis during Drosophila aging. Gerontol., 42, 123–136.

46. Stearns, S.C. and Kaiser, M. (1993) The effect of enhanced expression of elongation factor EF-1α on life span of Drosophila melanogaster. Genetica, 91, 158–166.

47. Stearns, S.C. and Kaiser, M. (1993) The effects of enhanced expression of elongation factor EF-1α on lifespan in Drosophila melanogaster. IV. A summary of three experiments. Genetica, 91, 167–182.

48. Dudas, S.P. and Arking, R. (1994) The expression of the EF1α genes of Drosophila is not associated with the extended longevity phenotype in a selected long-lived strain. Exp. Gerontol., 29, 645–657.

49. Silar, P. and Picard, M. (1994) Increased longevity of EF-1α high-fidelity mutants in Podospora anserina. J. Mol. Biol., 235, 231–236.

50. Takahashi, R., Mori, N. and Goto, S. (1985) Accumulation of heat-labile elongation factor 2 in the liver of mice and rats. Exp. Gerontol., 20, 325–331.

51. Riis, B., Rattan, S.I.S., Derventzi, A. and Clark, B.F.C. (1990) Reduced levels of ADP-ribosylatable elongation factor-2 in aged and SV40-transformed human cells. FEBS Lett., 266, 45–47.

52. Rattan, S.I.S., Ward, W.F., Glenting, M., Svendsen, L., Riis, B. and Clark, B.F.C. (1991) Dietary calorie restriction does not affect the levels of protein elongation factors in rat livers during ageing. Mech. Ageing Dev., 58, 85–91.

53. Redpath, N.T. and Proud, C.G. (1993) Purification and phosphorylation of elongation factor-2 kinase from rabbit reticulocytes. Eur. J. Biochem., 212, 511–520.

54. Celis, J.E., Madsen, P. and Ryazanov, A.G. (1990) Increased phosphorylation of elongation factor 2 during mitosis in transformed human amnion cells correlates with a decreased rate of protein synthesis. Proc. Natl. Acad. Sci. USA, 87, 4231–4235.

55. Riis, B., Rattan, S.I.S., Palmquist, K., Nilsson, A., Nygård, O. and Clark, B.F.C. (1993) Elongation factor 2-specific calcium and calmodulin dependent protein kinase III activity in rat livers varies with age and calorie restriction. Biochem. Biophys. Res. Commun., 192, 1210–1216.

56. Riis, B., Rattan, S.I.S., Palmquist, K., Clark, B.F.C. and Nygård, O. (1995) Dephosphorylation of the phosphorylated elongation factor-2 in the livers of calorie-restricted and freely-fed rats during ageing. Biochem. Mol. Biol. Int., 35, 855–859.

57. Nyborg, J., Nissen, P., Kjeldgaard, M., Thirup, S., Polekhina, G., Clark, B.F.C. and Reshetnikova, L. (1996) Structure of the ternary complex of EF-Tu: macromolecular mimicry in translation. Trends Biochem. Sci., 21, 81–82.

58. Kirkwood, T.B.L., Holliday, R. and Rosenberger, R.F. (1984) Stability of the cellular translation process. Int. Rev. Cytol., 92, 93–132.

59. Holliday, R. (1995) Understanding Ageing. Cambridge University Press, Cambridge.

60. Mori, N., Hiruta, K., Funatsu, Y. and Goto, S. (1983) Codon recognition fidelity of ribosomes at the first and second positions does not decrease during aging. Mech. Ageing Dev., 22, 1–10.

61. Luce, M.C. and Bunn, C.L. (1987) Altered sensitivity of protein synthesis to paromomycin in extracts from aging human diploid fibroblasts. Exp. Gerontol., 22, 165–177.

62. Luce, M.C. and Bunn, C.L. (1989) Decreased accuracy of protein synthesis in extracts from aging human diploid fibroblasts. Exp. Gerontol., 24, 113–125.

63. Holliday, R. and Rattan, S.I.S. (1984) Evidence that paromomycin induces premature ageing in human fibroblasts. Monogr. Devl. Biol., 17, 221–233.

64. Buchanan, J.H., Stevens, A. and Sidhu, J. (1987) Aminoglycoside antibiotic treatment of human fibroblasts: intracellular accumulation, molecular changes and the loss of ribosomal accuracy. Eur. J. Cell Biol., 43, 141–147.

65. Harley, C.B., Pollard, J.W., Chamberlain, J.W., Stanners, C.P. and Goldstein, S. (1980) Protein synthetic errors do not increase during the aging of cultured human fibroblasts. Proc. Natl. Acad. Sci. USA, 77, 1885–1889.

66. Johnson, T.E. and McCaffrey, G. (1985) Programmed aging or error catastrophe: an examination by two-dimensional polyacrylamide gel electrophoresis. Mech. Ageing Dev., 30, 285–287.

67. Kowald, A. and Kirkwood, T.B.L. (1993) Accuracy of tRNA charging and codon:anticodon recognition; relative importance for cellular stability. J. theor. Biol., 160, 493–508.

68. Bernier, L.W., Wang, E. (1996) A prospective view on phosphatases and replicative senescence. Exp. Gerontol., 31, 13–19.

69. Cristofalo, V.J. and Pignolo, R.J. (1996) Molecular markers of senescence in fibroblast-like cultures. Exp. Gerontol., 31, 111–123.

70. Campisi, J. (1996) Replicative senescence: an old lives' tale? Cell, 84, 497–500.

71. Stein, G.H., Besson, M. and Gordon, L. (1990) Failure to phosphorylate the retinoblastoma gene product in senescent human fibroblasts. Science, 249, 666–669.

72. Dulic´, V., Drullinger, L.F., Lees, E., Reed, S. and Stein, G.H. (1993) Altered regulation of G1 cyclins in senescent human diploid fibroblasts: accumulation of inactive cyclin E—Cdk2 and cyclin D1—Cdk2 complexes. Proc. Natl. Acad. Sci. USA, 90, 11034–11038.

73. Richter, K.H., Afshari, C.A., Annab, L.A., Burkhart, B.A., Owen, R.D., Boyd, J. and Barrett, J.C. (1991) Down-regulation of cdc2 in senescent human and hamster cells. Canc. Res., 51, 6010–6013.

74. Meinnel, T., Mechulam, Y. and Blanquet, S. (1995) Aminoacyl-tRNA synthetases: occurrence, structure, and function. In: tRNA: Structure, Biosynthesis, and Function. (Söll, D. and RajBhandary, U.L., Eds.), pp. 251–292, ASM Press, Washington D.C.

75. Kihara, F., Ninomyia-Tsuji, J., Ishibashi, S. and Ide, T. (1986) Failure in S6 protein phosphorylation by serum stimulation of senescent human diploid fibroblasts, TIG-1. Mech. Ageing Dev., 20, 305–313.

76. Blumenthal, E.J., Miller, A.C.K., Stein, G.H. and Malkinson, A.M. (1993) Serine/threonine protein kinases and calcium-dependent protease in senescent IMR-90 fibroblasts. Mech. Ageing Dev., 72, 13–24.

77. De Tata, V., Ptasznik, A. and Cristofalo, V.J. (1993) Effect of tumor promoter phorbol 12-myristate 13-acetate (PMA) on proliferation of young and senescent WI-38 human diploid fibroblasts. Exp. Cell Res., 205, 261–269.

78. Farber, A., Chang, C., Sell, C., Ptasznik, A., Cristofalo, V.J., Hubbard, K., Ozer, H.L., Adamo, M., Roberts, C.T., LeRoith, D., Dumenil, G. and Baserga, R. (1993) Failure of senescent human fibroblasts to express the insulin-like growth factor-1 gene. J. Biol. Chem., 268, 17883–17888.

79. Derventzi, A., Rattan, S.I.S. and Clark, B.F.C. (1993) Phorbol ester PMA stimulates protein synthesis and increases the levels of active elongation factors EF-1α and EF-2 in ageing human fibroblasts. Mech. Ageing Dev., 69, 193–205.

80. Miller, R.A. (1994) Aging and immune function: cellular and biochemical analyses. Exp. Gerontol., 29, 21–35.

81. Simbulan, C.M.G., Suzuki, M., Izuta, S., Sakurai, T., Savoysky, E., Kojima, K., Miyahara, K., Shizutsa, Y. and Yoshida, S. (1993) Poly(ADP-ribose) polymerase stimulates DNA polymerase alpha by physical association. J. Biol. Chem., 268, 93–99.

82. Shall, S. (1988) ADP-ribosylation of proteins: a ubiquitos cellular control mechanism. Adv. Exp. Med. Biol., 231, 597–611.

83. Dell'Orco, R.T. and Anderson, L.E. (1991) Decline of poly(ADP-ribosyl)ation during in vitro senescence in human diploid fibroblasts. J. Cell. Physiol., 146, 216–221.

84. Grube, K. and Bürkle, A. (1992) Poly(ADP-ribose) polymerase activity in mononuclear leukocytes of 13 mammalian species correlates with species-specific life span. Proc. Natl. Acad. Sci. USA, 89, 11759–11763.

85. Bürkle, A., Grube, K. and Küpper, J.-H. (1992) Poly(ADP-ribosyl)ation: its role in inducible DNA amplification, and its correlation with the longevity of mammalian species. Exp. Clin. Immunogenet., 9, 230–240.

86. Dell'Orco, R.T., McClung, J.K., Jupe, E.R. and Liu, X.-T. (1996) Prohibitin and the senescent phenotype. Exp. Gerontol., 31, 245–252.

87. Rattan, S.I.S., Derventzi, A. and Clark, B.F.C. (1992) Protein synthesis, post-translational modifications and aging. Ann. N.Y. Acad. Sci., 663, 48–62.

88. Mays-Hoopes, L.L. (1985) Macromolecular methylation during aging. Rev. Biol. Res. Aging, 2, 361–393.

89. McFadden, P.N. and Clarke, S. (1986) Protein carboxyl methyltransferase and methyl acceptor proteins in aging and cataractus tissue of the human eye lens. Mech. Ageing Dev., 34, 91–105.

90. Sellinger, O.Z., Kramer, C.M., Conger, A. and Duboff, G.S. (1988) The carboxylmethylation of cerebral membrane-bound proteins increases with age. Mech. Ageing Dev., 43, 161–173.

91. Stadtman, E.R. (1992) Protein oxidation and aging. Science, 257, 1220–1224.

92. Sohal, R.S., Agarwal, S., Dubey, A. and Orr, W.C. (1993) Protein oxidative damage is associated with life expectancy of houseflies. Proc. Natl. Acad. Sci. USA, 90, 7255–7259.

93. Sohal, R.S., Ku, H.-H. and Agarwal, S. (1993) Biochemical correlates of longevity in two closely related rodent species. Biochem. Biophys. Res. Commun., 196, 7–11.

94. Orr, W.C. and Sohal, R.S. (1994) Extension of life-span by overexpression of superoxide dismutase and catalase in Drosophila melanogaster. Science, 263, 1128–1130.

95. Spector, A. (1995) Oxidative stress-induced cataract: mechanism of action. FASEB J., 9, 1173–1182.

96. Lis, H. and Sharon, N. (1993) Protein glycosylation: structural and functional aspects. Eur. J. Biochem., 218, 1–27.

97. Oimomi, M., Maeda, Y., Hata, F., Kitamura, Y., Matsumoto, S., Hatanaka, H. and Baba, S. (1988) A study of the age-related acceleration of glycation of tissue proteins in rats. J. Gerontol., 43, B98–101.

98. Miksík, I. and Deyl, Z. (1991) Changes in the amount of ε-hexosyllysine, UV absorbance, and fluorescence of collagen with age in different animal species. J. Gerontol., 46, B111–116.

99. Sell, D.R. and Monnier, V.M. (1989) Structure elucidation of a senescence cross-link from human extracellular matrix. Implication of pentoses in the aging process. J. Biol. Chem., 264, 21597–21602.

100. Lee, A.T. and Cerami, A. (1992) Role of glycation in aging. Ann. N.Y. Acad. Sci., 663, 63–70.

101. Sell, D.R., Lane, M.A., Johnson, W.A., Masoro, E.J., Mock, O.B., Reiser, K.M., Fogarty, J.F., Ingram, D.K., Roth, G.S. and Monnier, V.M. (1996) Longevity and the genetic determination of collagen glycoxidation kinetics in mammalian senescence. Proc. natl. Acad. Sci. USA, 93, 485–490.

102. Makita, Z., Vlassara, H., Rayfield, E., Cartwright, K., Friedman, E., Rodby, R., Cerami, A. and Bucala, R. (1992) Hemoglobin-AGE: a circulating marker of advanced glycosylation. Science, 258, 651–653.

103. Gafni, A. (1990) Age-related effects in enzyme metabolism and catalysis. Rev. Biol. Res. Aging, 4, 315–336.

104. Gracy, R.W., Yüksel, K.Ü., Chapman, M.L., Cini, J.K., Jahani, M., Lu, H.S., Oray, B. and Talent, J.M. (1985) Impaired protein degradation may account for the accumulation of "abnormal" proteins in aging cells. In: Modifications of Proteins during Aging. (Adelman, R.C. and Dekker, E.E., Eds.), pp. 1–18. Alan R. Liss, New York.

105. Najbauer, J., Orpiszewski, J. and Aswad, D.W. (1996) Molecular aging of tubulin: accumulation of isopartyl sites in vitro and in vivo. Biochem., 35, 5183–5190.

106. Brunauer, L.S. and Clarke, S. (1986) Age-dependent accumulation of protein residues which can be hydrolyzed to D-aspartic acid in human erythrocytes. J. Biol. Chem., 261, 12538–12543.

107. Luthra, M., Ranganathan, D., Ranganathan, S. and Balasubramanian, D. (1994) Racemization of tyrosine in the insoluble protein fraction of brunescent aging human lenses. J. Biol. Chem., 269, 22678–22682.

108. Kay, M.M.B. (1990) Molecular aging of membrane molecules and cellular removal. In: Biomedical Advances in Aging. (Goldstein, A.L., Ed.), pp. 147–161, Plenum Press, New York.

109. Porter, M.B., Pereira-Smith, O.M. and Smith, J.R. (1992) Common senescent cell-specific antibody epitopes on fibronectin in species and cells of varied origin. J. Cell. Physiol., 150, 545–551.

110. Hébert, L., Pandey, S. and Wang, E. (1994) Commitment to cell death is signaled by the appearance of a terminin protein of 30 kDa. Exp. Cell Res., 210, 10–18.

111. Selkoe, D.J. (1992) Aging brain, aging mind. Sci. Amer., 267, 135–142.

112. DiPaolo, B.R., Pignolo, R.J. and Cristofalo, V.J. (1992) Overexpression of the two-chain form of cathepsin B in senescent WI-38 cells. Exp. Cell Res., 201, 500–505.

113. Baur, E.A., Kronberger, A., Stricklin, G.P., Smith, L.T. and Holbrook, K.A. (1985) Age-related changes in collagenase expression in cultured embryonic and fetal human skin fibroblasts. Exp. Cell Res., 161, 484–494.

114. Sottile, J., Mann, D.M., Diemer, V. and Millis, A.J.T. (1989) Regulation of collagenase and collagenase mRNA production in early- and late-passage human diploid fibroblasts. J. Cell. Physiol., 138, 281–290.

115. Park, M.H., Wolff, E.C. and Folk, J.E. (1993) Hypusine: its post-translational formation in eukaryotic initiation factor 5A and its potential role in cellular regulation. BioFactors, 4, 95–104.

116. Huttner, W.B. (1987) Protein tyrosine sulfation. Trends Biochem. Sci., 12, 361–363.

117. Marshall, C.J. (1993) Protein prenylation: a mediator of protein-protein interactions. Science, 259, 1865–1866.

118. Thelin, A., Runquist, M., Ericsson, J., Swiezewska, E. and Dallner, G. (1994) Age-dependent changes in rat liver prenyltransferases. Mech. Ageing Dev., 76, 165–176.

119. Bjorksten, J. and Tenhu, H. (1990) The cross linking theory of aging - added evidence. Exp. Gerontol., 25, 91–95.

120. Birckbichler, P.J., Anderson, L.E. and Dell'Orco, R.T. (1988) Transglutaminase, donor age, and in vitro cellular senescence. Adv. Exp. Med. Biol., 231, 109–117.

121. Nagaraj, R.H., Sell, D.R., Prabhakaram, M., Ortwerth, B.J. and Monnier, V.M. (1991) High correlation between pentosidine protein crosslinks and pigmentation implicates ascorbate oxidation in human lens senescence and cataractogenesis. Proc. Natl. Acad. Sci. USA, 88, 10257–10261.

122. Saunders, N.A., Smith, R.J. and Jetten, A.M. (1993) Regulation of proliferation-specific and differentiation-specific genes during senescence of human epidermal keratinocyte and mammary epithelial cells. Biochem. Biophys. Res. Commun., 197, 46–54.

123. Norsgaard, H., Clark, B.F.C. and Rattan, S.I.S. (1996) Distinction between differentiation and senescence and the absence of increased apoptosis in human keratinocytes undergoing cellular aging in vitro. Exp. Gerontol., 31, 563–570.

124. Pan, J.-X., Short, S.R., Goff, S.A. and Dice, J.F. (1993) Ubiquitin pools, ubiquitin mRNA levels, and ubiquitin-mediated proteolysis in aging human fibroblasts. Exp. Gerontol., 28, 39–49.

125. Ohtsuka, H., Takahashi, R. and Goto, S. (1995) Age-related accumulation of high-molecular-weight ubiquitin protein conjugates in mouse brains. J. Gerontol., 50A, B277-B281.

126. Edwards, D.R., Leco, K.J., Beaudry, P.P., Atadja, P.W., Veillette, C. and Riabowol, K.T. (1996) Differential effects of transforming groeth factor-β1 on the expression of matrix metalloproteinases and tissue inhibitors of metalloproteinases in young and old human fibroblasts. Exp. Gerontol., 31, 207–223.

127. West, M.D., Shay, J.W., Wright, W.E. and Linskens, M.H.K. (1996) Altered expression of plasminogen activator and plaminogen activator inhibitor during cellular senescence. Exp. Gerontol., 31, 175–193.

6

AGEING OF CELLS IN VITRO

Sydney Shall

Cell and Molecular Biology Laboratory
School of Biological Sciences
University of Sussex, Brighton
Sussex BN1 9Q, England

1. INTRODUCTION

The ageing of cells in vitro, that means in cell culture, has two rather different meanings. In one meaning, used in this article, we are referring only to the replicative limit which normal cells show when they are grown in vitro. The second meaning, which is of course quite different, refers to the loss of cellular functions which may occur with in vitro growth. It is quite reasonable to suppose that most of the loss of function that cells experience in culture occurs only after they have stopped dividing, and survive only as post-mitotic cells. We shall discuss in some detail the origin of these post-mitotic cells, but I suggest that during their post-mitotic life such cells in culture almost certainly experience gradual loss of function due to accumulation of unrepaired damage to the macromolecules, especially to their DNA. We also know that human neuronal cells are post-mitotic cells and they do survive, in the body, as post-mitotic cells for up to 100 years and many of them manage to retain most of their cellular functions for this length of time. So, we know that there is a phenomenon of post-mitotic survival of normal cells in culture, which would be an appropriate model for the study of ageing of post-mitotic cells in the body. In this discussion, however, I shall confine myself to the first and widely used meaning of cellular ageing, namely the gradual loss of replicative ability which is universally observed when animal cells are grown in vitro[1].

What then is the biological definition of the ageing process. I would say that it is the time-dependent deterioration in physiological response to stress of all sorts; this is associated of course with declining physiological function. It is the inability to respond adequately to stresses of all sorts that induce illness in older people. It seems to be the case that older people succumb with an illness when stresses are imposed on them, to which younger people can respond adequately. Why are older people less able to respond to stress? The answer ultimately seems to be that over time unrepaired damage to their tissues weakens their ability to respond. In the case of non-dividing tissues like neurones, one can easily exemplify this statement. When a neurone suffers damage, it is most likely that repair mechanisms will repair all or most of the damage. But in time, the small pro-

portion of unrepaired damage will accumulate and will eventually interfere with the ability of that cell to function and to respond to stress. This is the fundamental statement that
human ageing is at heart very much a problem of cell ageing. I do not ignore the changes
that occur in the integrative physiology of a whole organism, but it seems clear that the information that has accumulated so far, places great importance on the changes that occur
in individual cells with time. As I have done above, I would make a very sharp distinction
between the changes that occur in non-dividing post-mitotic cells like the neurones and in
dividing tissues. There are some people who would like to confine the word ageing to the
degenerative changes which occur in non-dividing, post-mitotic cells; I have no profound
problem with this, provided it is well recognised that most organs of the body consist of
tissues in which there is cell turnover. I shall argue that this cell turnover is crucial to how
we should approach the study of human ageing.

2. RELATIONSHIP BETWEEN CELLULAR SENESCENCE AND
   AGEING OF THE WHOLE ORGANISM

What then is the cellular basis of human ageing? In most tissues, there is tissue
turnover, and as a consequence the residence time of most cells is quite short. Cells are
produced usually from so-called stem cells, and acquire the distinctive properties of their
final existence. These cells in general have a rather short life time, and are then removed
sometimes after "marking," and sometimes after achieving a state of replicative sterility,
that is they become post-mitotic. The physical removal processes include mechanical removal in the gut and the skin, ingestion by other cells [macrophages] in many cases, and
very often by the process of apoptosis which also ends up with the now dead cell being
ingested by a macrophage. The "age" of the individual cells in a tissue depend on when
they were born from their respective stem cells. Most cells only have a short lifespan
from their birth to their death. Thus the accumulated damage in these cells cannot be of
profound importance; possibly much of the damage seen in these cells at any given time
was created in the recent past. Equally, however, these cells are destined to die and will
NOT pass on this damage to any descendants. Thus it is not the damage in these cells
that is crucial, but rather the ability of any given tissue to make new healthy cells to replace the older, damaged extant cells. As long as a tissue is able to give birth to new un-
damaged cells, then that tissue will be healthy. We must therefore direct our attention
much more to the ability of tissues to regenerate new cells during the lifespan of the individual. Thus cell proliferation seems to be required to maintain populations of cells
which are functionally efficient.

As we have said the cellular basis of ageing is the accumulation of defects in cells.
When there are too many defects in a cell, it will be unable to respond to stresses such as
damage, infection or tumours. It is NOT necessary for the damage to be so extensive that
the cells do not function at all. It is sufficient that a degree of damage should accumulate
which prevents the cells from responding adequately to stress. In such circumstances, the
individual will succumb to a disease; and in general people die of disease not of total
functional inability. In conclusion, the ability of the stem cells of a tissue to regenerate the
tissue by cell division seems to be an important component of keeping the tissue healthy.
A common experimental model for this regenerative ability is the growth of "normal"
cells in vitro, that is in cell culture in the laboratory. This model commonly uses mesodermal cells of many types, but there is also some reports with epithelial and endothelial

cells. These in vitro models have provided useful insights into what some animal cells can do and has been very widely studied.

There are several observations which suggest a correlation between the kinetics of cellular mortalisation in culture and ageing of the organism from which the cells are derived. Firstly, the proliferative potential of human fibroblast cells in culture correlates with the age of the donor from which the cells are derived[2-4]. There is a decrement in proliferative potential of about 0.2 population doublings per year of donor age. Röhme[5] reported a good correlation between species lifespan and proliferative capacity in culture for a variety of animals, from mouse with a maximum assumed lifespan of 2 years, to humans with a maximum assumed lifespan of about 110 years. The decrement of proliferative potential in these studies was also 0.2 population doublings per year of maximum lifespan.

The best evidence, in my view, for a correlation between cell culture dynamics and lifespan of the organism derives from the observations on fibroblasts from human individuals suffering from diseases which mimic accelerated ageing, such as progeria or Werner's syndrome and related conditions. Several investigators find that cells from patients with these "premature ageing" syndromes show very restricted proliferative potentials in culture[3,6-9]. I will discuss this connection between human genetic "premature ageing" diseases and ageing in culture in more detail later.

3. CELL AGEING IN THE LABORATORY

When embryonic or adult cells are placed in culture, a variety of cells grow out of the explant. Usually, only a very small fraction of the cells in the original sample successfully grow. This indicates that there has been very strong selection for the successful cells; the character of these cells is therefore of considerable interest. Regrettably, in general we cannot definitively identify these "founder" cells. It is commonly assumed that they may be "stem" cells, which upon transfer to the dish lose their character of being a stem cell, and become a cell descendant of the stem cell. The assumption is made that the stem cell requires a specific environment of either other cells, or of membrane attachment or of cytokines to maintain the character as stem cells. Consequently, upon transfer to the dish, these cells inevitably lose their stem cell properties. It is usually further assumed that non-stem cells in the explant, may also grow, but probably such cells only grow for a very limited number of generations, and that consequently the main features of the consequent culture are derived from the outgrowth of the small number of stem cells that succeed in gaining a foothold in the in vitro culture.

Usually the cultures grow sufficiently well so that after some time the cultures can be expanded into new dishes. Subsequently, the culture can be passaged successively many times. Initially, the culture is heterogeneous, with different cell types. The cells may be purified by cloning by limiting dilution, that is to disperse the cells into separate containers [usually little wells in a multi-well dish], so that on average there is less than 1 cell per well; this means that there is a high probability that any culture which grows will heave grown from a single cell. It is possible to do this with both fibroblasts and with lymphocyte T-cells derived from the blood circulation.

Given optimum cell culture conditions these cell culture populations can be repeatedly passaged for a fairly well-defined number of passages. To be exact, the lifespan of these cultures is precisely defined by the number of generations, that is the number of cell divisions, that these cells can achieve. It is common practice to describe this number of generations in the form of the number of population doublings achieved. It should be noted however, that since there is a declining growth fraction in these cultures, the number of population doublings is very much less than the number of cell generations achieved. The latter may be as many as three times the number of population doublings, because for much of the culture duration, the growth fraction is significantly less than one.

However, invariably these cultures will eventually cease to expand because the cells stop proliferating. The explanation for this is clearly that the cells irreversibly exit the cell cycle, and enter a post-mitotic phase where they metabolise and synthesise both RNA and protein, but do not synthesise DNA. Growth control in this experimental model is shown by the rapid expansion of cell numbers soon after explantation in the culture dish, followed by a gradual decline in the rate of expansion of the culture until it finally ceases expansion altogether. This definite limit to the in vitro culture of animal mesodermal cells was first reported by Swim and Parker[10] and by Hayflick and Moorehead[1]. It is customary to refer to the limited, finite lifespan of cells in culture as the "Hayflick limit," and to describe cultures which have ceased proliferation in this way as being "Hayfluck."

A variety of terms have been used to describe these observations and so far a consensus has not yet emerged. Consequently, it is necessary to explain the descriptive terms that are used in this area. The original explant and any cells that grow out from it in the original dish are referred to as the primary culture. After some variable time, the primary culture will fill the dish; the cells are then removed from this dish and a portion of the cell suspension is distributed to new dishes. The cells of this second culture which was obtained by "passing" the cells from the primary culture, are referred to as "Passage 1" cells. (But note that some investigators would call this Passage 0). The cells in Passage 1 will eventually become confluent and will then be transferred again, and become a "Passage 2" culture. [It is good practice to avoid confluence in this work; the cultures should be passaged in late logarithmic phase to a defined cell density to ensure a minimum lag time. In this way, the culture is always in a growing phase and the most accurate history of the cells can be kept]. This cyclical passaging can be continued until the culture ceases to expand further, even though there is adequate medium, space, and time allowed for their growth. It is often a subjective judgement in practice as to when a culture ceases to expand. Definitions vary from "no doubling in two weeks," to "fewer cells harvested than were seeded."

The cultures from the primary culture, through Passage 1 to Passage n, are referred to as "mortal" cultures. It is very important to remember that it is the population of cells that is mortal, not the individual cells. The cultures are described as mortal, because they will inevitably cease expanding. Clearly, all individual cells are always mortal. We call these cultures "mortal" consonant with the wide usage that permanent cell lines are always called "immortal" cultures. Other rather clumsy descriptions have also been applied to these cultures; they are sometimes referred to as non-established cultures or as early passage cultures. Both terms, although not really accurate, are synonymous with "mortal" cultures.

It is possible to derive from mortal, rodent cultures permanent, immortal lines of cells which arise from very rare, mutation events. These permanent cultures are generally termed "immortal" cultures, because they seem to show an unlimited reproductive potential. Some workers are not entirely happy with the use of the word "immortal," possibly because of its social overtones. However, it must be repeated that the description refers to the population of cells, to the whole culture and not to the individual cells. These immortal cultures are also sometimes referred to as permanent cultures, or as established cultures; neither of these terms is any more satisfactory than immortal, which is in common usage and is quite unambiguous.

Immortal cultures are those which apparently have an infinite reproductive capacity. But such immortal cultures could also have other properties: for example, they might otherwise be quite normal cells and I would then describe them as immortal, but untransformed cells. Such cultures can indeed be obtained from rodents, but not in general from humans. The cells might, however, have various combinations of altered phenotypes such as altered growth factor requirements, altered density dependent inhibition, ability to grow in soft agar, morphological changes, secretion of proteases, or the ability to form tumours in animals. All such cultures are referred to as immortal, transformed cultures. I would emphasise the point that immortalisation and malignant transformation are quite independent properties of cells.

4. DYNAMICS OF A MORTAL CELL POPULATION

We can now turn to a detailed analysis of the phenomenon of the limited lifespan of "normal" cells *in vitro* culture. The earliest observations by Hayflick and Moorehead[1] identified the central most important fact, namely that eventually the cell cultures ceased expanding. Subsequently, Smith and Hayflick[11] and Smith and Whitney[12] made the crucial observation that the cultures are reproductively heterogeneous. Some cells in these cultures have only a very limited reproductive potential; some cells show an intermediate ability to divide many times; and only a very few cells show the full, maximum reproductive capacity of the whole culture. It is not the case that all the cells achieve the maximum number of population doublings; on the contrary, only very few cells are able to divide the maximum number of times. Smith and Whitney[12] also showed that the ability to grow of any given cell was not a fixed characteristic of that cell lineage; because when they re-cloned a clone with a long lifespan they showed that they obtained clones covering a wide distribution; both long-lived and short-lived clones. The idea of specific clones succeeding each other in the culture is partially true only; the work of Smith and Whitney [12] clearly shows that there is a stochastic element in this mechanism. Thus, one cannot accurately predict the behaviour of individual cells in a clone, although one can accurately describe the average behaviour of the culture. These workers observed that at every age of the culture the cells displayed a wide distribution in replicative ability. This distribution was not stable, however. As the culture ages, the wide distribution moves to the left, that is to smaller and smaller clone sizes. That is, while the cells always show a wide range of reproductive ability, the modal clone size consistently becomes smaller as the culture ages. Thus, young cultures have a wide dispersion of reproductive abilities with a large fraction of cells having a very high replicative ability. Middle-aged cultures also have a wide dispersion of reproductive abilities but there are fewer cells which are able to form very large colonies, and many more cells which can only form small colonies. In older cultures the dispersion is again shifted towards smaller colonies. From these analyses Smith and Whitney[12] concluded that the reproductive ability of new-born cells was to some degree stochastically determined and that the dispersion of reproductive potentials was gradually moving towards smaller colonies and away from larger colonies.

These original observations have been confirmed and extended by Pontén et al[13]. It was originally reported by Macieiro-Coelho et al[14] that in older cultures the inter-mitotic times are very slightly longer; but not long enough to contribute to the slower culture growth which is observed. Consequently, it can be inferred that the growth fraction is declining. A direct *estimate* of the growth fraction by Pontén et al[13], indeed revealed a steady decline during the lifespan. It was observed that in the young cultures the growth fraction is very high, but at each succeeding passage the fraction of cells which cannot divide gradually increases[13]. This was a direct demonstration that these mortal cultures are self-limiting, because the fraction of dividing cells in the culture gradually declines until less than half of all new-born cells are capable of division. When this happens the culture must inevitably decline, because at each division now, the NUMBER of cells able to divide will decline to eventually reach zero.

This conclusion was confirmed by the examination of the distribution of colony sizes reached by individual cells. In this work the cloning was performed with a new technique that allowed cloning in the presence of a large number of living feeder cells of the same type as the cells being studied[13]. These observations confirmed the earlier observations by Smith and Whitney[12] made by the conventional cloning techniques. The distribution of colony sizes was observed to shift smoothly and continuously from the bigger to smaller sizes. Thus the overall picture is of a stochastic limitation of proliferative behaviour in individual cells, with the fraction of cells which are unable to divide gradually increasing from the beginning of the culture.

5. PROPERTIES OF SENESCENT CELLS

5.1 Morphology

Young cells are characteristically small and at confluence are packed closely together; as the culture ages and as senescent cells appear, one observes larger cells which at confluence occupy a much larger surface area. A detailed morphological description of the changes in human fibroblasts during cellular ageing has been published. It has been suggested that there are lineage relationships between the several morphologic types of cells observed, but this is still a very controversial conclusion.

The most characteristic feature of senescent cells—that is, cells that are no longer able to divide even in the presence of rich medium and adequate space—is that these cells do not any longer replicate their DNA. However, these cells are functionally competent to enter DNA synthesis because they can be induced to do so by fusion with cycling cells or by viral infection.

Although senescent cells do not replicate their DNA, and do not divide, they are metabolically very active. Senescent cells synthesise protein and RNA very actively for months after they can no longer enter the cell cycle.

6. KINETIC DESCRIPTION OF THE MORTALISATION PROCESS

It has been possible to make a simple kinetic description of the dynamic behaviour of both human and rodent mortal cultures[15,16]. This description is very simple and is only a first approximation to the dynamics of these cultures. This description emphasises the regularity and relative simplicity of the system and suggests a simple genetic basis for the phenomenon of cellular mortalisation.

This model defines mortalisation as the appearance of reproductive sterility in new-born cells. The model postulates that reproductive sterility appears randomly in the population of new-born cells and that the probability of mortalisation (P_m) in the culture increases with time. A relatively simple relationship is assumed between biological time and the probability of mortalisation, namely that

$$P_m = t / (\gamma + t) \tag{1}$$

where t is time in generations and γ is that number of generations at which the probability of mortalisation reaches 0.5. It is noted that the form of the equation is hyperbolic: as $t \rightarrow 0$, P_m also approaches 0, and as t becomes very large compared to γ, then P_m tends to approach 1.0. However, when $P_m = 0.5$, each dividing cell gives rise to two cells, only one of which will itself divide. Thus the population only increases by one at each division; it does not double at each division as in exponential growth. Indeed, when $P_m = 0.5$ the population grows arithmetically (one unit at each generation) not exponentially (doubling at each generation). Moreover, when $P_m > 0.5$, at each division there will be fewer dividing cells than at the previous division, and hence the population inherently and inevitably will cease to expand, and the number of dividing cells will move to zero. Finally, if the value of P_m is less than 0.5 and constant, we would still have an exponentially growing population, but the population doubling rate would be a constant fraction of the cell generation rate. It is useful to emphasise the simplicity of this description and also the paucity of the assumptions. This condition is needed to make the kinetic description experimentally testable. The available experimental data show a very reasonable fit to this simple model. Further amplification and refinement of the model may now be more useful.

A related but much more complex kinetic model has been published by James Smith and coworkers[17,18]. This model also allows for the detailed prediction of colony sizes that are observed during the lifespan, and shows a good concordance with experimental data.

Very recently Kitano and Imai [Kitano, H. and Imai, S-I (1997) Two distinct intrinsic mechanisms regulate the stochastic and catastrophic phases in cellular senescence. J. Theor. Biol., in Press] have evolved a sophisticated kinetic model which starts from assumptions of the behaviour of relevant genes. This model proposes that there are two parallel, independent but time-aligned processes which regulate mortalisation. One they call stochastic growth arrest and the second they call catastrophic senescence. Both mechanisms are assumed to be transcriptional mechanisms coupled to DNA replication. The first mechanism is supposed to be generated by local growth inhibitory cell-to-cell interactions, and the second mechanism is thought to be under the control of the highly ordered chromatin structure. Computer simulation experiments, not computer mathematical modelling, of cellular growth kinetics and senescence-associated gene expression successfully generated experimental data using the combined effects of these two mechanisms. From the results of their computer simulation they propose a new theory that so far successfully describes a comprehensive set of experimental data presently available. This new theory is very interesting and will be worth close study.

7. KINETICS OF CELL AGEING IN WERNER'S SYNDROME CELLS

In human populations there occur naturally somewhat rare genetic diseases of ageing. Werner's syndrome is such a rare, autosomal recessive disease, which is characterised by short stature, a hoarse high-pitched voice, juvenile bilateral cataracts, premature greying of the hair, skin changes, diabetes, atherosclerosis, neoplasms, and other diseases characteristic of the elderly[19,20]. It is noteworthy that the behavioural changes and Alzheimer-like neuropathologies, which are usually associated with old age, are only seen very infrequently. This contrast with natural ageing emphasises the multiple causes of the several features of ageing in human populations. It suggests that this genetic condition may be a very useful model of some aspects of human ageing; Werner's syndrome simplifies our analysis of the process of human ageing and allows us to examine some aspects of the process without the neurodegenerative processes which presumably have separate causes. Werner's syndrome is frequently studied as a model of some aspects of human ageing.

Generally, the condition only becomes apparent between 20 and 30 years of age. Progressive atrophy of the skin and subcutaneous tissue of the distal limbs precedes characteristic foot ulceration and eventual gangrene. It is very striking that persons with Werner's syndrome usually die of a variety of diseases characteristic of the elderly in normal populations. These people have an average age at death of about 47 years, which is about one half that of the normal population. Perhaps one can provisionally postulate that these people "age" twice as fast as normal people do. We have observed earlier in this chapter that there is some correlation between the lifespan of cells in vitro and the lifespan of the organism from which the cells were taken. This correlation is very dramatically exemplified with cells taken from patients with Werner's syndrome. Fibroblast cultures from Werner's syndrome patients show a dramatic shortening of the lifespan in culture[3,6-9]. Normal human fibroblasts achieve about 60 population doublings in culture, while Werner's syndrome cells usually only achieve a maximum of about 20 population doublings. Presumably this large decrease in cellular lifespan is due to the same gene mutation responsible for the other features of the Werner's syndrome, since this condition is known to be due to a mutation at a single locus, giving rise to a rare recessive, autosomal disease. Because a single gene mutation seems to be responsible for both the human premature ageing phenotype and for the large decrease in the lifespan of the fibroblast cells in culture, it is important to try to understand why the cells from Werner's syndrome patients are unable to divide as many times as normal cells.

There are two alternative kinetic explanations for the decreased lifespan of Werner's syndrome cells. First, it may be that the fraction of cycling cells in the original explant may be approximately the same as in an explant derived from a normal subject, but the rate of loss of reproductive ability may be much higher in Werner cells. Alternatively, the Werner cells when freshly explanted may contain a much smaller fraction of cycling cells, which lose their reproductive potential at a normal rate. Of course, a combination of the two mechanisms may be found.

By measuring the fraction of S-phase cells, identified by incorporation of the thymidine analogue bromodeoxyuridine, Faragher et al[21] have shown that cells in these mutant cultures usually exit, apparent irreversibly, from the cell cycle at a faster rate than do normal cells, although they mostly start off with a good replicative ability. The inference of these results is that the Werner's syndrome gene seems to control the number of times that human cells are able to divide before becoming senescent.

A useful outcome of the study of the Werner's syndrome cells has been to develop a reliable way to estimate the in vitro "age" of a human cell culture. Two parameters are needed to identify a meaningful biological age of cells in culture; the number of cell generations the culture has survived and the fraction of cells in the culture that are still capable of proliferation. Passage number and accumulated population doublings achieved, provide only a very rough estimate of the ageing of cell cultures. The most commonly used method to determine the age of a population of cells, has been to estimate by autoradiography the fraction of cells incorporating radioactive nucleosides after a long exposure[22]. The estimates obtained could then be compared with previous estimates made at each passage level, and from this an estimate of the population age could be deduced. This method has been cogently criticised by Macieiro-Coelho[23], because this method is acutely dependent upon unknown factors like the duration of the cell cycle phases and the variances of their distributions. This method is very sensitive to the period used for the labelling compared to the cell cycle phases; too short a label fails to label all the cycling cells, while too long a label overestimates cycling cells because some of the labelled cells divide and are counted twice. An alternative method is to estimate the fraction of the lifespan completed, but this requires a knowledge of the complete lifespan kinetic record.

An alternative is now available. Immunocytochemical markers of cellular proliferation have been widely used to analyse the proportion of proliferating cells in cell cultures and in histological sections. A simple, reliable method for the determination of cell age in culture has been devised by Kill et al[24]. The method is based upon the estimation of the fraction of cells expressing antigens associated with cell proliferation such as Ki-67, PCNA which is part of a DNA polymerase, and topoisomerase II, in an indirect immunofluorescence [or enzyme] assay performed at each passage. The results with this method show a smooth decline in the fraction of cells expressing each of the cell cycle-dependent antigens, while the fraction of cells expressing antigens which are independent of the cell cycle are always constant at 100% of all the cells. By measuring the fraction of cells expressing the cell cycle-associated antigens at three time points one can reconstruct the line indicating the ageing of the cell population.

Thus, estimation of the fraction of cells expressing cell cycle-associated antigens throughout the lifespan of a culture provides a useful means for the estimation of the age of the culture and of the rate of ageing of the culture. When such an assay was performed with Werner's syndrome cells[24], a smooth decline in the fraction of cells displaying the proliferation-dependent antigens was observed. This is similar to normal cells. However, in the case of the Werner's syndrome cells the rate of decline in the fraction of proliferating cells was accelerated by between 5- and 6-fold compared with normal fibroblasts. Once again, it is observed that Werner's cells seem to age a lot faster than do normal cells.

Although a growing body of evidence confirms that the ageing of normal mammalian cultures is characterised by a smooth and gradual decline in the fraction of proliferating cells, this has not been adequately incorporated into our understanding of cell ageing in culture. It is not uncommon to find reference to the age of cultures in terms of phases I, II and III as defined by Hayflick[2]. I suggest that this classification is inappropriate, since it implies that there are intrinsic differences between young, middle-aged and old cultures. It is now clear that the difference between these cultures at different in vitro ages is the relative proportions of proliferating and non-proliferating cells, post-mitotic cells.

8. MOLECULAR GENETICS OF CELLULAR SENESCENCE

The first evidence for a molecular genetic explanation of cell ageing came from experiments in which different types of cells were fused together, and then it was asked what ageing properties did the fusion products show. When clonal populations of various human, immortal cell types were fused with normal, mortal human cells the hybrid cells were mortal, that is they had a limited, finite proliferative potential[25-28]. The mortal hybrids that are obtained in these experiments have a definite tendency to spontaneously immortalise at a low frequency; this indicates that the immortalisation process involves at least two steps. The fusion process enforces mortality on the culture, but probably spontaneous mutation at the usual frequency is sufficient to re-immortalise at a low frequency. These results demonstrate very clearly that cellular mortalisation is dominant, and that immortalisation is achieved by inactivating gene functions. Fusion of normal human lympho-cyte T cells or endothelial cells with immortal cells also yielded hybrids with a limited lifespan[29].

In a very elegant series of experiments Pereira-Smith and coworkers[30] determined the apparent number of complementation groups which determine cellular mortalisation. The design of the experiments were as follows; She assumed that different immortal clones may be mutated in different genes. If this were true then by fusing two cells each with a mutation in a different gene, she might observe complementation, because each cell would contribute a wild-type copy of one gene and she had previously shown that the wild-type alleles of these genes were dominant. Complementation would be shown when two immortal cells produced a hybrid that showed mortalisation. The results clearly showed that all the cells could be placed into at least 4 different complementation groups, indicating that there might be at least 4 different genes required to achieve cellular mortal-isation. She found that each cell type only assigned to one complementation group, and that the complementation groups did not correspond to cell type, tumour type, embryonic layer of origin. The identification of only four groups indicates that the fairly limited number of genes or gene pathways are altered when immortalisation occurs. Therefore, only a relatively small number of genes are involved in the control of the mortalisation process.

9. MOLECULAR MECHANISM OF CELLULAR AGEING

A full examination of the possible molecular mechanisms of cellular mortalisation requires a complete review of the molecular biology of this process. Although our knowl-edge is very incomplete, two rather different types of hypotheses may be plausibly enter-tained. It is now quite clear that there are genes required for the process of cellular mortalisation. However, we do not have enough experimental information to decide what constitutes the molecular basis of cellular mortalisation. The two most favoured views are either that the process of loss of reproductive ability is due to accumulating, unrepaired damage in the cells or alternatively that there is a genetic programme that is biologically driven and that induces cells to differentiate into post-mitotic, metabolically active cells.

REFERENCES

1. Hayflick, L. Moorehead, PS. (1961) The serial cultivation of human diploid cell strains. Exp. Cell Res. 25, 585–621.
2. Hayflick, L. (1965)The limited in vitro lifetime of human diploid cell strains. Exp cell. Res. 37, 614–636.
3. Martin GM, Sprague CA, Epstein CJ. (1970) Replicative lifespan of cultured human cells: effects of donor age, tissue, and genotype. Lab. Invest. 23, 86–92.
4. Schneider EL, Mitsui Y. (1976) The relationship between in vitro cellular aging and in vivo human age. Proc. Natl. Acad. Sci. U.S.A. 73, 3584–3588.
5. Röhme, D. (1981) Evidence for a relationship between longevity of mammalian species and lifespans of normal fibroblasts in vitro and erythrocytes in vivo. Proc. Natl. Acad. Sci. U.S.A. 78, 5009–5013.
6. Goldstein S. (1979) Studies on age-related diseases in cultured skin fibroblasts. J. Invest. Dermatol. 73, 19–23.
7. De Busk. The Hutchinson-Gifford progeria syndrome. J. Pediatrics. 80, 697–724.
8. Salk D, Bryant E, Au k, Hoehn H, Martin GM. (1981) Systematic growth studies, cocultivation and cell hybridisation studies of Werner syndrome cultured skin fibroblasts. Human Genetics. 58, 310–316.

9. Salk D. (1982) Werneris syndrome: A review of recent research with an analysis of connective tissue meta-bolism, growth control of culture cells, and chromosomal aberrations. Human Genetics. 62, 1–15.

10. Swim HE, Parker RF. (1957) Culture characteristics of human fibroblasts propagated serially. Amer. J. Hygiene. 66, 235–243.

11. Smith JR, Hayflick L. (1974) Variation in the lifespan of clones derived from human diploid cell strains. J. Cell Biol. 62, 48–53.

12. Smith JR, Whitney RG. Intraclonal variation in proliferative potential of human diploid fibroblasts: stochastic mechanism for cellular aging. Science. 207, 82–84.

13. Pontén J, Stein WD, Shall S. (1983) A quantitative analysis of the ageing of human glial cells in culture. J. Cell. Physiol. 117, 342–352.

14. Macieira-Coelho A, Pontén J, Phillipson L. (1966) The division cycle and RNA synthesis in diploid human cells at different passage levels in vitro. Exp. Cell res. 42, 673–684.

15. Stein WD, Ellis D, Shall S. (1978) A mortalisation theory for the control of cell proliferation and for the origin of immortal cell lines. in Cell Reproduction. (eds Dirksen ER, Prescott DM, Fox CF.) pp. 147–154, Academic Press, New York.

16. Shall S, Stein WD. (1979) A mortalisation theory for the control of cell proliferation and for the origin of immortal cells lines. J. Theoret. Biol. 76, 219–231.

17. Smith JR, Lumpkin CK. (1980) Stochastic model of cellular senescence. I. Theory. Mech. Age. Dev. 13, 387–397.

18. Jones RB, Smith JR. (1982) A Stochastic model of cellular senescence. II. Concordance with experimental data. J. Theor. Biol. 96, 443–460.

19. Epstein CJ, Martin GM, Schultz AL, Motulsky AG. (1966) Werneris syndrome. A review of its symptomatology, natural history, pathological features, genetics and relationships to the natural ageing process. Medicine. 45, 177–221.

20. Herd RM, Faragher RGA, Shall S, Hunter JAA. (1993) Werner's syndrome: a review of the clinical and pathological features and pathogenesis. Eur. J. Dermatol. 3, 425–432.

21. Faragher RFA, Kill IR, Hunter JAA, Pope FM, Tannock C, Shall S. (1993) The gene responsible for Werner syndrome may be a cell division "counting" gene. Proc. Natl. Acad. Sci. U.S.A. 90, 12030–12034.

22. Cristofalo VJ, Sharf B. (1973) Cellular senescence and DNA synthesis. Thymidine incorporation as a measure of population age in human diploid cells. Exp. Cell. Res. 76, 419–427.

23. Macieiro-Coelho A. (1974) Are non-dividing cells present in ageing cell cultures? Nature. 248, 421–422.

24. Kill IR, Faragher RFA, Lawrence K, Shall S. (1994) The expression of proliferation-dependent antigens during the lifespan of normal and progeroid human fibroblasts in culture. J. Cell Sci. 107, 571–579.

25. Bunn CL, Tarrant GM. (1980) Limited lifespan in somatic cell hybrids and cybrids. Exp. cell Res. 37, 385–396.

26. Muggleton-Harris A, De Simone D. (1980) Replicative potentials of various fusion rproducts between WI-38 and SV40 transformed WI-38 cells and their components. Somatic Cell Genetics 6, 689–698.

27. Pereira-Smith OM, Smith JR. (1981) Expression of SV40 T antigen in finite lifespan hybrids of normal-SV40 transformed fibroblasts. Somatic Cell Genetics. 7, 411–421.

28. Pereira-Smith OM, Smith JR. (1983) Evidence for the recessive nature of cellular immorrtlaity. Science. 221, 964–966.

29. Pereira-Smith OM, Robetorye S, Ning Y, Orson FM. (1990) Hybrids from fusion of normal human T lymphocytes with immortal human cells exhibit limited lifespan. J. Cell. Physiol. 144, 546–549.

30. Pereira-Smith OM, Smith JR. (1988) Genetic analysis of indefinite division in human cells: Identification of four complementation groups. Proc. Natl. Acad. Sci. U. S. A. 85, 6042–6046.

7

STRESS AND ENERGY METABOLISM IN AGE-RELATED PROCESSES

O. Toussaint and J. Remacle

Laboratoire de biochimie cellulaire
FUNDP
Rue de Bruxelles, 61 B-5000, Namur, Belgium
Fax: 32 81 724135, E-mail : oltou, jrem@biocell.fundp.ac.be

1. INTRODUCTION

The hypothesis that any type of stress can provoke accelerated ageing of cells is difficult to understand from an analytical point of view but is easily understood from a theoretical point of view which considers the cell as a global system. Thermodynamics of far from equilibrium open systems can be used as a theoretical framework to understand how stresses considered as fluctuations of the inner or external cellular environment which can alter any highly organized biological system. The nature of stresses and of their mechanisms of action may change, but the fate of the biological system may be the same : acceleration of ageing or cell death[1]. In this article, we will first review data obtained on cultivated human fibroblasts showing evidence that cell ageing can be accelerated under the action of intense but non lethal stresses. We will then examine the complementary role of the cellular defense systems and of the energetic metabolism in modulating these effects of stresses. Finally, we will show how these conclusions obtained from *in vitro* experiments can be applied to various stressing and pathological conditions.

2. EFFECTS OF STRESSES ON FIBROBLAST AGEING AND DEATH

Human WI-38 and AG04432 embryonic lung fibroblasts, and HH-8 skin fibroblasts were used as *in vitro* ageing models to test the effects of stresses on cellular ageing. These fibroblasts have a finite life-span calculated in population doublings as originally described[2]. More recently, their progressive *in vitro* ageing has been proposed to result from a sequential and irreversible shift through 7 morphotypes, the 4 latest being post-mitotic cells. These morphotypes have been biochemically and morphologically discriminated by their specific polypeptide patterns observed in 2-dimensional electrophoresis and their characteristic morphological appearance observed or by optic microscopy after staining with Coomassie blue[3-8]. Since the passage through these 7 morphotypes is a natural and progressive process occuring along with the sub-cultures, these different steps can be considered as steady states from the general theory of thermodynamics of open system, for reasons already explained in details [1] (Figure 1), which enabled to predict that non lethal but intense stresses could acceler-

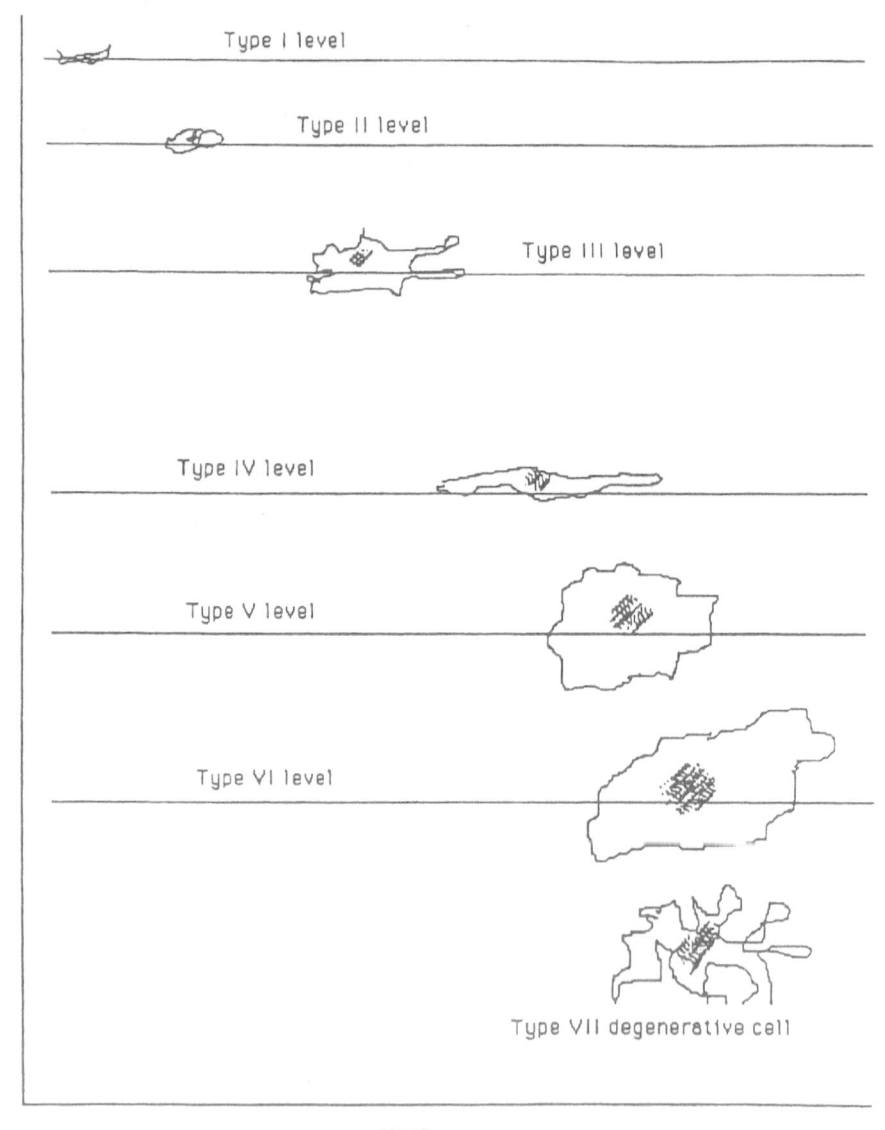

Figure 1. Schematic presentation of the fibroblast stem cell system. Cells evolve through a succession of 7 morphotypes, each characterized by a stable steady state of entropy production (metabolic level). This metabolic level decreases from morphotype I to the degenerative morphotype VII (reproduced with permission from Experimental Gerontology, 30, 1–22, 1995[6]).

ate cellular ageing. Experiments were deviced in order to test if stresses would speed up the fibroblast passage through their 7 morphotypes.

First, a qualitative study was performed by observing the evolution of the cells after a stress performed either under various ethanol concentrations or in the presence of a molecule generating oxygen-derived free radicals : tert-butylhydroperoxide (TBHP). Results are shown in Figure 2 which gives a typical evolution of the cell morphotypes after a 2h stress under 10^{-4} M TBHP diluted in phosphate buffer saline at pH 7.4 . In the first example (lane 1), cells of type I and II proliferated from day 0 (day of the stress) to day 28 and most of them remained in morphotypes I and II. In the other one (lane 2), morphotypes I and II became morphotypes III and IV after 28 days. In the first case, morphotypes I and II actively divided to make confluent clones, and in the second case, cells I and II shifted to morphotypes III, which can still divide but much more slowly, and can also produce post-mitotic morphotypes IV. When stresses were performed under ethanol, similar results were obtained as examplified in Figure 3. Cells were observed at day 1, 4 and 7 after the stress. After stresses of 2 h with 0, 2, 3 and 4% ethanol (v:v) diluted in PBS. The increase in morphotypes III and IV and the decreased cell proliferation is clearly seen as a function of ethanol concentration 7 days after the stress.

The proportions of each morphotype were then determined before and after a single stress under ethanol or TBHP for a quantitative analysis of the effect of the stress on the morphotype shift. In both cases, a modification in the pattern of morphotypes was obtained, which was strongly linked to the stress intensity: morphotypes I evolved into II and later III, which became much more abundant. Morphotypes IV also increased, with morphotypes V and VI completely disappearing, being shifted into degenerative VII morphotypes. If the stress was performed with the same cells being subcultivated and re-stressed at each successive passage, increased proportions of "old" morphotypes were obtained. The percentages of morphotypes III and IV obtained after these successive stresses were also dependent on the stress intensity[8]. Similar results were obtained after successive stresses of other kinds : U.V. light, mitomycin C, strong electromagnetic fields[5,9,10]. Honda and Matsuo performed successive oxidative stresses on cultivated fibroblasts[11,12] and found a decrease in the maximum population doublings. Cheng and Ames stressed F65 human diploid fibroblasts under 2.10^{-4} M hydrogen peroxide for 2 h, which did not affect the viability of the cells, and observed that the cells subsequently failed to respond to a stimulus of serum, platelet-derived growth factor, basic fibroblast growth factor, or epidermal growth factor for DNA synthesis. Serial subcultivation of cells stressed with hydrogen peroxide showed that they had a lower number of doublings compared to non-stressed cells[13], as observed by[8] who performed TBHP stresses. Moreover these stressed cells, like senescent cells, were less able to activate ornithine decarboxylase and thymidine kinase[13]. Von Zglinicki et al.[14] have shown that exposure of WI-38 fibroblasts to 40 % oxygen partial pressure blocks proliferation irreversibly after one to three population doublings. Cells acquire a senescent morphology, present lipofuscin accumulation, are arrested in G1 phase as senescent cells, and display an increase in the rate of telomere shortening from 90 bp per population doubling in controls to 500 bp per population doublings in hyperoxia. In every case, proliferation is blocked if a telomere cutoff length of about 4 kb is arrived at. Ammendola et al.[15] examined the changes in cell transcription patterns after oxidative stress induced by diethylmaleate. Using the differential display technique they identified several differentially expressed sequence tags, three of which are identical or highly homologous to sequences contained in the human cDNAs encoding vimentin, c-fos, and cytochrome oxidase IV, which are differentially expressed in old fibroblasts,

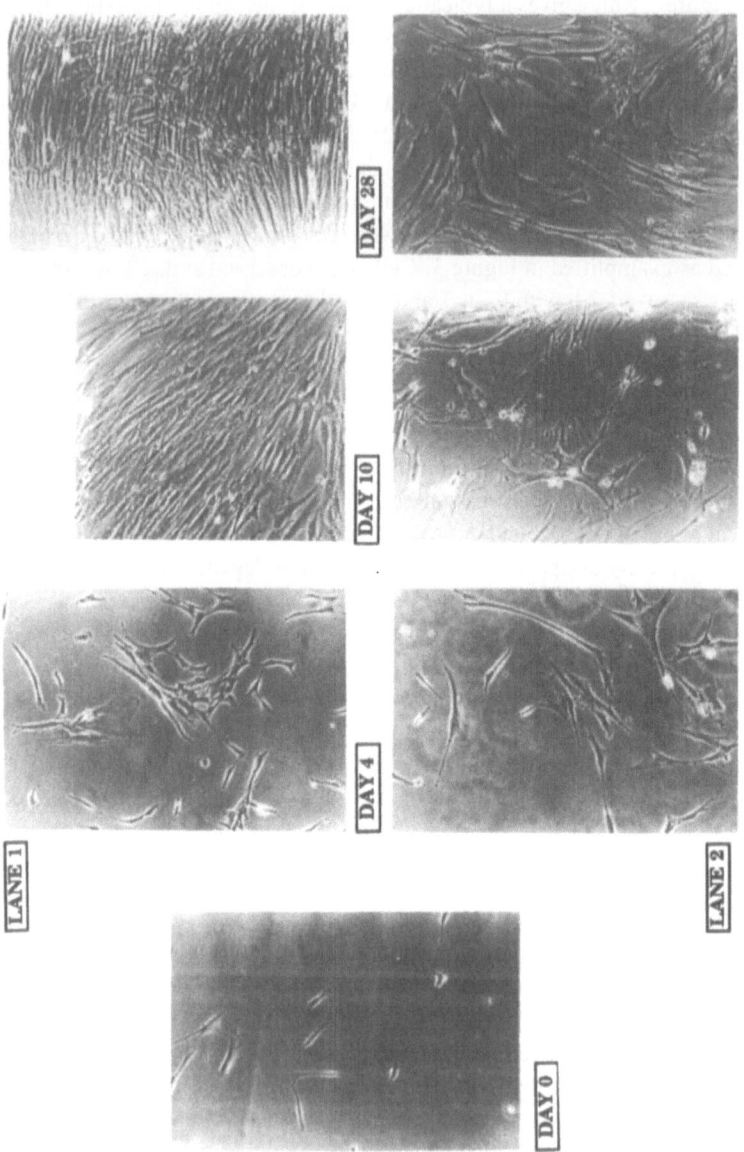

Figure 2. Evolution of the morphotypes of WI-38 fibroblasts after a 2h stress under 10^{-4} M TBHP. Before the stress, cells were cultivated classicaly in BME medium + 10 % foetal calf serum as described in[8]. After the stress performed under 10^{-4} M TBHP in phosphate buffer saline (PBS, pH 7.4, 10 mM) as described in[6-8], cells were rinsed twice in culture medium and reincubated in normal culture medium + 10 % serum. Cells were then followed and randomly photographed at different times after the stress. In the first lane 1, cells proliferated from day 0 (day of the stress) to day 28 and most of them remained in morphotypes I and II. In the other one (lane 2), morphotypes I and II became morphotypes III and IV after 28 days.

How can stresses so different in nature lead to the same global effect at the cellular level? A single deterministic differentiation process can not be responsible for such an effect, because of the multiple mechanisms involved in the cellular damages and in the various cellular responses to such different stresses. Thermodynamics of far from equilibrium open systems could be used to explain these effects of accelerated ageing. First, it is theoretically demonstrated that even very low stresses such as the ubiquitous fluctuations in temperature, pH, free radical concentrations can participate to the increase in the level of intracellular irreversible damages with time[1]. Secondly, thermodynamics allow to consider intermediary stresses, defined as non-lethal and non ubiquitous stresses, as fluctuations which can destabilize the steady state of the cell. In these conditions, the perturbations caused by the stress may be so important that cells will not be able to come back to their original steady state of free energy production and utilization, and they will have to find a lower steady state of free energy production and utilization[1].

If the shift from one morphotype to the other, and finally cell death, after a stress are the result of a destabilization process which is energy dependent, it is theoretically possible to modulate the shift from one morphotype to the other or cell death by acting on the cell energetic level[1,6]. This prediction was tested by stressing cells in conditions of partial mitochondria uncoupling, thus lowering the ATP turn-over compared to the numbers of electron given by NADH and $FADH_2$. In these conditions, stresses with TBHP or with ethanol gave a synergistic effect with the mitochondria uncoupling for cell mortality. Therefore a decrease in the capacity to effectively regenerate ATP leads to an increase in the cellular susceptibility to these stresses[16]. Since TBHP and ethanol act on different molecular targets, these results suggest that this synergism betwwen stress and energy depletion could be a general phenomenon. On the other hand, energetic substrates were found protective for cell survival in these conditions. The molecules tested were D-glucose, pyruvate, glutamate, and malate. Naftidrofuryl oxalate and bilobalide, two pharmacological stimulators of the energetic metabolism, were also found to be protective[7,16]. These energetic substrates and molecules effectively protected cell survival by increasing the energetic potential of the cell as they kept the intracellular amount of ATP at a high level during the stresses. A parallel evolution was obtained between the ATP content present just after the stress and the proportion of surviving cells found after the stress[16]. The presence of these energetic substrates and molecules not only protected cells against death but also against the shift of young morphotypes towards late morphotypes observed after non lethal stresses[6].

Under stress, an additional energy utilization is required for the defense systems, for the induction of the synthesis of new molecules like all the stress protein families, for the replacement of damaged cell components, or for the repair of cellular components like DNA, membranes, etc. For instance, in response to DNA strand breaks induced by H_2O_2, poly-(ADP-ribose) polymerase is activated, leading to depletion of cellular NADH and ATP, and loss in cell viability[17]. The needs for free energy availability under stress can be fulfilled for example by upregulating the glucose uptake[18]. It has been demonstrated that oxidant, heavy metals and heat shock stress stimuli enhance protein tyrosine phosphorylation, modulate protein tyrosine phosphatase activity, activate the src family protein tyrosine kinase, p56cl, and also enhance glucose uptake in peripheral blood mononuclear cells. It was also shown that the ubiquitous glucose transporters GLUT-1 belongs to the glucose-regulated protein family of stress-inducible protein[19]. Glucose has also been found to protect normal thymocytes against TBHP stresses by preventing ATP depletion which is followed by intracytoplasmic Ca^{++} ($[Ca^{++}]_i$) elevation[20,21]. Rat blood cells exposed to TBHP show a rapid decrease in reduced glutathione and an increase in oxidized

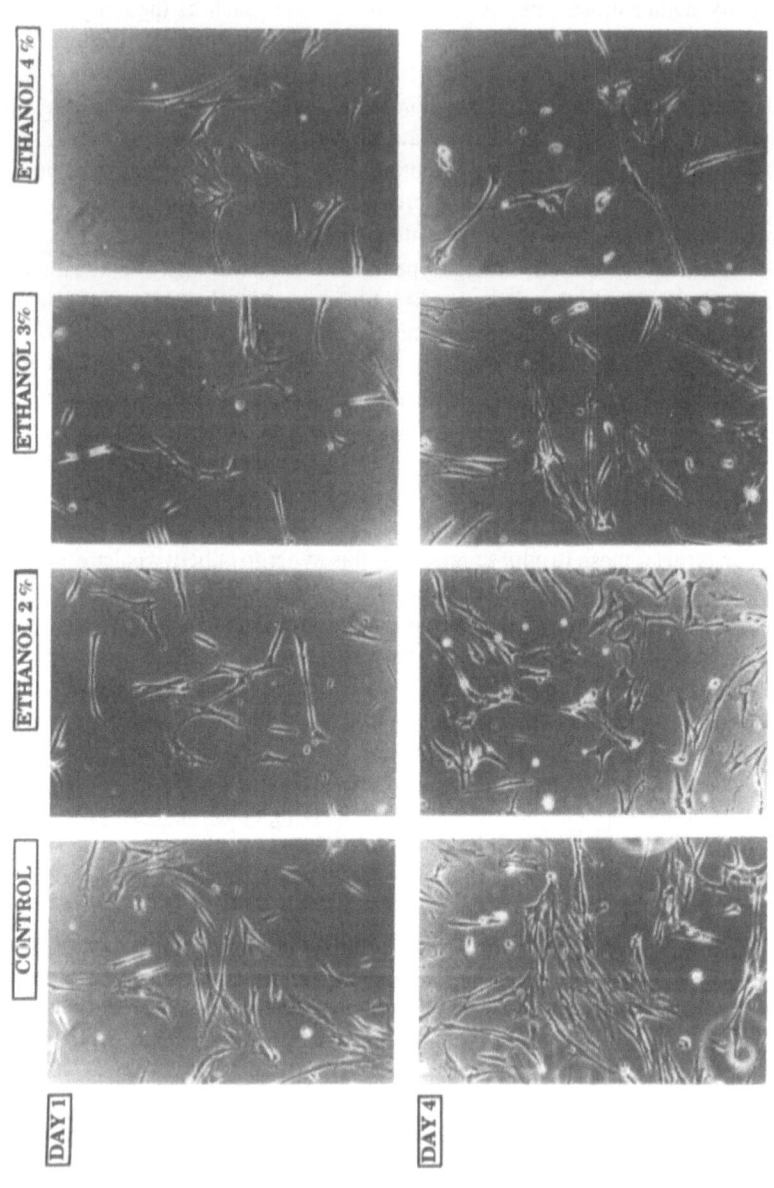

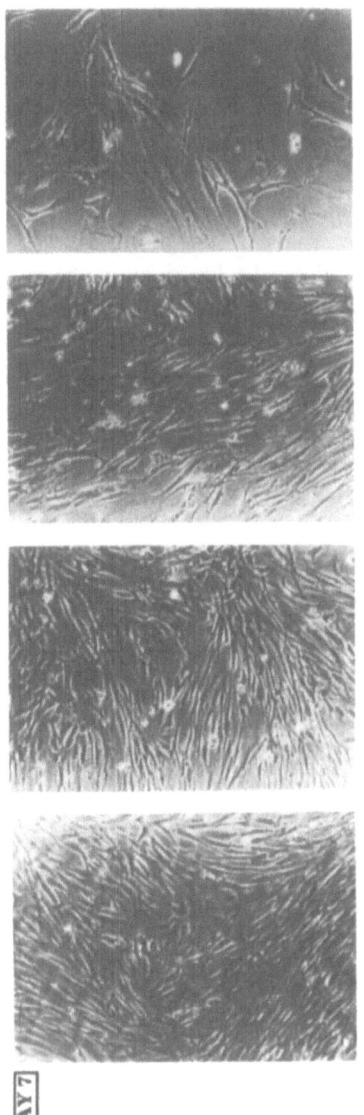

Figure 3. Evolution of the morphotypes of **WI-38** fibroblasts after 2h stresses under 0, 2, 3 and 4% ethanol (v:v). Before the stress, cells were cultivated classically in **BME** medium + 10 % foetal calf serum. After the stress performed under ethanol diluted in phosphate buffer saline (PBS, pH 7.4, 10 mM) as described in [6–8], cells were rinsed twice in culture medium and reincubated in normal culture medium + 10 % serum. Cells were randomly followed at day 1, 4 and 7 after the stress. The increase in morphotypes III and IV is clearly seen as a function of ethanol concentration 7 days after the stress.

glutathione together with an increase in glutathione-protein mixed disulfides. Glucose can also protect against this increase in glutathione-protein mixed disulfides[22]. By following the fluorescence of two probes imaged simultaneously using laser-scanning confocal microscopy, Nieminen et al. have observed the onset of mitochondrial permeability transition pores after exposure of hepatocytes to TBHP. Trifluoperazine, a phospholipase inhibitor which inhibits the formation of these pores in isolated mitochondria prevents mitochondrial depolarization, ATP depletion and cell death[23].

Glucose regulated protein GRP 78 and heat shock protein HSP 70 are induced in rat brain by a short focal ischaemia. However, once the duration of the ischaemia increases, the availability of ATP decreases leading to an arrest of the transcription of the concerned genes[24]. Induction of GRP 78 in hypoxic conditions is essential for the cell survival since inhibition of its induction in chronic hypoxic cells, such as found in solid tumors, is cytotoxic[25]. The induction of GRP 78, and also of GRP 94 and GRP 170, are not only triggered by hypoxia or glucose starvation, but also by calcium ionophore A23187, glucosamine and several other conditions that adversely affect the functions of the endoplasmic reticulum[26]. Seven endoplasmic reticulum proteins were shown to bind selectively to denatured proteins, including histone, gelatin, alpha fetoprotein, thyroglobulin, lysozyme, casein, and IgG. Among these proteins were GRP 94, protein-disulfide isomerase, ERp72, calreticulin, and p50, which were released after Ca^{++} stimulation. These proteins appear to function as Ca^{++}-dependent chaperones, and also require ATP to restore protein folding[27]. Other stresses such as metabolic acidosis concomitant with the release of glucocorticoids[28], or cerebral ischaemia[29] activate ubiquitin and proteasomes ATP-dependent degradation systems.

All these results are in agreement wih the theoretical model of cells considered from a global point of view[1] : a high level in the energetic metabolism ameliorates the cell response to stresses, by giving the possibility to increase the maintenance-related energy consumption. A mathematical model of ageing, which combines the free radical theory and the protein error theory, also supports the positive correlation between maintenance-related energy consumption and life-span[30]. In conclusion, different steps can be proposed for the fibroblast accelerated ageing or death in stressful conditions. The first step is the cell damages generated by the stress, then increase in energy utilization and repair functions take place as schematized in Figure 4. A lack of energy may then occur, which can be reinforced by alterations of the energy-producing systems. This energy lack leads to a decrease in the efficiency of defence, repair and regulation systems, which can lead to accelerated cell ageing or cell death. This model was developed on cultivated fibroblasts but other cell types could follow the same pathway. Since a lot of data have accumulated on the ageing brain either in normal or pathological conditions, we'll now examine some data on the effect of stress in the brain and on the importance of the cellular energetic metabolism.

3. CELL DEATH IN THE BRAIN IN STRESSFUL SITUATIONS AND IN AGEING

The pattern of brain ageing is not uniform[31]. The greatest cell loss occurs in those regions which are most affected by the excitatory neurotransmitter glutamate. Moreover, it is well known that stresses enhance the rate of neuronal loss in these areas, and that both ageing and stress are linked to substantial increases in glutamate release in these regions[32].

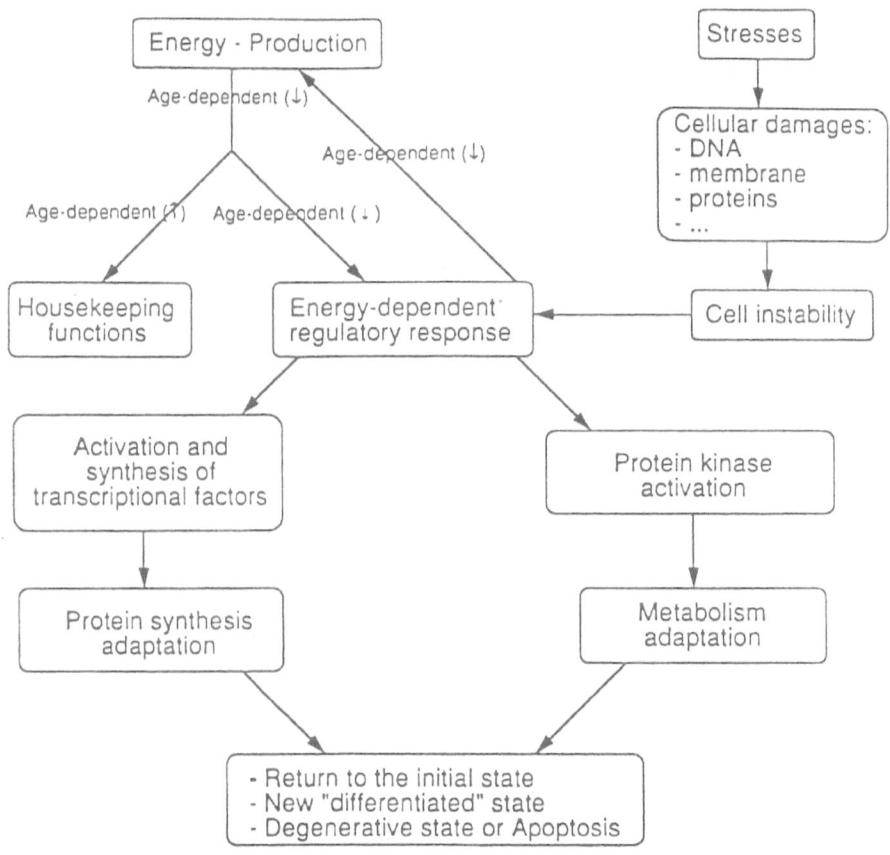

Figure 4. Schematic representation of the cellular responses to stress and how the decrease in energy affects the cellular resistance to stress. Cell damages caused by stressfull conditions lead to energy consumption by defence and repair systems and/or to alterations of the energy production systems, both situations leading to energy lack. This energy lack is accentuated in situations where there is a shortage in energetic substrates, and lead to a decrease in defence, repair and regulation systems, increasing again cell damage and causing accelerated cell ageing or even cell death (reproduced with permission from Experimental Gerontology, 30, 1–22, 1995[6]).

Of great importance is the selective age-related neuronal death of specific hypothalamic nuclei. The suprachiasmatic nucleus which is mainly innervated by glutamatergic neurons originating from the retina, exhibits severe cellular loss during the normal ageing process and during accelerated ageing associated with Alzheimer's disease[33-35]. Cell loss occuring during normal ageing and pathological ageing has also been demonstrated in hippocampal glutamatergic interneurons, which receive strong input from cortical glutamatergic projections. Some cortical glutamatergic neurons also kill themselves during normal ageing. It is striking that the deficits in memory and learning existing in Alzheimer's disease and in very old individuals are associated with to the autodestruction of these glutamatergic neurons[36]. Before looking for the complementation of their effects, let us examine first the effect of the release of glucocorticoid and situations with decrease in the cellular energy potential.

3.1. Glucocorticoids and Glutamate Release

What are the effect of the glutamate increase provoked by the glucocorticoids, released in stressing situations ? Glucocorticoids increase hippocampal vulnerability to metabolic insults and potentiate the deleterious action of glutamate and of the other excitatory amino acids. It was shown recently that glucocorticoids do not depress baseline ATP levels but accelerate the rate of the decline in ATP concentrations observed during metabolic insults such as glucose starvation or cyanide exposition, leading to calcium overload and free radical over-production[37]. It has been conclusively demonstrated that glutamate administration sharply increases hydroxyl radical generation in the brain[38]. The severity of the irreversible brain lesions is directly proportional to the rate of hydroxyl radical formation in these brain structures[32,39–43]. The hydroxyl radical is an extremely reactive oxygen-derived species which provokes hydroxylation, hydrogen atom abstraction and electron transfer of cellular constituants[44]. These radicals initiate and propagate radical chain reactions leading to cross-linkage of fragmented and peroxidized biomolecules[44].

3.1.1. Glutamate and Cell Toxicity. Normally, the glutamate release acts as a normal signal between neurons, triggering the brief opening of N-methyl D aspartate (NMDA) receptor-coupled channels. Activation of the NMDA receptor, under partially depolarizing conditions, permits Na^+ and Ca^{++} to flow through their channels. Depolarization activates voltage-dependent Ca^{++} channels, permitting the influx of Ca^{++}. Depolarization also increases ATP consumption by the Na/K^+ ATPase, with increased oxidative phosphorylation, with production of superoxide anions as a by-product [For a review, see[45]]. A huge influx of calcium can also decrease the energetic metabolism[46,47] by uncoupling the mitochondrial respiration chain[32,36,38] and generating free radicals[48,49]. Stimulation of phospholipase A_2 by Ca^{++} and the subsequent release of arachidonic acid leads to the generation of oxygen free radicals and to the activation of nitric oxide synthetase. Arachidonic acid and free radicals enhance the release of glutamate and inhibits its uptake inactivation by neuronal and glial transporter, thus promoting a vicious circle [For reviews, see[45,46,49,50]]. Elevated intraneural Ca^{++} also activates peptidases, such as calpain I, catalyzing the enzymatic conversion of xanthine dehydrogenase to xanthine oxidase, yielding superoxide anions by metabolizing purines. Together with hydroxyl radicals, calcium stimulates many proteases and nucleases, leading finally to cell degeneration and death[32,51].

3.1.2. Protective Effects of Growth Factors, Cytokines and Other Peptides on the Influx of Ca^{++}. Fortunately, some growth factors, cytokines and other peptides have been shown to protect cells against such an influx of Ca^{++}. Nerve-growth factor (NGF) prevent loss of both $[Ca^{++}]_i$ homeostasis and mitochondrial transmembrane potential, and protect hippocampal neurons against hypoglycemic injury. Loss of $[Ca^{++}]_i$ homeostasis may be a critical event leading to mitochondrial damage and further cell loss resulting from energy failure, even after the glucose deprivation is[52]. NGF protect hippocampal and cortical neurons against free radical generation induced by iron[53]. A protective effect of recombinant human NGF and mouse NGF against glutamate-induced neurotoxicity was found in cultured rat cortical neurones[54].

Basic fibroblast growth factor (bFGF) protects rat central nervous system neurons against hypoxia[55]. bFGF also promoted the survival of cultured rat hippocampal neurons exposed to phorbol ester[56], or glutamate[56,57]; bFGF also protected against the latter when released by glucose deprivation[52,58], hypoxia or ischaemic-like conditions[56,59]. As NGF, bFGF protects hippocampal and cortical neurons against free radical generation induced by iron[53]. BFGF also enhances the survival of transplanted rat septal neurons, if incubated with bFGF before transplantation[60]. A possible mechanism whereby bFGF reduces neuron damage, caused by the protein kinase C (PKC)-activating phorbol ester 12-O-tetrade-canoylphorbol 13-acetate (TPA), glutamate and ischaemia-like conditions, could be by counteracting the excessive activation of PKC and secondly by preventing the loss of PKC occuring after prolonged exposure to TPA or ischaemia-like conditions[56]. bFGF also suppresses the expression of a functional 71 kDa NMDA receptor protein that mediates calcium influx and neurotoxicity occuring during excitotoxic/ischaemic damage[61]. Insulin-like growth factor-II (IGF-II) protect hippocampal neurons against glucose deprivation; IGF-I and II protect hippocampal and cortical neurons against iron-induced toxicity[53,58]. PDGF-AA or PDGF-BB (platelet-derived growth factor) protect hippocampal neurons against energy deprivation and oxidative injury[62]. A protection against the deleterious effects of ethanol on WI-38 fibroblasts was also obtained, in non proliferative conditions, with PDGF[63].

Brain-derived neurotrophic factor (BDNF) and neurotrophin-3 were shown to attenuate glucose deprivation-induced and glutamate-induced damage in rat hippocampal, septal and cortical neurons in culture by attenuating the intracellular calcium elevation[64]. Moreover, BDNF increases neurotrophin-3 expression in cerebellar granule neurons[65]. BDNF protected rat hippocampal neurons against ischaemia[66] and cultured rat cerebellar granule neurons against glutamate-induced cytotoxicity[67].

Tumor necrosis factors (TNF) -α and -β also protect cultured rat embryonic hippocampal, septal and cortical neurons against metabolic-excitotoxic insults (glucose deprivation, glutamate exposure) by promoting the maintenance of calcium homeostasis[68]. Very low concentrations of staurosporine, a kinase inhibitor, also prolong the survival of these cells by attenuating the increase in $[Ca^{++}]_i$ following glucose deprivation. The tyrosine kinase inhibitor genistein reduced this protective effect, indicating that tyrosine phosphorylation was required for the neuroprotection by staurosporine[69]. The mechanism whereby TNF-α stabilizes the intracellular calcium concentration may involve the expression of the calcium-binding protein-D28k[68].

Secreted forms of β-amyloid protein have also been shown to protect hippocampal neurons against amyloid β-peptide-induced oxidative injury. In fact, alternative processing of the β-amyloid precursor protein (βAPP) results in liberation of either secreted forms of βAPPs, or amyloid β-peptide (AβP). In rat hippocampal cell cultures, AβP caused a reduction in neuronal survival. The secreted forms of βAPP 695 and βAPP 751 reduced this AβP-induced cytotoxicity. AβP caused an increase in intracellular calcium levels which is attenuated by the secreted forms of βAPPs. Moreover, AβP-induced toxicity was reduced by vitamin E, showing the involvement of free radical in this toxic process. Furthermore, the secreted forms of βAPPs protected neurons against free radical-induced injury caused by iron[70]. The secreted forms of βAPPs also protect neurons against excitotoxic or ischaemic insult by stabilizing the intracellular $[Ca^{++}]$[71]. It has been proposed that certain βAPP mutations, Down syndrome and age-related changes in brain metabolism like reduced energy availability or increased oxidative stress may favour accumulation of AβP and destabilization of calcium homeostasis[71].

3.1.3. Protective Effects of Antioxidant Molecules. Glutathione (GSH) is part of the anti-oxidant protection acting as a substrate of glutathione peroxidase and is by itself a radical scavenger[72]. The rate-limiting precursor for GSH synthesis is cystine. Cystine is transported by the anionic amino acid transport system, which is highly specific for cystine but also for glutamate, and this explains the glutamate inhibitory activity on the cystine uptake[73]. The consequence of this lower cystine uptake is a decrease in the cellular antioxidative capacity, eventually inducing cell death if an excess of free radicals is produced. Like cystine, ascorbate ion, a hydrophilic reducing agent, is also transported via the glutamate uptake carrier, by an antiport mechanism, and exposure to high levels of glutamate also depletes the intracellular ascorbate store. More dangerous is the increase in extracellular ascorbate, since in conjuction with the presence of transition metals such as iron or copper, it induces redox cycling and hydroxyl radical generation[74–76]. In aged brains, where the deficit in antioxidants becomes important, the activity of glutamine synthetase, which is localized in astrocytes, and detoxifies glutamate, is decreased due to the prooxidant medium. The decrease in this activity is specific to the brain regions highly innervated by glutamatergic neurons and is much more important in patients with Alzheimer's disease[36,46,47].

With ageing, modifications in the amounts of reducing equivalents such as NADPH, NADH, or GSH could also contribute to the increase in the prooxidant state of the cell[77–79]. For instance, in the rat cerebral cortex, reduced glutathione, total glutathione (reduced + oxidized), the glutathione redox index (reduced/oxidized) and the activity of γ-glutamylcysteine synthetase, involved in glutathione synthesis, decrease during ageing[80]. This probably explains why the vulnerability to excitatory amino acids and to exogenous neurotoxins, both generating hydroxyl radicals, is increased in old animals[81,82].

3.2. Decrease in Energy Potential and Glutamate Release in Ageing, Alzheimer's Disease and Brain Ischaemia

Glucose metabolism in the brain is of central signifiance since it contributes to the synthesis of the neurotransmitters acetylcholine, glutamate, aspartate, gamma-aminobutyric acid (GABA) and glycine, and yields ATP as driving force of almost all molecular work. Subsequently to changes in the glucose metabolism and energy production, as observed in ageing, variations occur in acetylcholine synthesis and release, extracellular concentration and binding of glutamate and cytosolic Ca++ homeostasis, additionally to increases in free radical production and membrane structure changes[83]. A synergistic effect on cell death can be obtained experimentally between an exposition to hydroxyl radicals, and a decrease in the energetic potential. Transient ischaemia, not only decreases the availability in cellular energy as ATP, but also allows the elevation in endogenous excitatory amino acids, which leads to neuronal loss in the hippocampus. This effect is again exacerbated by the corticosterone elevation and retarded by bilateral adrenoectomization[84,85].

A slight but persistent reduction in energy formation is observed in brain ageing so that an increase in energy demand, or stress conditions increase energy shortage particularly in old age. Mitochondrial dysfunction in both ATP formation and ATP release may be assumed to be causative for the reduced availability of energy in cerebral cells leading to diminished cellular work. For instance, significant decreases in complex IV activities are found in synaptic mitochondria and in complexes I, II + III and IV in non synaptic mitochondria taken from the brain of old mice compared to those observed for young ones[86]. As a consequence, the impaired energy metabolism reduces the activity of ATP-ases, which normally consume as much as 40% of the cellular energy production in the neurons[79].

Ageing is considered as a risk factor for sporadic late-onset dementia of the Alzheimer type. Metabolically, this disorder seems to be characterized by an early energy shortage in cerebral cells with the same consequences for the disturbance of cellular work. Using a non invasive method based on nuclear magnetic resonance in the brain of mildly Alzheimer patients, an increase in the levels of phosphomonoesters, a decrease in the levels of phosphocreatine, and an increased oxidative metabolic rate were observed. This situation is typical of cells which resist a stress by increasing their energy production and utilization[87]. The causative abnormality may also be seen in a perturbation of the control of cerebral glucose metabolism[88].

Brain glucose utilization and ATP formation were found to be reduced to 54% and 81%, respectively, compared to control values in sporadic dementia of Alzheimer type, causing reduced availability of the glucose-derived neurotransmitter acetylcholine. With respect to energy shortage, impacts on energy-dependent processes such as synaptic transmission, ion homeostasis, protein processing and degradation, extracellular transmission, and extracellular phosphorylation may be expected[89]. Acetyl-L-carnitine has been considered of potential use in senile dementia of the Alzheimer type because of its ability to serve as a precursor for acetylcholine. Moreover, as acetyl-L-carnitine and L-carnitine are "shuttles" of long chain fatty acids between the cytosol and the mitochondria to undergo β-oxidation, they can maximize energy production, thereby supporting the physiological functions of the mitochondria. This remark must be linked to recent observations of significant decrease in carnitine acetyltransferase, in autopsied Alzheimer brains[90].

In diseases like Huntington's disease, Parkinson's disease, amyotrophic lateral sclerosis, and Alzheimer's disease, the sudden release of glutamate is not the only factor for the neuronal cell death. Other causes and mechanisms are involved, and a defect in mitochondrial energetic metabolism is often found. For instance, several reports showed a defect of mitochondrial complex I[91] and alpha-ketoglutarate dehydrogenase complex[92] in the substancia nigra of the parkinsonian brain. A defect in mitochondrial energetic metabolism can secondarily increase excitotoxic neuronal death, by making neurons more vulnerable to endogenous glutamate (Figure 5). With a reduced oxidative metabolism and a partial cell membrane depolarization, voltage-dependent NMDA receptor ion channels would be more easily activated [for a review see[91]].

3.3. Brain Ischaemia

Ischaemia injury involves death mostly of hippocampal glutamatergic neurons and of some cortical glutamatergic neurons, as observed in victims of the Alzheimer's disease. More evidence in support of this effect on the energetic metabolism, is that metabolizable sugars attenuate the glucocoticoid effect on neuronal destruction through the release of excitatory amino acids[84]. More generally, energy-failure due to hypoxia or hypoglycemia results in a marked efflux of glutamate in the extracellular space high enough to generate neurotoxic effects[93].

Concerning the specific cases of focal ischaemia, it has recently been proposed to revise the excitotoxic hypothesis in favor of the energetic hypothesis. Critical analysis of published data on glutamate toxicity *in vitro*, by comparison with *in vivo* release of glutamate and the therapeutic effect of glutamate antagonists, suggest an alternative explanation for the glutamate-mediated injury by hypoxia due to peri-infarct spreading depression-like depolarizations. These depolarizations are triggered in the core of the ischaemic infarct and spread at irregular intervals into the peri-infarct surrounding. In the penumbra region of focal ischaemia, the hemodynamic constraints of collateral blood cir-

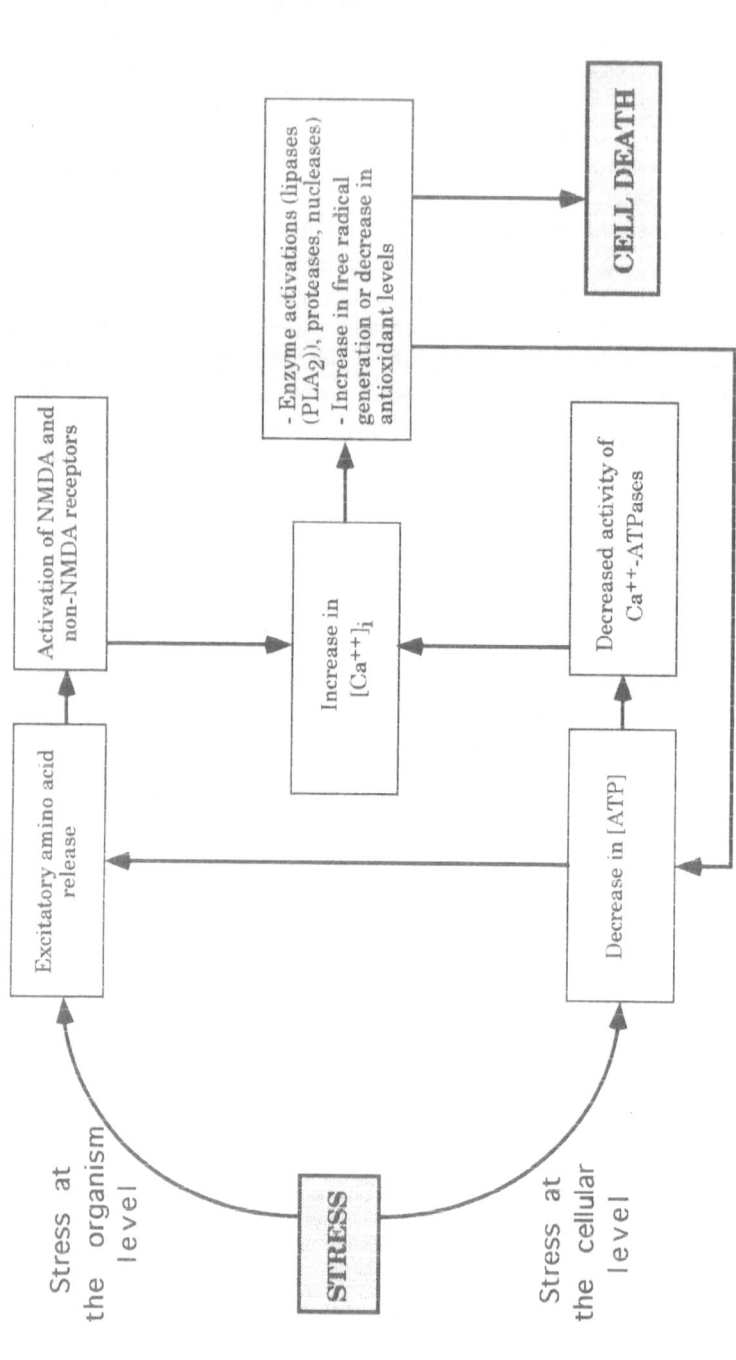

Figure 5. Schematic influence of stresses on neuronal death. Stresses at the level of the organism provoke a release of glucocorticoids and of excitatory amino acids which lead to cell death by increasing the cytosolic calcium concentration, activating enzymes, and generating free radicals. Stresses at the level of the neurons provoke a decrease in the intracellular ATP level, which lead to excitatory amino acid release or to an increase in the cytosolic calcium concentration, the activation of enzymes and an increase in the free radical concentration, leading in turn to cell death.

culation prevail the adequate adjustment of oxygen delivery, leading to transient episodes of relative tissue hypoxia. These transient hypoxic episodes cause a suppression of protein synthesis, a gradual deterioration of energy metabolism and a progression of irreversibly damaged tissue into the penumbra zone. The generation of peri-infarct spreading depressions and the associated metabolic workload can be suppressed by NMDA and non-NMDA antagonists. As a result, the penumbral inhibition of protein synthesis and the progressing energy failure is prevented, so that the volume of ischaemic infarct decreases. Interventions to improve ischaemic resistance should therefore aim at improving the oxygen supply or reducing the metabolic workload in the penumbra region[94].

Taken together, all these data suggest that in order to obtain protection against stresses, one must stabilize the intracellular neuronal ATP concentration, and therefore $[Ca^{++}]_i$. Data suggest to perform a very early intervention because the extracellular ATP, released by cell lysis, can negatively affect the calcium metabolism of the neighbouring cells by stimulating calcium influx in ATP-hydrolysis -dependent or -independent processes[95–97] which can be receptor-mediated or not.

In the last part of this review, we propose to examine the role of the energetic metabolism to other pathobiological situations occuring in vivo.

4. OTHER EXAMPLES OF DECREASES IN THE ENERGETIC POTENTIAL

4.1. Ischaemia

Ischaemia can not only be found in the brain, but also in other tissues. Blood stasis in leg veins is a situation commonly linked to the development of veinous diseases such as varicoses. Such a stasis provokes ischaemia. It has been proposed that the effect of oxygen deprivation on the functional state of the endothelium is the starting point of a cascade of events leading to the disorganization of the vessel wall typical of these pathologies, in addition to genetic, hormonal and mechanical factors[98]. Here also, the decrease in free energy availability during hypoxia was correlated with modifications in the intracellular calcium concentrations[99]. Experimentally, human endothelial cells obtained from umbilical veins are activated in hypoxic conditions leading to the synthesis of proinflammatory molecules like platelet activating factor or prostaglandins. The adhesion of human polymorphonuclear neutrophils on these endothelial cells mediated by the CD18/CD11b integrins molecules are markedly increased during hypoxia incubation[100]. Energetic substrates were found protective as well as two pharmacological stimulators of the energetic metabolism like for instance naftidrofuryl oxalate[101] or extracts of Gingko biloba[102].

In anoxia or during preservation of donor tissues, where an elevated glycolysis metabolism can lead to a lactate accumulation and pH decrease, other energetic substrates than glucose like β-hydroxybutyrate can be used as protectors. This is a ketone body which was found to be protective for the retention of the metabolic activity of human and rabbit corneas under storage, for the integrity of the endothelium and for the physiological function of the cornea in vivo[103].

L-acetyl-carnitine, which is a transmitochondrial carrier of acetyl and long-chain acyl groups, facilitates the transport of fatty acids into the mitochondria where they are used for the formation of high-energy phosphate bonds during oxidative phosphorylation[104]. L-carnitine and its derivatives have been shown to protect ischaemic myocardium[105,106], to improve cardiac performances[105,107-109] and to restore high-energy phosphate pools (ATP, creatine phosphate) in myocardial cells[108,109]. It has been shown that compounds which decrease the cellular energy status may significantly inhibit DNA repair[110,111]. In this respect, treatment with L-carnitine accelerates the disappearance of single-strand breaks induced by oxygen radicals and alkylating agents in human peripheral blood lymphocytes, in a process where antioxidant action of L-carnitine seems to be ruled out[104].

Glycolytic ATP production protects the myocardium against ischemic contracture and falicitates post-ischaemic functional recovery[112,113]. CoQ which protects the mitochondrial activity, enhances cardiac functional and allows a metabolic recovery during post-ischaemic reperfusion of rat hearts. These experiments also demonstrated that the mechanism of action of CoQ was not scavenging the primary burst of superoxide anions or hydroxy radicals but rather enhancing the recovery of high-energy phosphate, thereby preventing calcium overload and subsequent toxic events[114]. Co-administration of L-carnitine and coenzyme Q10 (CoQ) reduces damage induced by chronic alcohol poisoning and hyperbaric oxygen[115]. Inhibitors of adenosine deaminase attenuate myocardial ischaemic injury and improve postischaemic recovery of contractile function and metabolism through endogenous myocardial adenosine enhancement and ATP preservation[116].

4.2. Mitochondrial Alterations

Decreases in energy production can be a direct result of the lower efficiency of the mitochondria. Deletions in the mitochondrial DNA (mt DNA) were observed in aged tissues of mice[117], rats[118], in rhesus monkeys[119], or human beings[120-123], and in sicknesses like the Parkinson's disease, several myopathies, Leber's optic neuropathy, Kearns-Sayre syndrome, chronic progressive external ophtalmoplegia, Pearson's marrow/pancreas syndrome[124], myoclonic epilepsy and ragged-red fiber disease (MERRF), neurogenic muscle weakness, ataxia and retinous pigmentosa (NARP)[125]. It becomes clear that a study of the factors which are responsible for the appearance of these mt DNA alterations with ageing is of primary importance. A first progress has been achieved by distinguishing 2 classes of mt DNA deletions, A and B. Class A carries a commonly detected 4,977-bp deletion occuring at a pair of 13-bp directly repeated sequences. Class B comprises a family of at least nine closely related 8.04-kb deletions at the same pair same 5-bp direct repeats, where the breakpoints differ at the base-pair level[126]. In the skin tissues, the amount of the 4,977-bp-deleted mtDNA is associated with the physiologic conditions of the skin tissues : the sun-exposed skin tissues harbored higher level of the 4,977-bp-deleted mtDNA than the nonexposed normal aged skin tissues. This observation provides an argument in favor of photooxidation-induced mtDNA mutations[127]. Moreover, the mtDNA deletions occur more frequently and abundantly in high energy-demanding tissues during the ageing process of the human[128].

Other mitochondrial genetic alterations like point mutations are also observed in many pathological or aged tissues[129-133]. Other mitochondrial alterations exist like morphological, enzymatic, membrane, physiological and genetic modifications[120]. So mtDNA deletion is probably not the only factor responsible for mitochondrial alterations. It has been demonstrated that the accumulation of nuclear recessive somatic mutations can also be partly responsible for the in vivo age-related mitochondrial dysfunctions observed in human skin fibroblasts[134]. At the transcriptional level, the age-related decrease in the DNA-binding efficiency of transcription factors, such as shown for SP1 transcriptional factor[135,136], could be responsible for a decreased expression of the mitochondrial energy genes by a decreased binding capacity to their promoters, like OXBOX and REBOX overlapping promoter elements of the mitochondrial FoF1-ATP synthetase β subunit gene[137], activated by SP1. If too much altered, mitochondria will be inactive and eliminated; but if not, they could replicate at the same rate than the non-altered mitochondria so that, with time and subsequent mitochondria and cell divisions, less efficient mitochondria will accumulate and the energy production will be reduced in these cells[138].

Whatever the mechanisms, the accumulation of bioenergetically defective cells is anyway the physiological consequence of these mitochondrial alterations. The lower efficiency of the cells is probably not detrimental in the beginning since the cells and the tissues as a whole will adjust their metabolism accordingly. However old cells possess mitochondria that are less efficient and dissipate more energy so that the ATP regeneration is lowered. In normal conditions, such lowered efficiency is probably not harmful to the cells since mitochondria do not function at their highest rate. However, in a less favourable environment, their efficiency to counteract stresses, such as oxidative stresses, is lower since their general metabolism, their adapatability and their defence or repair systems are lowered[139,140], resulting in severe impairments of cell functions.

5. CONCLUSION

As a general mechanism, we propose that, despite the use of free energy and defense systems to resist stressful situations, the cellular fluctuations encountered may be so important that the protective systems are not sufficient anymore, or even overwhelmed, so that the energy production is also impaired, leading to the impossibility to counteract the increasing fluctuations. If this phenomenon overpasses a critical threshold, the cells are committed to "differentiate" to older cell types or to die[141].

Concerning the implication of the energetic metabolism, we have seen that the multiple defense or repair systems (chemical and enzymatic antioxidants, altered fatty acid or protein destruction and replacement, DNA repair, induced or not) against stresses need energy in various forms, such as NADPH, NADPH, $FADH_2$, ATP and other phosphorylated nucleosides. A "re-energization therapy" has been shown to be highly effective in the case of certain mitochondrial-related pathologies such as mitochondrial myopathy, encephalomyopathies, lactic acidosis, stroke-like episodes and Kearns-Sayre syndrome[125]. Replacement therapy and pharmacological support could also be tested in age-related pathologies by a similar mechanism using appropriate energetic and redox compounds. New compounds could be rationally developed in clinical use for the development of "re-energization" strategies[120] using as already shown in vitro for instance co-enzyme Q 10[114], ketone bodies[103], acetyl-L-carnitine[90,142], bioavailable Mg^{++} 120 or drugs acting on mitochondrial activity. For instance, Arenas et al.[143] have shown that daily oral uptake of L-carnitine resulted in dramatic increase in pyruvate dehydrogenase complex activities in Vastus lateralis muscles of long-distance runners, together with an increase in complex I, II and IV of the respiratory chain, leading to improvements in maximal oxygen consumption.

The research for genes involved in the cellular resistance to stresses, or definitively activated after stresses will also greatly help in the development of adequate therapeutical strategies, as the study of the cellular regulatory processes taking place in the stressed cells.

These different research strategies cited here above should be a priority for a rationale research on ageing and age-related diseases.

ACKNOWLEDGMENTS

The writing of this article was made possible thanks the EU Biomed and Health Research Programme, Concerted Action Programme on Molecular Gerontology, BMH1 CT94 1710. We also thank Prof. Sapolsky, University of California at Los Angeles (UCLA), Ca, USA for critical reading of the manuscript.

REFERENCES

1. Toussaint O., Raes M., Remacle J. (1991) Aging as a multi-step process characterized by a lowering of entropy production leading the cell to a sequence of defined stages. *Mech Ageing Dev* , 61, 45–64.
2. Hayflick L., Moorehead P. S. (1961) The serial cultivation of human diploid cell strains. *Exp Cell Res* , 25, 585–621.
3. Hamels H., Toussaint O., Le L., Houbion A., Eliaers F., Remacle R. (1995) AGO4432 fibroblasts also shift through 7 successive morphotypes during in vitro ageing. Comparison with WI-38 human foetal lung fibroblasts and HH-8 human skin fibroblasts. *Arch Int Physiol Biochem Biophys* , 103, B 16.
4. Bayreuther K., Rodeman H. P., Hommel R., Dittman K., Albiez M., Francz P. (1988) Human skin fibroblasts in vitro differentiate along a terminal cell lineage. *Proc Natl Acad Sci USA* , 85, 5112–5116.
5. Rodeman H. P., Bayreuther K., Francz P. I., Dittman F., Albiez M. (1989) Selective enrichment and biochemical characterization of seven fibroblasts cell types in vitro. *Exp Cell Res* , 180, 84–93.
6. Toussaint O., Michiels C., Raes M., Remacle J. (1995) Cellular ageing and the importance of the energetic factors. *Exp Gerontol* , 30, 1–22.
7. Toussaint O., Eliaers F., Houbion A., Remacle J., Drieu K. (1995) In *Advances in Ginkgo biboba Extract Research, vol. 4. Effect of Ginkgo biloba Extracts (EGb 761) on Aging and Age-related Disorders.* Elsevier publications: Paris, Vol. ; pp 1–16.
8. Toussaint O., Houbion A., Remacle J. (1992) Aging as a multi-step process characterized by a lowering of entropy production leading the cell to a sequence of defined stages. II. Testing some predictions on aging human fibroblasts in culture. *Mech Ageing Dev*, 65, 65–83.
9. Rodeman H. P. (1989) Differential degradation of intracellular proteins in human skin fibroblasts of mitotic and mitomycin-C (MMC)-induced postmitotic differentiation states in vitro. *Differentiation*, 42, 37–43.
10. Rodeman H. P., Bayreuther K., Pfleiderer G. (1989) The differentiation of normal and transformed fibroblasts in vitro is influenced by electromagnetic fields. *Exp Cell Res* , 182, 610–621.
11. Honda S., Matsuo M. (1983) Shortening of the *in vitro* lifespan of human diploid fibroblasts exposed to hyperbaric oxygen. *Exp Gerontol* , 18, 339–345.
12. Honda S., Matsuo M. (1988) Relationships between the cellular glutathione levels and *in vitro* life span of human diploid fibroblasts. *Exp Gerontol* , 23, 81–86.
13. Chen Q., Ames B. N. (1994) Senescence-like growth arrest induced by hydrogen peroxide in human diploid fibroblast F65 cells. *Proc Natl Acad Sci U S A* , 91, 4130–4134.
14. Von-Zglinicki T., Saretzki G., Docke W., Lotze C. (1995) Mild hyperoxia shortens telomeres and inhibits proliferation of fibroblasts: a model for senescence? *Exp Cell Res* , 220, 186–93.
15. Ammendola R., Fiore F., Esposito F., Caserta G., Mesuraca M., Russo T., Cimino F. (1995) Differentially expressed mRNAs as a consequence of oxidative stress in intact cells.,
16. Toussaint O., Houbion A., Remacle J. (1994) Effects of modulations of the energetical metabolism on the mortality and ageing of cultured cells. *Biochim Biophys Acta* , 1186, 209–220.
17. Kirkland J. B. (1991) Lipid peroxidation, protein thiol oxidation and DNA damage in hydrogen peroxide-induced injury to endothelial cells: role of activation of poly(ADP-ribose)polymerase. *Biochim Biophys Acta*, 1092, 319–25.
18. Pasternak C. A., Aiyathurai J. E., Makinde V., Davies A., Baldwin S. A., Konieczko E. M., Widnell C. C. (1991) Regulation of glucose uptake by stressed cells. *J Cell Physiol* , 149, 324–331.
19. Wertheimer E., Sasson S., Cerasi E., Ben-neriah Y. (1991) The ubiquitous glucose transporter GLUT-1 belongs to the glucose-related protein family of stress-inducible proteins. *Proc Natl Acad Sci USA*, 88, 2525–2529.
20. Bartoli G. M., Piccioni E., Agostara G., Calviello G., Palozza P. (1994) Different mechanisms of tert-butyl hydroperoxide-induced lethal injury in normal and tumor thymocytes. *Arch Biochem Biophys*, 312, 81–87.

21. Calviello G., Ricci P., Bartoli G. M. (1993) tert-butyl hydroperoxide induced [Ca2+]i increase in thymus and thymoma cells. *Biochem Biophys Res Commun*, 197, 859–68.

22. Di-Simplicio P., Rossi R. (1994) The time-course of mixed disulfide formation between GSH and proteins in rat blood after oxidative stress with tert-butyl hydroperoxide. *Biochim Biophys Acta*, 1199, 245–52.

23. Nieminen A. L., Saylor A. K., Tesfai S. A., Herman B., Lemasters J. J. (1995) Contribution of the mitochondrial permeability transition to lethal injury after exposure of hepatocytes to t-butylhydroperoxide. *Biochem J*, 307, 99–106.

24. Wang S., Longo F. M., Chen J., Butman M., Graham S. H., Haglid K. G., Sharp F. R. (1993) Induction of glucose regulated protein (grp78) and inducible heat shock protein (hsp70) mRNAs in rat brain after kainic acid seizures and focal ischemia. *Neurochem Int*, 23, 575–582.

25. Koong A. C., Chen E. Y., Lee A. S., Brown J. M., Giaccia A. J. (1994) Increased cytotoxicity of chronic hypoxic cells by molecular inhibition of GRP78 induction. *Int J Radiat Oncol Biol Phys*, 28, 661–666.

26. Lin H. Y., Masso-Welch P., Di Y. P., Cai J. W., Shen J. W., Subjeck J. R. (1993) The 170-kDa glucose-regulated stress protein is an endoplasmic reticulum protein that binds immunoglobulin. *Mol Biol Cell*, 4, 1109–1119.

27. Nigam S. K., Goldberg A. L., Ho S., Rohde M. F., Bush K. T., Sherman M. (1994) A set of endoplasmic reticulum proteins possessing properties of molecular chaperones includes Ca(2+)-binding proteins and members of the thioredoxin superfamily. *J Biol Chem*, 269, 1744–1749.

28. Price S. R., England B. K., Bailey J. L., Van-Vreede K., Mitch W. E. (1994) Acidosis and glucocorticoids concomitantly increase ubiquitin and proteasome subunit mRNAs in rat muscle. *Am J Physiol*, 267, C955-C960.

29. Dewar D., Graham D. I., Teasdale G. M., McCulloch J. (1994) Cerebral ischemia induces alterations in tau and ubiquitin proteins. *Dementia*, 5, 168–73.

30. Kowald A., Kirkwood T. B. (1994) Towards a network theory of ageing: a model combining the free radical theory and the protein error theory. *J Theor Biol*, 168, 75–94.

31. Mann D. A. (1991) Is the pattern of nerve cell loss in aging and Alzheimer's disease a real, or an apparent selectivity ? *Neurobiol Aging*, 12, 340–343.

32. Hall E. D., Andrus P. K., Althaus J. S., VonVoigtlander P. F. (1993) Hydroxyl radical production and lipid peroxidation parallels selective post-ischemic vulnerability in gerbil brain. *J Neurosci Res*, 34, 107–112.

33. Miller J. D. (1992) Pharmacological intervention in sleep and circadian processes. *Ann Rep Med Chem*, 27, 11–19.

34. Meijer J. H., Albus H., Weidema F., Ravesloot J. (1993) The effects of glutamate on membrane potential and discharge rate of suprachiasmatic neurons. *Brain Res*, 603, 284–288.

35. Poeggeler B., Balzer I., Hardeland R., Lerchl A. (1991) Pineal hormone melatonin oscillates also in the dinoflagellate Gonyaulax polyedra. *Naturwissenshaften*, 78, 268–269.

36. Smith C. D., Carney J. M., Tatsumo T., Stadtman E. R., Floyd R. A., Markesbery W. R. (1992) Protein oxidation in aging brain. *Ann NY Acad Sci*, 663, 110–119.

37. Lawrence M. S., Sapolsky R. M. (1994) Glucocorticoids accelerate ATP loss following metabolic insults in cultured hippocampal neurons. *Brain Res*, 646, 303–306.

38. Boisvert D. P. (1992) In *The role of neurotransmitters in brain injury*; M. Globus and W. D. Dietrich, Ed.; Plenum Press: New-York, Vol. ; pp 361–366.

39. Joseph J. A., Gupta M., Han Z., Roth G. S. (1991) The deleterious effects of aging and kainic acid may be selective for similar striatal neuronal populations. *Aging*, 3, 361–371.

40. Kitamura Y., Zhao X.-H., Ohnuki T., Takei M., Nomura Y. (1992) Age-related changes in transmitter glutamate and NMDA receptor/channels in the brain of senescence accelerated mouse. *Neurosci Let*, 137, 169–172.

41. May P. C., Kohama S. G., Finch C. E. (1989) N-methyl-aspartic acid lesions of the arcuate nucleus in C57BL:6J mice : a new model for the age-related lenghtening of the estrous cycle. *Neuroendocrinol*, 50, 605–612.

42. Palkovits M., Lang T., Patthy A., Elekes I. (1986) Distribution and stress-inducd increase of glutamate and aspartate levels in discrete brain nuclei of rats. *Brain Res*, 373, 252–257.

43. Zawia N., Arendash G. W., Wecker L. (1992) Basal forebrain cholinergic neurons in aged rat brain are more susceptible to ibotenate-induced degeneration than neurons in young adult rat brain. *Brain Res*, 589, 333–377.

44. Halliwell B., Gutteridge J. M. C. (1992) Biologically relevant metal ion-dependent hydroxyl radical generation. *FEBS Lett*, 307, 108–112.

45. Coyle J. T., Puttfarcken P. (1993) Oxidative stress, glutamate, and neurodegenerative disorders. *Science*, 262, 689–695.

46. Peruche B., Kriegelstein J. (1993) Mechanisms of drug action against neuronal damage caused by ischemia - an overview. *Prog Neuropsychopharmacol Biol Psychiatry* , 17, 21–70.

47. McIntosch T. K., Vink R., Soares H., Hayes R., Simon R. (1990) Effect on non competitive blockade of N-methyl-D-aspartate receptors on the neurochemical sequelae of experimental brain injury. *J Neurochem* , 55, 1170–1179.

48. Harley A., Cooper J. M., Schapira A. H. (1993) Iron induced oxidative stress and mitochondrial dysfunction: relevance to Parkinson's disease. *Brain Res* , 627, 349–353.

49. Dickens B. F., Weglicki W. B., Li Y.-S., Mak I. T. (1992) Magnesium deficiency in vitro enhances free radical-induced intracellular oxidation and cytotoxicity in endothelial cells. *FEBS Lett* , 311, 187–191.

50. Hermes-Lima M., Castilho R. F., Valle V. G. R., Bechara E. J. H., Vercesi A. E. (1992) Calcium-dependent mitochondrial oxidative damage promoted by 5-aminovulinic acid. *Biochim Biophys Acta* , 1180, 201–206.

51. Reiter R. J., Poeggeler B., Tan D.-X., Chen L.-D., Manchester L. C., Guerrero J. M. (1993) Antioxidant capacity of melatonin : a novel action not requiring a receptor. *Neuroendocrinol Lett* , 15, 103–116.

52. Cheng B., McMahon D. G., Mattson M. P. (1993) Modulation of calcium current, intracellular calcium levels and cell survival by glucose deprivation and growth factors in hippocampal neurons. *Brain Res* , 607, 275–285.

53. Zhang Y., Tatsuno T., Carney J. M., Mattson M. P. (1993) Basic FGF, NGF, and IGFs protect hippocampal and cortical neurons against iron-induced degeneration. *J Cereb Blood Flow Metab*, 13, 378–388.

54. Shimohama S., Ogawa N., Tamura Y., Akaike A., Tsukahara T., Iwata H., Kimura J. (1993) Protective effect of nerve growth factor against glutamate-induced neurotoxicity in cultured cortical neurons. *Brain Res* , 632, 296–302.

55. Akaneya Y., Takahashi M., Hatanaka H. (1994) Death of cultured postnatal rat CNS neurons by in vitro hypoxia with special reference to N-methyl-D-aspartate-related toxicity. *Neurosci Res* , 19, 279–285.

56. Louis J. C., Magal E., Gerdes W., Seifert W. (1993) Survival-promoting and protein kinase C-regulating roles of basic FGF for hippocampal neurons exposed to phorbol ester, glutamate and ischaemia-like conditions. *Eur J Neurosci* , 5, 1610–1621.

57. Fernandez-Sanchez M. T., Novelli A. (1993) Basic fibroblast growth factor protects cerebellar neurons in primary culture from NMDA and non-NMDA receptor mediated neurotoxicity. *FEBS Lett* , 335, 124–131.

58. Mattson M. P., Zhang Y., Bose S. (1993) Growth factors prevent mitochondrial dysfunction, loss of calcium homeostasis, and cell injury, but not ATP depletion in hippocampal neurons deprived of glucose. *Exp Neurol* , 121, 1–13.

59. Mattson M. P., Cheng B. (1993) Growth factors protect neurons against excitotoxic/ischemic damage by stabilizing calcium homeostasis. *Stroke* , 24, 1144–1145.

60. Shitaka Y., Saito H. (1994) The effect of basic fibroblast growth factor (bFGF) and nerve growth factor (NGF) on the survival of septal neurons transplanted into the third ventricle in rats. *Jpn J Pharmacol*, 64, 27–33.

61. Mattson M. P., Kumar K. N., Wang H., Cheng B., Michaelis E. K. (1993) Basic FGF regulates the expression of a functional 71 kDa NMDA receptor protein that mediates calcium influx and neurotoxicity in hippocampal neurons. *J Neurosci* , 13, 4575–4588.

62. Cheng B., Mattson M. P. (1995) PDGFs protect hippocampal neurons against energy deprivation and oxidative injury: evidence for induction of antioxidant pathways. *J Neurosci*, 15, 7095–7104.

63. Toussaint O., Houbion A., Remacle J. (1992) Effect of cellular stress on the ageing of cultured cells. *Arch Int Physiol Biochem Biophys*, 100, B91.

64. Cheng B., Mattson M. P. (1994) NT-3 and BDNF protect CNS neurons against metabolic/excitotoxic insults. *Brain Res* , 640, 56–67.

65. Leingartner A., Heisenberg C. P., Kolbeck R., Thoenen H., Lindholm D. (1994) Brain-derived neurotrophic factor increases neurotrophin-3 expression in cerebellar granule neurons. *J Biol Chem*, 269, 828–830.

66. Beck T., Lindholm D., Castren E., Wree A. (1994) Brain-derived neurotrophic factor protects against ischemic cell damage in rat hippocampus. *J Cereb Blood Flow Metab*, 14, 689–692.

67. Lindholm D., Dechant G., Heisenberg C. P., Thoenen H. (1993) Brain-derived neurotrophic factor is a survival factor for cultured rat cerebellar granule neurons and protects them against glutamate-induced neurotoxicity. *Eur J Neurosci* , 5, 1455–1464.

68. Cheng B., Christakos S., Mattson M. P. (1994) Tumor necrosis factors protect neurons against metabolic-excitotoxic insults and promote maintenance of calcium homeostasis. *Neuron*, 12, 139–153.

69. Cheng B., Barger S. W., Mattson M. P. (1994) Staurosporine, K-252a, and K-252b stabilize calcium homeostasis and promote survival of CNS neurons in the absence of glucose. *J Neurochem* , 62, 1319–1329.

70. Goodman Y., Mattson M. P. (1994) Secreted forms of beta-amyloid precursor protein protect hippocampal neurons against amyloid beta-peptide-induced oxidative injury. *Exp Neurol* , 128, 1–12.

71. Mattson M. P., Barger S. W., Cheng B., Lieberburg I., Smith-Swintosky V. L., Rydel R. E. (1993) beta-Amyloid precursor protein metabolites and loss of neuronal Ca2+ homeostasis in Alzheimer's disease. *Trends Neurosci* , 16, 409–414.

72. Ritchie J. P. (1992) The role of glutathione in aging and cancer. *Exp Gerontol* , 27. 615–626.

73. Murphy T. H., Miyamota M., Sastre A., Schnaar R. L., Coyle J. T. (1989) Glutamate toxicity in a neuronal cell line involves inhibition of cystine transport leading to oxidative stress. *Neuron* , 2. 1547–1558.

74. Ramassamy C., Naudin B., Christen Y., Clostre F., Costentin J. (1992) Prevention by Gingko biloba extracts (EGb 761) and trolox C of the decrease in synaptosomal dopamine or serotonin uptake following incubation. *Biochem Pharmacol* , 44. 2395–2401.

75. Pazdernik T. L., Layton M., Nelson S. R., Samson F. E. (1992) The osmotic/calcium stress theory of brain damage : are free radicals involved? *Neurochem Res* , 17. 11–21.

76. Fischer-Nielsen A., Poulsen H. E., Loft S. (1992) 8-hydroxydeoxyguanosine in vitro : effets of glutathione, ascorbate, and 5-aminosalicylic acid. *Free Rad Biol Med* , 13. 121–126.

77. Chau R. M. W., Skaper S. D., Varon S. (1988) Peroxidative block of glucose utilisation and survival in CNS neuronal cultures. *Neurochem Res* , 13. 611–616.

78. Storey E., Hyman B. T., Jemkins B., Brouillet E., Miller J. M., Rosen B. R., Beal M. F. (1992) 1-methyl-4-phenylpiridinium produces excitotoxic lesions in rat striatum as a results of impairment of oxidative metabolism. *J Neurochem* , 58. 1975–1978.

79. Viani P., Cervato G., Fiorilli A., Cestaro B. (1991) Age-related differences in synaptosomal peroxidative damage and membrane properties. *J Neurochem* , 56. 253–258.

80. Iantomasi T., Favilli F., Marraccini P., Stio M., Treves C., Quatrone A., Capaccioli S., Vincenzini M. T. (1993) Age and GSH metabolism in rat cerebral cortex, as related to oxidative and energy parameters. *Mech Ageing Dev* , 70. 65–82.

81. Dawson R., Wallace D. R. (1992) Kainic acid-induced seizures in aged-rats : neurochemical correlates. *Brain Res Bull* , 29. 459–468.

82. Wozniak D. F., Steward G. R., Miller J. P., Olney J. W. (1991) Age-related sensitivity to kainate neurotoxicity. *Exp Neurol* , 114. 250–253.

83. Hoyer S. (1995) Age-related changes in cerebral oxidative metabolism. Implications for drug therapy. *Drugs Aging* , 6. 210–218.

84. Sapolski R. (1986) Glucocorticoid toxicity in the hippocampus : reversal by supplementation with brain fuels. *J Neurosci* , 6. 2240.

85. Sapolski R. (1986) Glucocorticoid toxicity in the hippocampus : temporal aspects of synergy with kainic acid. *Neuroendo* , 43. 440.

86. Ferrandiz M. L., Martinez M., De-Juan E., Diez A., Bustos G., Miquel J. (1994) Impairment of mitochondrial oxidative phosphorylation in the brain of aged mice. *Brain Res* , 644. 335–338.

87. Pettegrew J. W., Panchalingam K., Klunk W. E., McClure R. J., Muenz L. R. (1994) Alterations of cerebral metabolism in probable Alzheimer's disease: a preliminary study. *Neurobiol Aging* , 15. 117–32.

88. Hoyer S. (1994) Age as risk factor for sporadic dementia of the Alzheimer type? *Ann N Y Acad Sci* , 719. 248–256.

89. Hoyer S. (1993) Abnormalities in brain glucose utilization and its impact on cellular and molecular mechanisms in sporadic dementia of Alzheimer type. *Ann N Y Acad Sci* , 695. 77–80.

90. Carta A., Calvani M., Bravi D., Bhuachalla S. N. (1993) Acetyl-L-carnitine and Alzheimer's disease: pharmacological considerations beyond the cholinergic sphere. *Ann N Y Acad Sci* , 695. 324–326.

91. Beal M. F. (1992) Does impairment of energy metabolism result in excitotoxic neuronal death in neurodegenerative illnesses? *Ann Neurol* , 31. 119–130.

92. Mizuno Y., Ikebe S., Hattori N., Nakagawa-Hottori Y., Mochizuki H., Tanaka M., Ozawa T. (1995) *Biochem Biophys Acta* , 1271,

93. Katchman A., Hershkowitz N. (1993) Early anoxia-induced vesicular glutamate release results from mobilization of calcium from intracellular stores. *J Neurophysiol* , 70. 1–7.

94. Hossmann K. A. (1994) Glutamate-mediated injury in focal cerebral ischemia: the excitotoxin hypothesis revised. *Brain Pathol* , 4. 23–36.

95. Chueh S. H., Hsu L. S., Song S. L. (1994) Two distinct ATP signaling mechanisms in differentiated neuroblastoma x glioma hybrid NG108–15 cells. *Mol Pharmacol* , 45. 532–539.

96. Chueh S. H., Kao L. S. (1993) Extracellular ATP stimulates calcium influx in neuroblastoma x glioma hybrid NG108–15 cells. *J Neurochem* , 61. 1782–1788.

97. Chen Z. P., Levy A., Lightman S. L. (1994) Activation of specific ATP receptors induces a rapid increase in intracellular calcium ions in rat hypothalamic neurons. *Brain Res* , 641. 249–256.

98. Michiels C., Arnould T., Remacle J. (1993) Hypoxia-induced activation of endothelial cells as a possible cause of veinous diseases. An hypothesis. *Angiology* , 639–646.

99. Arnould T., Michiels C., Alexandre I., Remacle J. (1992) Effect of hypoxia upon intracellular calcium concentration in human endothelial cells. *J Cell Physiol* , 152, 215–221.

100. Michiels C., Arnould T., Knott I., Dieu M., Remacle J. (1993) Stimulation of prostaglandin synthesis by human endothelail cells exposed to hypoxia. *Am J Physiol* , 264, C866-C874.

101. Michiels C., Arnould T., Alexandre I., Houbion A., Remacle J. (1993) Effect of naftidrofuryl on hypoxia-induced activation and mortality of human endothelial cells. *J Pharmacol Exp ther* , 267, 904–911.

102. Arnould T., Michiels C., Remacle J. (1992) Effet du Ginkor fort sur l'activation des cellules endothéliales induite par une hypoxie sévère. *Act Vasc Intern* , 4, 40–47.

103. Chen C. H., Chen S. C. (1994) The efficacy of non-lactate-generating metabolites as substrates for maintaining donor tissues. *Transplantation* , 57, 1778–1785.

104. Boerrigter M. E., Franceschi C., Arrigoni-Martelli E., Wei J. Y., Vijg J. (1993) The effect of L-carnitine and acetyl-L-carnitine on the disappearance of DNA single-strand breaks in human peripheral blood lymphocytes. *Carcinogenesis* , 14, 2131–6.

105. McFalls E. O., Paulson D. J., Gilbert E. F., Shug A. L. (1986) Carnitine protection against adriamycin-induced cardiomyopathy in rats. *Life Science* , 38, 497–505.

106. Paulson D. J., Traxler J., Schmidt M., Noonan J., Shug A. L. (1986) Protection of ischaemic myocardium by L-propionylcarnitine, effects on the recovery of cardiac output after ischemia and reperfusion, carnitine transport and fatty acid oxidation. *Cardiovasc Res* , 20, 536–541.

107. Duan J. M., Karmazyn M. (1989) Effect of D,L-carnitine on the response of the isolated heart of the rat to ischemia and reperfusion : relation to mitochondrial function. *Br J Pharmacol* , 98, 1319–1327.

108. Whitmer J. T. (1987) L-carnitine treatment improves cardiac performances and restores high-energy phosphate pools in cardiomyopathic Syrian hamster. *Circulation Res* , 61, 369–408.

109. Ferrari R., Ceconi C., Cargnoni A., Pasini E., Boffa G. M., Curello S., Visioli O. (1991) The effect of propionyl-L-carnitine on the ischemic and reperfused intact myocardium and on their derived mitochondria. *Cardiovasc Drug Therap* , 5, 57–65.

110. Seki S., Mori S., Nakashima A., Oda T. (1987) Effect of ATP and other nucleotides on DNA repair synthesis in bleomycin-pretreated permeable mouse sarcoma cells. *Carcinogenesis* , 8, 1391–1394.

111. Dwarkanath B. S., Jain V. K. (1989) Energy linked modifications of the radiation response in a human cerebral glioma cell line. *Int J Radiat Oncol Biol Phys* , 17, 1033–1040.

112. Opie L. H. (1990) Myocardial ischemia - metabolic pathways and implications of increased glycolysis. *Cardiovasc Drug Ther* , 4, 777–790.

113. Jeremy R. W., Ambrosio G., Pike M. M., Jacobus W. E., Becker L. C. (1993) The functional recovery of post-ischemic myocardium requires glycolysis during early reperfusion. *J Mol Cell Cardiol* , 25, 261–276.

114. Hano O., Thompson-Gorman S. L., Zweier J. L., Lakatta E. G. (1994) Coenzyme Q10 enhances cardiac functional and metabolic recovery and reduces Ca2+ overload during postischemic reperfusion. *Am J Physiol* , 266, H2174-H2181.

115. Bertelli A., Cerrati A., Giovannini L., Mian M., Spaggiari P., Bertelli A. A. (1993) Protective action of L-carnitine and coenzyme Q10 against hepatic triglyceride infiltration induced by hyperbaric oxygen and ethanol. *Drugs Exp Clin Res* , 19, 65–8.

116. Sandhu G. S., Burrier A. C., Janero D. R. (1993) Adenosine deaminase inhibitors attenuate ischemic injury and preserve energy balance in isolated guinea pig heart. *Am J Physiol* , 265, H1249-H1256.

117. Brossas J. Y., Barreau E., Courtois Y., Treton J. (1994) Multiple deletions in mitochondrial DNA are present in senescent mouse brain. *Biochem Biophys Res Commun* , 202, 654–659.

118. Gadaleta M. N., Rainaldi G., Lezza A. M., Milella F., Fracasso F., Cantatore P. (1992) Mitochondrial DNA copy number and mitochondrial DNA deletion in adult and senescent rats. *Mutat Res*, 275, 181–193.

119. Lee C. M., Chung S. S., Kaczkowski J. M., Weindruch R., Aiken J. M. (1993) Multiple mitochondrial DNA deletions associated with age in skeletal muscle of rhesus monkeys. *J Gerontol* , 48, B201-B205.

120. Linnane A. W., Zhang C., Baumer A., Nagley P. (1992) Mitochondrial DNA mutation and the ageing process; bioenergy and pharmacological intervention. *Mutation Res* , 275, 195–208.

121. Yen T. C., King K. L., Lee H. C., Yeh S. H., Wei Y. H. (1994) Age-dependent increase of mitochondrial DNA deletions together with lipid peroxides and superoxide dismutase in human liver mitochondria. *Free Radic Biol Med* , 16, 207–214.

122. Yen T. C., Pang C. Y., Hsieh R. H., Su C. H., King K. L., Wei Y. H. (1992) Age-dependent 6kb deletion in human liver mitochondrial DNA. *Biochem Int* , 26, 457–68.

123. Yen T. C., Su J. H., King K. L., Wei Y. H. (1991) Ageing-associated 5 kb deletion in human liver mitochondrial DNA. *Biochem Biophys Res Commun* , 178, 124–131.

124. Linnane A. W., Ozawa T., Marzuki S., Tanaka M. (1992) Mitochondrial DNA mutations as an important contributor to ageing and degenerative diseases. *Lancet* , March 25, 642–645.

125. Wallace D. C. (1992) Mitochondrial genetics: A paradigm for aging and degenerative diseases? *Science* , 256, 628–632.
126. Baumer A., Zhang C., Linnane A. W., Nagley P. (1994) Age-related human mtDNA deletions: a heterogeneous set of deletions arising at a single pair of directly repeated sequences. *Am J Hum Genet* , 54, 618–630.
127. Pang C., Lee H., Yang J., Wei Y. (1994) Human skin mitochondrial DNA deletions associated with light exposure. *Arch Biochem Biophys* , 312, 534–538.
128. Lee H. C., Pang C. Y., Hsu H. S., Wei Y. H. (1994) Differential accumulations of 4,977 bp deletion in mitochondrial DNA of various tissues in human ageing. *Biochim Biophys Acta* . 1226, 37–43.
129. Hayashi J., Ohta S., Kagawa Y., Takai D., Miyabayashi S., Tada K., Fukushima H., Inui K., Okada S., Goto Y. (1994) Functional and morphological abnormalities of mitochondria in human cells containing mitochondrial DNA with pathogenic point mutations in tRNA genes. *J Biol Chem*, 269, 19060–19066.
130. Munscher C., Muller-Hocker J., Kadenbach B. (1993) Human aging is associated with various point mutations in tRNA genes of mitochondrial DNA. *Biol Chem Hoppe Seyle* , 374, 1099–1104.
131. Munscher C., Rieger T., Muller-Hocker J., Kadenbach B. (1993) The point mutation of mitochondrial DNA characteristic for MERRF disease is found also in healthy people of different ages. *FEBS Lett* , 317, 27–30.
132. Huang C. C., Chen R. S., Chen C. M., Wang H. S., Lee C. C., Pang C. Y., Hsu H. S., Lee H. C., Wei Y. H. (1994) MELAS syndrome with mitochondrial tRNA(Leu(UUR)) gene mutation in a Chinese family. *J Neurol Neurosurg Psychiatry* , 57, 586–589.
133. Zhang C., Linnane A. W., Nagley P. (1993) Occurrence of a particular base substitution (3243 A to G) in mitochondrial DNA of tissues of ageing humans. *Biochem Biophys Res Commun* , 195, 1104–1110.
134. Hayashi J., Ohta S., Kagawa Y., Kondo H., Kaneda H., Yonekawa H., Takai D., Miyabayashi S. (1994) Nuclear but not mitochondrial genome involvement in human age-related mitochondrial dysfunction. Functional integrity of mitochondrial DNA from aged subjects. *J Biol Chem* , 269, 6878–6883.
135. Ammendola R., Mesuraca M., Russo T., Cimino F. (1992) Sp1 DNA binding efficiency is highly reduced in nuclear extracts from aged rat tissues. *J Biol Chem* , 267, 17944–17948.
136. Ammendola R., Mesuraca M., Russo T., Cimino F. (1994) The DNA-binding efficiency of Sp1 is affected by redox changes. *Eur J Biochem* . 225, 483–489.
137. Haraguchi Y., Chung A. B., Neill S., Wallace D. C. (1994) OXBOX and REBOX. overlapping promoter elements of the mitochondrial F0F1-ATP synthase beta subunit gene. OXBOX/REBOX in the ATPsyn beta promoter. *J Biol Chem* , 269, 9330–9334.
138. Bandy B., Davison A. J. (1990) Mitochondrial mutation may increase oxidative stress: Implications for carcinogenesis and aging? *Free Rad Biol Med* , 8, 523–539.
139. Corbisier P., Raes M., Michiels C., Pigeolet E., Houbion A., Delaive E., Remacle J. (1990) Influence of the energetic pattern of mitochondria in cell ageing. *Mech Ageing Dev* , 51, 249–262.
140. Corbisier P., Toussaint O., Remacle J. (1995) In *Thirty years of progress in mitochondrial bioenergetics and molecular biology*; F. Palmieri, S. Papa, C. Saccone and M. N. Gadaleta. Ed.: Elsevier: Amsterdam. Vol. : pp 31–35.
141. Remacle J., Michiels C., Raes M. (1992) The importance of antioxidant enzymes in cellular ageing and degeneration. *In "Free radicals and ageing"* I Emerit and B Chance Eds. Birkhäuser Verlag AG. Basel, pp. 99–108.
142. Calvani M., Caruso G., Iannuccelli M., Carta A. (1990) In *Stress and the aging brain*; G. Nappi, E. Martignoni, A. R. Genazzani and F. Petraglia, Ed.; Raven Press: New York. Vol. 37; pp 131–137.
143. Arenas J., Huertas R., Campos Y., Diaz A. E., Villalon J. M., Vilas E. (1994) Effects of L-carnitine on the pyruvate dehydrogenase complex and carnitine palmitoyl transferase activities in muscles of endurance athletes. *FEBS let* . 341, 91–93.

8

GROWTH FACTORS AND AGEING

The Case of Wound Healing

Dimitri Stathakos, Dimitris Kletsas, and Stelios Psarras

Laboratory for Enzyme Research
Institute of Biology
National Centre for Scientific Research "Demokritos"
153 10 Athens, Greece

1. INTRODUCTION[*]

The structural and functional integrity of a living organism is impaired in tissue lesion. In this case, homeostasis serves as a universal mechanism which restores the tissue continuity by a specific sequence of events, collectively termed "wound healing". This process is regulated by a complex network of growth factors and cytokines acting on a wide range of target-cell types[1]. In ageing, where by definition a decline in homeostatic capability is observed, wound healing is significantly deterred, as has been widely confirmed by clinical studies.

The complexity of parameters involved in these processes in the intact organism, has rendered the *in vitro* systems into basic tools of absolute importance for the respective studies. Especially for the studies on ageing, cell assay systems have been proved extremely valuable for understanding and exploring the underlying mechanisms, the reason for this being the fact that isolated cells cultured *in vitro* can be grown under fully controlled conditions. Therefore, they offer a well defined system of interactions, whose parameters can be relatively easily monitored by modern biochemical techniques. In this aspect, the findings of Leonard Hayflick in the 1960's—that cultured normal human fibroblasts exhibit a finite life span, at the end of which, the cell population is no more able to undergo cellular division[2,3]—opened a whole new area of investigation. Indeed, several lines

---

* Abbreviations used in the text: Cdks: cyclin-dependent kinases; CPD: cell population doubling; DAG: diacylglycerol; ECM: extracellular matrix; EGF: epidermal growth factor; ERK : extracellular signal-regulated kinase; FGF: fibroblast growth factor; gas genes: growth-arrest-specific genes; IGF-I: insulin-like growth factor-I; MAPK: mitogen-activated protein kinase; NGF: nerve growth factor; PDGF: platelet-derived growth factor; PDGFR : PDGF receptor; PI-3K: phosphoinositide 3-kinase; PLC-γ: phospholipase C-γ: PLD: phospholipase D; PKC: protein kinase C; TGF-α : transforming growth factor-alpha; TGF-β: transforming growth factor-beta; TNF-α: tumour necrosis factor-alpha.

Molecular Gerontology, edited by Rattan and Toussaint
Plenum Press, New York, 1996

of evidence suggest that *in vitro* cellular ageing is analogous to *in vivo* ageing (reviewed in Refs. 4–6).

The characteristics of cellular ageing *in vitro* have been described in a number of normal cell systems of different origins, as, for example, keratinocytes, chondrocytes, epithelial cells, smooth muscle cells, glial cells, vascular endothelial cells, hepatocytes, spleen leucocytes and lymphocytes[7]. However, as the most thoroughly studied examples remain the ageing diploid fibroblast strains, they will be extensively covered in this chapter.

During the last years, we have investigated some important aspects of cellular ageing in this system, and in the present article, we will discuss some of the recent data and their respective consequences for the study of ageing.

2. THE PROLIFERATIVE CAPACITY OF THE CELLS DECLINES WITH AGEING

2.1. Growth Factors Are Involved in Cellular Ageing

One of the main changes exhibited by cell populations during their *in vitro* ageing is the gradual loss of their replicative capacity. A senescent culture of normal fibroblasts, i.e. a culture consisting of cells close to the end of their life span, is no longer able to proliferate under the influence of any endogenous or exogenous stimuli. As characteristically shown in Fig. 1, the proliferative response of human embryonic fibroblasts to human serum, decreases dramatically during *in vitro* ageing.

The loss of cell capacity to divide *in vitro* is a very important issue, since it may reflect an *in vivo* cellular feature, critical for the ageing of an organism. Therefore, a considerable amount of research effort in gerontology has been aimed at investigating the possible mechanisms of this loss of responsiveness.

In the following, we will summarize some of the more recent results which try to elucidate the molecular aspects of this phenomenon.

Cellular proliferation is regulated positively or negatively by a set of growth regulatory molecules, e.g. growth factors, cytokines and hormones, secreted by a variety of cell types and tissues in the organism. "Traditional" growth factors, such as platelet-derived growth factor (PDGF), epidermal growth factor (EGF), fibroblast growth factor (FGF) etc., are polypeptides that can act as mitogens, i.e. they can trigger the proliferation of various cell types. Serum for example, a well known mitogenic stimulus, is actually a

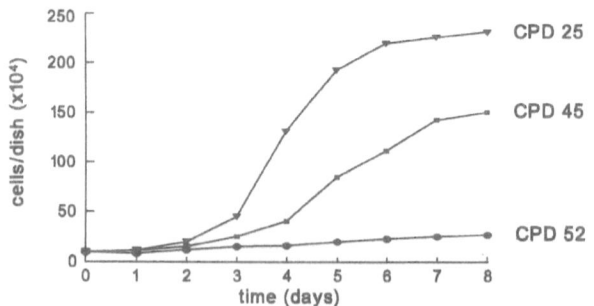

Figure 1. Age-related decline in the responsiveness of fibroblasts to external growth stimuli. After plated in 50-mm dishes at a density of 10^5 cells/dish (approx. 5×10^3 cells/cm^2), human embryonic fibroblasts of different cell population doubling (CPD) levels (25, 45, 52, respectively) were grown in MEM containing 10% human serum. Cells were harvested and counted daily in duplicates (adapted from Ref. 11).

mixture of growth factors and other growth regulatory molecules. As expected, the loss of cellular replicative capacity during ageing (see Fig. 1), reflects a diminished response of the cells to individual growth factors and hormones[8,9].

In particular, age-related decrease in the proliferative response of senescent cells has been reported, among other mitogens, for: PDGF[10-12], EGF[10,12], FGF[11,12], insulin-like growth factor-I (IGF-I)[10,13], insulin[10] and dexamethasone[14].

Moreover, we have recently found that human embryonic fibroblasts arrested in the G_0 phase of the cell cycle, when cultured at sparsity in homologous environment, enter S phase after treatment with the multifunctional agent tranforming growth factor-beta (TGF-β)[15]. This proliferative response also declines gradually with cellular *in vitro* ageing.

Why are then the exogenous mitogens unable to subvert the senescent state?

Extracellular signals such as hormones, growth factors and cytokines are acting at the cellular level by binding to specific cytoplasmic or membrane receptors, which are then activated and through a series of complex enzymic[+] reactions, occuring inside the cells, regulate the action of several transcription factors. Like many cellular responses, the mitogenic response is also the final result of a specific programme of transcriptional events activated by exogenous stimuli. The first step in this process is the binding of the growth regulatory molecules to their cellular receptors.

2.2. Growth Factor Binding during Cellular Ageing

From the above, it is plausible to suspect that specific ageing-related defects in the signal transduction pathways leading to transcriptional activation might result in the loss of proliferative response observed during ageing. There are some recent data that reinforce this hypothesis; in order to understand them, we have first to summarize briefly the main features of the signal transduction pathways of "traditional" growth factors, which are the issue of this article. All other growth regulatory molecules, as for example polypeptides involving G-proteins in their signal transduction, have a rather moderate impact on the proliferative response, especially of connective tissue cells, and will not be handled here.

Growth factors, such as PDGF, EGF, FGF etc., exert their cellular action by binding to specific membrane receptors whith intrinsic protein tyrosine kinase activity. After ligand binding, receptor dimerization occurs, leading to activation of the tyrosine kinase activity, as well as autophosphorylation of the receptor molecules (for review see Ref. 16). The phosphorylated receptors are then able to recruit and bind specific cytoplasmic molecules (see Fig. 2 and respective legend), which contain SH2-regions, i.e. sequences homologous to a specific part of the Rous sarcoma virus oncogene, src. The recruited molecules are then activated, triggering a cascade of enzymic reactions, which further transduce the mitogenic signal within the cell.

The major signal transduction pathway within this superfamily of growth factors is the so-called Ras/Raf/MEK/ERK pathway, which leads to trancriptional activation of a set of genes (Fig. 2). Some of these genes (e.g. c-fos, c-jun) control in turn, among other cellular responses, the traverse of the G1/S border of the cell cycle and subsequent mitosis (for recent reviews see Refs. 17, 18). This is a typical MAPK (mitogen-activated protein kinase)-pathway, consisting of a set of serine/threonine and tyrosine phosphorylations. In

---

+ Note that the term "enzymic" is preferred here, instead of the erroneous form "enzymatic". "Enzyme" from the Greek word ένζυμον, originating from the prefix εν= in and ζύμη = yeast, i.e. literally "in yeast", forms its correct derivatives from the stem word "enzym-" and the appropriate ending, such as enzymic, enzymically, enzymology, etc.

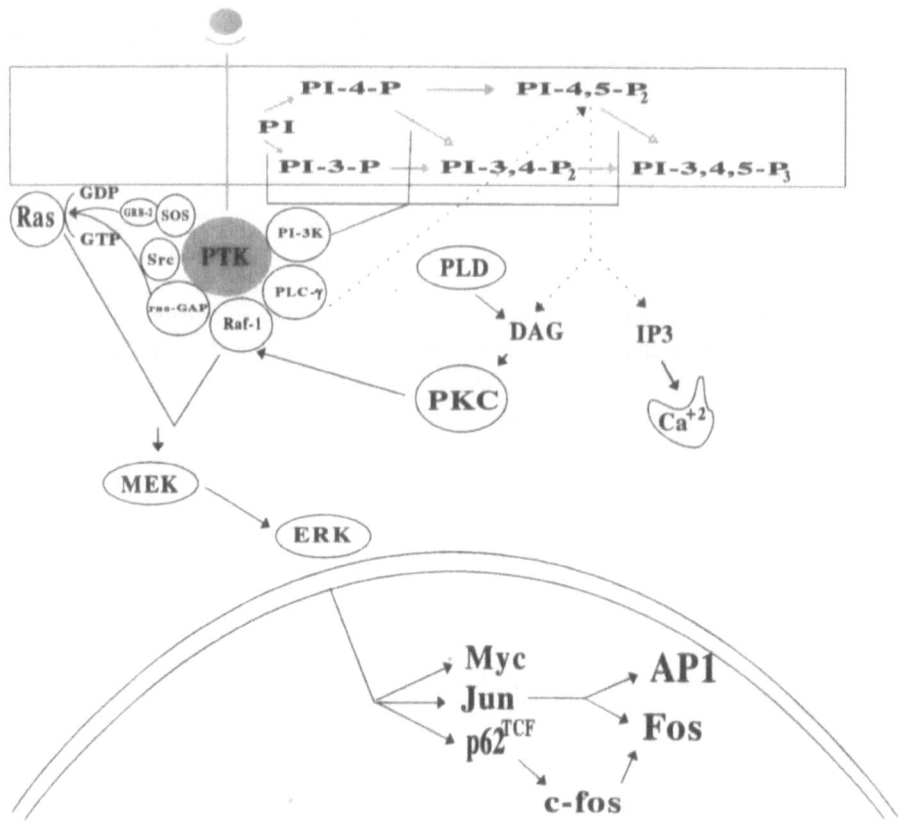

Figure 2. Signal tranduction pathways of growth factor receptors with protein tyrosine kinase activity. After ligand (e.g. PDGF) binding at the cell surface receptor, the intrinsic protein kinase (PTK) activity of the receptor is
activated. As a result, several cytoplasmic molecules containing SH2-regions (see text) are recruited and activated
by the autophosphorylated receptor dimer (the dimerization is not depicted in the figure for reasons of simplicity).
Among them are the products of the family of src oncogenes and ras-GAP, which stimulates the intrinsic GTPase
activity of the Ras protein. Subsequently, a cascade of phosphorylations is initiated, transducing the mitogenic signal to the nucleus. Here is depicted the major signal transduction pathway, the so called "Ras/Raf/MEK/ERK"
pathway; initially, the GRB-2/SOS complex activates Ras protein, which then associates with the product of the c-
raf oncogene (also recruited by the PTK receptor), leading to activation of the dual specificity protein kinase (mitogen activated protein kinase kinase, MAPKK or MEK). The latter phosphorylates and activates the
serine/threonine kinase ERK (extracellular signal- regulated kinase), leading finally to transcriptional activation of
specific genes in the nucleus. Among them, very important for the G1→S transition, are: c-fos, which is activated
by the ternary complex factor p62^{TCF}, c-jun, the product of which forms, as a homodimer or as a heterodimer with
the product of the c-fos, the very important transcription factor AP1, and c-myc. Apart from the
Ras/Raf/MEK/ERK pathway, phospholipase C-γ (PLC-γ) causes the hydrolysis of phosphatidylinositol 4,5 biphos-
phate (PI 4,5 P_2) to inositol triphosphate (IP$_3$) and diacylglycerol (DAG). IP$_3$ triggers the release of the very important second messenger Ca^{+2} from its intacellular deposits. On the other hand, DAG, also activated by
phospholipase D (PLD), activates protein kinase C (PKC), a serine/threonine kinase also implicated in the
Ras/Raf/MEK/ERK pathway. Finally, phosphoinositide 3-kinase (PI3K) generates a set of 3-phosphorylated phosphatidylinositol (PI) products, which are believed to be second messengers.

particular, Raf-1 (which by exception is recruited by the phosphorylated receptor, although it does not contain any SH2 region) and ERK (extracellular signal-regulated kinase) are serine/threonine kinases, whereas MEK is a "dual specificity" kinase. Apart from the Ras/Raf/MEK/ERK pathway, other activated enzymes lead to the formation of second messengers of special importance for cellular stimulation (Fig. 2); these include:

- phosphoinositide 3-kinase (PI-3K), which generates 3-phosphorylated inositol lipids
- phospholipases C-γ (PLC-γ) and D (PLD), which generate diacylglycerol (DAG). DAG activates then protein kinase C (PKC), which in turn leads again to activation of the Ras/Raf/MEK/ERK pathway (see Fig. 2). PKC is a serine/threonine kinase, and its activation is a key event in the mitogenic signal transduction (for a recent review see Ref. 19).

Since the first step of this complex scheme of mitogenic signal transduction is the binding of the growth factor to its membrane receptors (Fig. 2), some laboratories have asked whether the diminished proliferative response during ageing reflects a relevant defect at the binding level. We focused our interest on PDGF, since this growth factor is the most potent mitogen for connective tissue cells.

To our surprise, we and others have found that PDGF binding per cell is significantly increased during cellular *in vitro* ageing, mainly due to an increase in the number of PDGF receptors, whose affinity for the ligand remains though unaltered[8,20-22]. A similar age-related increase has been reported for EGF[23] and IGF-I[24] binding, again with no change in the apparent dissociation constant (K_D). However, by taking into account the well-known increase in cell size during *in vitro* ageing, it has been shown that the number of EGF- or IGF-I receptors per square micrometer of cell surface area remains practically unchanged[23,24]. Thus, the increase in growth factor binding during cellular ageing could be simply attributed to differences in cell size. The fact remains however, that the overall PDGF binding to cells is increased during ageing, although the proliferative response is reduced.

However, it should be mentioned that, in contrast to the results obtained during *in vitro* ageing of embryonic fibroblasts[23], the situation in fibroblasts of early *in vitro* passages, but derived from newborn or young adults or old donors, i.e. of cells aged *in vivo*, is as follows: the numbers of EGF receptors per cell, as well as their internalization rates are reduced during *in vivo* ageing[25]. This, suggests that even at the level of membrane events, there is a marked difference between embryonic cells aged *in vitro* and adult cells aged *in vivo*. This will be further analyzed in the last section of this article.

Furthermore, we have recently observed a very interesting age-related change concerning the ability of the cells to overcome a crisis at the receptor level, e.g. their ability to restore PDGF binding, after down-regulation of PDGF receptors. Down-regulation is a common phenomenon for various growth factor receptors, such as those of EGF [26] and PDGF[27]. After ligand binding at 37°C, the complex of the growth factor with its receptor is endocytosed via clathrin-coated pits. Following internalization the receptor is found in endosomes, where it can either enter the degradative pathway to the lysosomes, or recycle back to the cell surface. After binding with excess PDGF, the cell surface is depleted from PDGF receptors as a consequence of their degradation. This depletion is termed down-regulation of PDGF receptors. We examined the restoration rates of PDGF binding after down-regulation and found that aged fibroblasts achieve higher restoration levels than younger cells[22]. Moreover, late-passage cells treated with TGF-β (which induces the restoration of down-regulated PDGF receptors[22]), completely recover their initial binding capacity for PDGF within 24 hours, namely much earlier than early-passage fibroblasts.

In summary, by comparing the characteristics of PDGF-binding at various ageing levels, we observed that a set of its parameters (K_D, internalization and degradation rates of PDGF, extent of PDGF receptor down-regulation) remains unchanged during *in vitro* ageing, whereas late passage fibroblasts express more PDGF receptors and exhibit more efficient restoration mechanisms than early passage cells.

Additionally, no change has been observed in the autophosphorylation of EGF- and PDGF-receptors after ligand binding, in membrane preparations (for PDGF receptors) or intact cells (for both receptors), between senescent and young fibroblasts[21].

The above indicate that the decrease in proliferative response can not be attributed to a defect of the growth factor binding system. Therefore, we have to look for an explanation downstream along the signal transduction pathway to the cell nucleus.

2.3. Alterations in the Signal Transduction from the Cell Surface to the Nucleus during Ageing

Some evidence that there are distinct age-related defects in the tyrosine kinase receptor pathways came recently from studies on phospholipase D (PLD). PLD, a DAG generating enzyme (Fig. 2), is not activated in senescent cells, possibly due to abundance of its inhibitor, ceramide[29]. The inactivation of PLD results also in defective generation of phosphatidic acid (a potent mitogen for fibroblasts[30]), as well as, of course, DAG. DAG generation is an intermediate step in the activation of PKC, one of the key enzymes in the mitogenic signal transduction pathway (Fig. 2). An important step in PKC activation is its translocation to the plasma membrane. PKC translocation is significantly reduced in senescent cells after stimulation by serum, but not by phorbol esters (e.g. PMA)[31]. The tumour promoting agent PMA activates PKC directly by binding, bypassing PLD and DAG. Since PKC activation by serum is mediated through PLD, the above are consistent with a defect in PLD activation in senescent cells.

This defect could explain the respective diminished proliferative response of senescent cells to growth factors. However, this is not enough, since PLD is responsible for only a part of the mitogenic signal transduction (see also Fig. 2). Therefore we have to focus on the targets of the mitogenic signal transduction, i.e. genes expressed after growth factor stimulation.

2.4. Altered Gene Expression during Cellular Ageing

Cytoplasmic signal transduction leads to activation of specific genes that are responsible for cell proliferation. Senescent cells are incapable of novel DNA synthesis (i.e. entry into S phase) after mitogen treatment. Accordingly, research on cellular ageing has focused on the defects of the expression of those "cell cycle marker" genes, which are involved in the G1 → S transition.

Some of the properties of senescent human fibroblasts coincide with those of their young counterparts, arrested in the G_0 phase of the cell cycle by serum deprivation or contact inhibition, or both[32]. However, senescent and arrested young cells do not represent the same state, as it has been shown by using flow cytometric analysis that senescent diploid fibroblasts are not blocked at the G_0 phase, but rather at the G1/S boundary[33]. Furthermore, their main characteristic difference is that while serum- or growth factor-treated young arrested cells initiate a signal transduction cascade that leads to entry into S phase and concomitant DNA synthesis, senescent cells are unable to do so.

So, much effort has been invested to identify the defects of the signal transduction pathway in stimulated senescent cells. It has been found that a large percentage of cell cycle regulated genes, which are expressed during the G1 → S transition, are still expressed by senescent fibroblasts (Fig. 3). These are early-G1 genes (such as c-myc, c-jun, junB, c-erbB, 4F1, JE-3) or mid-G1 genes (such as c-H-ras, p53, odc, 2F1, 2A9, eIF-5A) (for a review see Ref. 32).

On the other hand, some important age-related defects in the expression of a number of other G1 → S genes have been found by several laboratories. In particular:

There is an inability of senescent fibrobasts to express the immediate early gene c-fos after stimulation by serum[34]. As already reported (see section 2.2.), the product of c-fos represents part of the AP-1 transcription factor, whose activity is required for the proliferation of cells after stimulation by serum or individual mitogens. The DNA binding activity of AP-1 is substantially decreased as fibroblasts age[35]; it has also been observed that the serum response factor (SRF), bound to the serum response element (SRE) of the c-fos promoter and acting as a trancriptional activator for this gene, is hyperphosphorylated in senescent human diploid fibroblasts. Hyperphosphorylated SRF seems to be unable to bind to SRE, leading possibly to

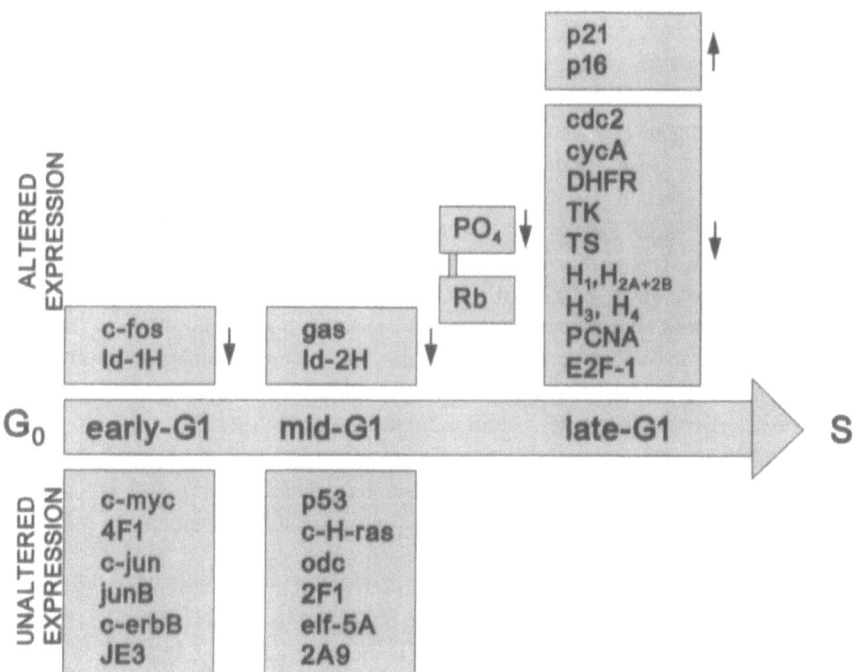

Figure 3. Altered gene expression during cellular ageing. The major genes expressed during early-, mid- and late-G1 are demonstrated here. Listed in the rectangles above the horizontal G1 → S arrow are the genes, whose expression is altered, i.e., enhanced (↑) or repressed (↓) during cellular ageing. In the case of the retinoblastoma protein (Rb), its hypophosphorylated state is also represented by a small vertical arrow (↓). The rectangles below the G1 → S arrow contain those genes, whose expression remains unaltered during cellular ageing. DHFR: dihydrofolate reductase; PCNA: proliferating cell nuclear antigen; H: histone; TK: thymidine kinase; TS: thymidilate synthetase; cyc: cyclin.

diminished c-fos expression[36]. However, as Rose et al.[37] have shown, re-stimulation of c-fos expression and functional AP-1 transcription activity—obtained through microinjection of c-H-ras protein—in senescent fibroblasts, does not lead to DNA synthesis. This suggests that the senescent phenotype cannot be solely attributed to impaired c-fos expression and AP-1 activity, but may involve other defects too.

A second case is a group of six genes, the growth-arrest-specific (gas) genes, all well known to be highly expressed in arrested fibroblasts and down-regulated after mitogen treatment during the mid-G1[38,39]. Cowled et al.[40] have shown that when murine fibroblasts become senescent *in vitro*, they express dramatically reduced amounts of gas1 and gas6 compared to young cells, but only barely detectable levels of gas2, gas3 and gas5. Furthermore, while gas1 was clearly down-regulated after serum treatment, no down-regulation was observed for gas6, confirming a general difference in the regulation of expression of the different gas genes in senescent murine fibroblasts (Fig. 3). However, in human fibroblasts, we have found that at least three of the gas genes, i.e. gas1, gas3 and gas6, show the same pattern of expression in young, as well as in senescent cells[28].

Another set of genes, the Id genes, which encode helix-loop-helix (HLH) proteins are involved in the G1 $\rightarrow$ S transition[41,42]. The expression of the Id-1H and Id-2H genes is induced after serum stimulation of young human fibroblasts with transient peaks at 1–2h and 10–12h, respectively[43]. Accordingly, antisense Id oligomers inhibit the entry of human fibroblasts into S phase, suggesting that Id proteins inhibit one or more growth-suppressive HLH transcription factors present in quiescent cells. Id-1H and Id-2H mRNAs were barely induced in senescent human fibroblasts (Fig. 3), supporting the idea that their suppression may be responsible for the senescent phenotype[43]. Nevertheless, as was also the case of the c-fos gene (see above), even when these genes were forced into expression, they failed to induce DNA synthesis. Therefore, their expression seems to be unable to prevent cellular senescence.

Even more extended age-related changes have been observed at the G1/S boundary. This was expected as it is at this point of the cell cycle that senescent cells are arrested[33]. Pang and Chen[45] have observed a "global" change, as they termed it, in gene expression at this boundary. In particular, they have found in senescent human fibroblasts a decrease in the mRNA production of a group of "G1/S" genes, i.e. thymidine kinase, thymidylate synthetase, dihydrofolate reductase, ribonucleotide reductase, proliferating cell nuclear antigen, histone H1, histone H2A+2B, histone H3 and histone H4 (Fig. 3).

A very important and possibly key event observed in senescent human fibroblasts is their inability to fully phosphorylate the retinoblastoma protein, pRb[46] (Fig.3). Hypophosphorylated pRb inhibits the cells' progression from G1 into S, possibly by binding to transcription factors, such as E2F, which are necessary for the expression of late-G1 genes. Furthermore, the E2F-1 gene mRNA expression is repressed in senescent human fibroblasts. On the other hand, the restoration of E2F-1 expression cannot induce DNA synthesis in senescent cells[47]. Another gene involved in pRb phosphorylation is the cdc2 gene. The product of this gene, $p34^{cdc2}$, in association with cyclins (cyc), plays a key role in the entry of cells into the S phase, possibly by phosphorylating the retinoblastoma protein. Furthermore, Stein et al.[44] have shown, that senescent cells fail to express cdc2, cyc A and cycB after mitogen treatment (Fig. 3). However, nuclear microinjection of cdc2 DNA into senescent cells, although it causes rounding of the cells, it is unable to restore DNA synthesis. pRb phosphorylation is achieved by the kinase activity of cyc D/Cdk4 and cyc E/Cdk2 complexes[48]. Senescent cells have low cyc E/Cdk2 activity and as a consequence fail to phosphorylate the pRb protein. This is probably due to the accumulation of large amounts of an inhibitor of cyclin-dependent kinases

(Cdks)[49–52], known as p21 [SDI1, CIP1, WAF1], which probably binds to cyclin E/Cdk2 complexes, rendering them inactive in senescent cells[32].

Finally, p16[ink4] is another gene product with properties of a G1-cyclin-Cdk inhibitor. Accordingly, mutational inactivation of p16[ink4] can result in uncontrolled cell growth[53]. In senescent human fibroblasts, elevated levels of p16[ink4] have been observed, as in the case of p21, suggesting that these two cyclin-Cdk inhibitors may be significant for the inability of senescent cells to enter the S phase of the cell cycle[4].

3. HOMEOSTASIS, APOPTOSIS, GROWTH FACTORS AND AGEING

A major homeostatic force regulating tissue mass and architecture in embryonic development, as well as in the adult tissue, is cell death by apoptosis[54]. It is characterized by specific morphological and biochemical changes that distinguish this type of cell death from necrosis. During apoptosis, chromatin condensation is observed followed by DNA fragmentation and eventually formation of membrane-bound apoptotic bodies. These cell fragments are engulfed by neighboring cells and macrophages *in vivo*, showing that apoptosis is a controlled manner of eliminating unwanted cells without immunological reactions. This process can be triggered in various tissues and cell types, *in vivo* and *in vitro*, by external noxes, such as cytotoxic drugs. Particularly *in vitro*, apoptosis is commonly achieved by serum deprivation [55,56]. In the process of determining whether defined serum components can modulate cell death induced by serum-removal, it has been shown that apoptosis can be suppressed by certain growth factors, meaning that the latter can be considered also as survival factors (see also Table 1). Even more interesting, growth factor-

Table 1. Effect of growth factors on apoptosis

Treatment	Cell type
Induction of apoptosis	
Deprivation of	Normal cells
Serum	Human fibroblasts,
	Rat pheochromocytoma
FGF	Human endothelial cells
EGF	Rat kidney
NGF	Rat neurons
Addition of	Tumour cells
TNF-α	Human prostatic cells
Activin	Murine myeloma
EGF	Human breast carcinoma
TGF-β	Human cancer ovarian cells
	Human gastric carcinoma
	Human transformed fibroblasts
	Rat hepatoma
Rescue from apoptosis	
Addition of	Normal cells
IGF-I	Rat neurons
EGF/FGF	Mouse astrocyte progenitor cells
PDGF	Human fibroblasts
FGF	Bovine endothelial cells
IGF-I/IGF-IR	Mouse fibroblasts

mediated protection from apoptosis is also evident in the post-commitment phases of the cell cycle, i.e. S, G2 and M phases, the progression of which is growth-factor independent, showing that the antiapoptotic function of growth factors is not directly linked to their proliferative activity[57].

In this context, PDGF, the major serum mitogen, as well as IGF-I, protect quiescent rat fibroblasts with constitutively expressed, i.e. deregulated, c-myc expression from undergoing apoptosis[57]. Similar protective effects have been reported for the action of PDGF on glial cells[58] and for IGF-I and its receptor (IGF-IR) on various normal and tumour cells[59–61]. In the previous example of rat fibroblasts driven to apoptosis by deregulated c-myc expression—and in contrast to PDGF and IGF-I—EGF and FGF do not appear to act as survival factors. On the other hand, their presence can inhibit apoptosis in other types of cells, like mouse astrocyte progenitor cells[62], bovine endothelial cells[63], human umbilical vein endothelial cells[64] and rat kidney cells[65]. The above indicate that the ability of each growth factor to act as survival factor is cell-type specific (see also Table 1).

Even more interesting is the case of the factors that induce apoptosis in malignant cells or tissues (Table 1). Such factors are tumour necrosis factor-α (TNF-α)[66], activin[67], EGF[68] and TGF-β[69–71]. Notably, in co-cultures of normal and transformed fibroblasts, TGF-β triggers the elimination of transformed cells by induction of apoptosis[72]. The above demonstrate a general protective effect by growth factors for the normal tissue. Their role as "survival" factors appears of equal importance towards tissue homeostasis, as their ability to promote cell proliferation.

Recently, Wang[56] reported that serum deprivation induces apoptotic death in quiescent young human fibroblasts, as well as in senescent cells, though not to the same extent, suggesting that *in vitro* aged human fibroblasts are resistant to undergoing programmed cell death. Furthermore, an unusually high concentration of the product of the "antiapoptotic" bcl-2 oncogene has been found in senescent cells. This finding, along with the repression of c-fos expression and pRb phosphorylation, as well as the increased presence of p21 (see above, section 2.4.), could be viewed as a mechanism by which senescent cells inhibit their own cell cycle traverse, that may be abortive, resulting in their cell death[56]. From this, one could deduce that *in vitro* cellular senescence introduces a selection process which may favour survivors, i.e. cells that are less dependent to exogenously added growth factors[56].

4. THE CASE OF WOUND HEALING

4.1. The Wound-Healing Process

As stated in the beginning, the most characteristic demonstration of tissue homeostasis in the adult organism is the process of wound healing. For better understanding, wound repair can be divided in three consecutive phases: inflammation, granulation and remodelling[1]. As in every unified *in vivo* process, these phases exhibit a considerable overlapping: After wound formation and blood vessel disruption, blood platelets aggregate to effect blood coagulation and haemostasis. Furthermore, they release a large number of growth factors that are chemotactic and induce cell proliferation. Most important among the attracted cells are fibroblasts and monocytes/macrophages. Macrophages in turn, beyond phagocytosing bacteria and scavenging cell debris, can also release chemotactic agents and growth factors. Consequently, granulation tissue is formed, containing mainly fibroblasts, macrophages, and a loose extracellular matrix (ECM), composed of collagen, fi-

Table 2. Growth factors speed wound healing

PDGF:
- is a chemoattractant for fibroblasts, smooth muscle cells and inflammatory cells
- induces proliferation of fibroblasts and smooth muscle cells
- activates macrophages for the production and secretion of growth factors and cytokines
- stimulates production and secretion of extracellular matrix proteins.

EGF:
- primarily stimulates the growth of epithelial cells and is involved in angiogenesis.

FGF:
- stimulates proliferation of endothelial cells and is also involved in angiogenesis.

TGF-α:
- stimulates endothelial cell proliferation
- stimulates collagen synthesis.

IGF-I:
- regulates connective tissue formation.

TGF-β:
- induces synthesis and deposition of extracellular matrix components
- regulates the proliferation of numerous cell types
- regulates the immune responses in the wound healing area.

bronectin and hyaluronic acid. ECM proteins are deposited mainly by the attracted fibroblasts and provide a final substratum for the migration and proliferation of cells. At the same time, angiogenesis sets in, in order to establish tissue integrity. During the subsequent phase of the healing process, matrix remodelling occurs by the elimination of most fibronectin and the accumulation of type I collagen to provide the regenerating tissue with increased strength. The final result is a gradual substitution of the granulation tissue by the newly formed connective tissue. By that time, re-epithelialization is also completed, providing the wound surface with a new epithelium, consisting mainly from keratinocytes. Finally, myofibroblasts effect the contraction of the wound edges and the healing process is completed.

During the last stages of wound repair, characteristic changes can be observed concerning the cell composition of the new tissue: During the elimination of the granulation tissue there is a significant decrease in cellularity; in particular, the population of fibroblasts and epithelial cells is reduced. Furthermore, as the wound closes and evolves into a scar, myofibroblasts disappear. This temporally regulated disappearance of the various cell types, such as myofibroblasts, seems to be mediated by apoptosis[73].

Since the process of wound healing involves cell migration, proliferation and apoptotic death, it is plausible to expect that growth factors will have an important role in the process. Indeed, growth factors actually regulate all stages of wound healing. The major growth factors that are involved in this process are listed in Table 2. The table also summarizes briefly their wound-healing profiles, which show a considerable overlapping[74,75].

4.2. Effect of Ageing on Wound Healing

Clinical experience with aged individuals has shown that ageing introduces some intrinsic physiological changes that result in impaired wound healing. The parallel appearance of age-related diseases (cardiovascular diseases, diabetes, cancer, atherosclerosis, dementia, etc.) obscures the statistical analysis of these data. The above indicate the need and importance of *in vitro* models in the study of the effect of ageing in wound repair. Until recently *in vitro* studies were mainly focused on histological examination, growth properties, mechanical studies and epithelialization rates[76–79], often with variable and contradictory results. Hence, the need for cell assay systems simulating *in vitro* the conditions of the wound became of absolute importance.

Foetal tissue repair is markedly different from that in the adults. Recently, research on wound healing has paid great attention to the results obtained from embryonic models. Between foetal and adult repair there are a few similarities, but also well described differences. In the adult, the classical sequence inflammation, granulation and remodelling does not lead to a recapitulation of the normal tissue architecture after injury. The usual situation is the production of a collagenous scar to maintain the integrity of the remaining tissue[80]. In contrast, response of the foetal tissue proceeds in the absence of an inflammatory reaction and the whole process resembles tissue regeneration. So, it is conceptually more appropriate to refer to adult repair as "healing" and to embryonic repair as "regeneration"[80]. These differences have been attributed to the specific environment of the foetus, i.e. the amniotic fluid, which is sterile and its composition of polypeptide growth factors may be different of that of the adult wound fluid[81].

We have tried to investigate the differences between embryonic tissue regeneration and adult wound healing, as well as the effect of ageing on wound healing at the cellular level.

To that end, we have developed a novel specific cell assay system, roughly approximating *in vitro* the conditions of the open wound *in vivo* [12]. This system is completely homologous in its constituents, i.e. cells, growth factors and milieu, all of human origin. It is based on a human embryonic lung fibroblast cell strain, which is normal, diploid and with a limited *in vitro* life span (approx. 60 cell population doublings, CPDs). During sequential passages, these cells exhibit ageing characteristics, i.e. increased population doubling time, diminished responsiveness to mitogenic stimuli, increased cell surface, etc. In the presence of minute amounts of human plasma-derived serum, the cells can be synchronized at sparsity, arrested in the G_0 phase of the cell cycle[12]. As this system simulates very effectively the initial stages of the healing wound, it proved to be a very good tool for studying the actions of growth factors released from platelets on fibroblasts, which are the main cell type involved in tissue regeneration.

We studied first the concerted action of the two major growth regulatory polypeptides contained in human platelets, i.e. PDGF and TGF-β (see Table 2) and found that in sparse cultures both factors are mitogenic and, furthermore, when added together they act in synergism. Interestingly, these results remain qualitatively unaltered during cellular *in vitro* ageing, although, as expected, the overall cellular response decreases. We then studied this synergism at the level of PDGF binding. Both types of the PDGF receptors (PDGFRs) are expressed in these cells, with type β being the predominant species at a ratio of [type β/type α]=4: 1 (Psarras et al., unpublished results). The kinetics of the restoration of PDGF binding after down-regulation of PDGF receptors (see section 2.2) showed that it is a time-consuming process (see Fig. 4); 48 hours after the depletion of cell-surface PDGFRs, young fibroblasts cannot restore more than 50% of their initial binding capacity[22]. Interestingly, TGF-β significantly enhances the restoration (Fig. 4). This is in agreement with the synergistic action of these two factors, observed at the proliferative level.

We next found that TGF-β affects subsequent PDGF binding in a time-dependent manner: 12 hours after TGF-β treatment PDGF binding is transiently and moderately reduced, whereas after longer incubation periods it is completely restored, and moreover, it exceeds the initial binbing capacity[15]. This dual effect of TGF-β has also been studied at the level of mRNA expression. Here, TGF-β initially up-regulates the expression of PDGF A chain

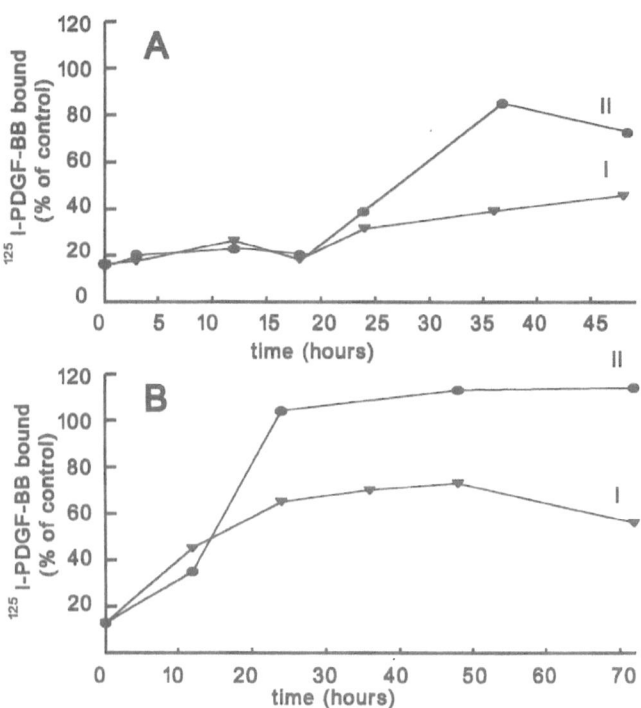

Figure 4. Age-related changes in restoration kinetics of down-regulated PDGF receptors.A. Human embryonic fibroblasts of CPD 34 were plated at 6.9×10^3 cells/cm^2. After 5 days, they were incubated with 20 ng/ml PDGF-BB for 1 hour at 37°C, in order to achieve down-regulation of PDGFRs, followed by an incubation for the indicated periods (II) with, or (I) without 2 ng/ml TGF-β. Subsequently, the binding of 1 ng/ml [^{125}I]PDGF-BB was measured as described in Ref. 22. B. Human embryonic fibroblasts of CPD 48 were plated at the same density and further treated as in A. Adapted from Ref. 22.

mRNA and down-regulates PDGFR type α mRNA expression. Subsequently, a significant induction of PDGFR type β mRNA is observed[28]. These studies, both on binding and mRNA expression, led us to propose the following model, explaining the synergism between PDGF and TGF-β in these cells (Fig. 5). According to this, TGF-β induces initially (Fig. 5, I) the production and secretion of PDGF-AA, which then binds to PDGF receptors type α in an autocrine manner and induces their down-regulation. This, in concert with the reduced mRNA expression of PDGF receptor type α, results in the moderate decrease of PDGF binding. Subsequently, after longer incubation periods (Fig. 5, II), TGF-β induces the production of PDGF receptor type β, rendering the cells more responsive to PDGF.

The restoration studies have also been conducted in cultures of aged fibroblasts. More intriguingly, we have found that, after down-regulation, the restoration mechanism of PDGF binding is much more efficient in aged cells. Especially in the presence of TGF-β, cells of high CPD levels completely restore PDGF binding within 24 hours[22]. Since we still observed some restoration in the presence of cycloheximide, suggesting that this process includes also PDGFR recycling (Psarras et al., unpublished results), it is plausible to assume that this phenomenon could be due to altered recycling rates during ageing.

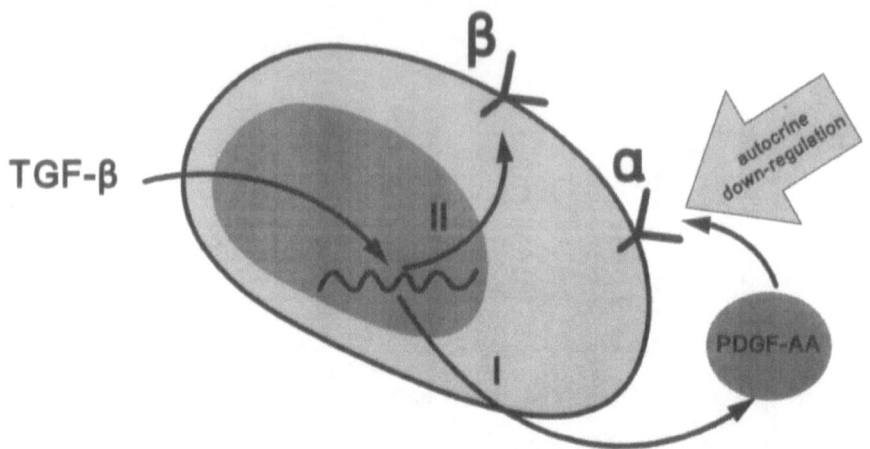

Figure 5. Model for the action of TGF-β in human embryonic fibroblasts. TGF-β incubated with the cells for less than 12 hours (I) induces the expression of mRNA of PDGF A chain. The newly synthesized PDGF-AA can be secreted by the cells and down-regulate the PDGF receptors type α in an autocrine manner. The decrease in the number of PDGF receptors type α is further supported by the fact that TGF-β also down-regulates their mRNA expression. After 12 hours of incubation (II), TGF-β induces the mRNA expression of PDGF receptor type β, resulting in the recovery of the initial capacity of the cells to bind and to respond to PDGF.

Trying to follow *in vitro* the subsequent stages of tissue regeneration, we have used cultures of increasing cell densities ranging from sparsity to confluency. We observed that PDGF acts as a potent mitogen at all densities, whereas TGF-β, which is stimulatory at sparsity, becomes inhibitory in confluent cultures. Furthermore, it even inhibits the action of PDGF itself[15]. So, it seems that TGF-β can be considered as the effective "healing"— and a true regulatory factor during embryonic tissue regeneration.

The *in vivo* results with wound healing models point also to the same direction. In particular, Pierce et al. [82] conducted wound healing studies in rats, showing that TGF-β becomes inactive as a proliferative agent two weeks after application, whereas PDGF continues to do so even after 7 weeks. It should be noted here, that in confluent cultures of senescent human fibroblasts, TGF-β remains inhibitory, as in young cultures, proving that *in vitro* ageing does not alter qualititavely this density-dependent action of TGF-β (Kletsas et al., manuscript in preparation)

As mentioned above, there are fundamental differences in the tissue repair processes between the foetus and the adult individual.

We studied this phenomenon at the level of the action of platelet-derived mitogens at increasing cell culture densities. By using human adult lung fibroblasts we have found—in contrast to the results from embryonic cells—that in both sparse and confluent cultures, PDGF as well as TGF-β remain mitogenic. In contrast, TGF-β is inhibitory in confluent cultures of human embryonic dermal fibroblasts, as is also the case for human embryonic lung fibroblasts. Furthermore, in adult dermal fibroblasts from donors of various *in vivo* ages (from 7 yr old to centenarians), we have observed that, although the overall cellular response to growth factors declines with ageing, the action of TGF-β remains stimulatory irrespective of the cell density used (Kletsas et al., manuscript in preparation). In general, the pattern of the proliferative response of dermal cells to PDGF and TGF-β regarding cell density is similar to that of lung cells, showing that this phenomenon is not cell type-specific.

The above data suggest that during the foetal to adult transition there is a clear alteration in the responsiveness of the cells to the growth factors that are critical for wound repair. This is in accordance with the observed differences between repair processes of foetal and adult tissues. Furthermore, it indicates that these differences are not only due to the specific environment of the foetus—the amniotic fluid—but also to developmental changes at the cellular level[80].

5. CURRENT PERSPECTIVES

The aspects on growth factors and cellular ageing presented in this chapter are clearly part of a much broader field of interest : The control of homeostasis by the regulation of cell proliferation and apoptosis throughout ontogenesis—from the embryo to the elderly.

Accordingly, from cell level to tissue level, several questions arise which direct our present—and, possibly, future—studies. Here are some examples:

- Are there changes in the endocytosis, trafficking, recycling, degradation and function of the specific receptors for growth factors during ageing?
- Human somatic cells secrete a large number of growth factors that regulate their proliferation—an autocrine mechanism especially important for homeostasis at tissue level. Are there specific changes during development and ageing? Our preliminary observations point in that direction.
- Extracellular matrix (ECM) plays a significant role during the wound healing process. Its production is regulated by the action of growth factors; ECM, in turn, regulates the latter's proliferative action. The altered production of ECM components during the foetal-to-adult transition and at the various stages of *in vivo* and *in vitro* ageing, as well as the consequences on cell proliferation, are already under investigation in our laboratory.
- How significant are the changes in the expression of "cell cycle marker" genes, and how close are we to the detection of gerontogenes, i.e. genes expressed at senescence that may dominate the overall gene expression at this developmental stage?

Beyond these and similar questions within our immediate interests, there is, of course, the urgently needed, but immense in its complexity, information concerning the entire cascade of the signal transduction events, from ligand binding at the cell surface to transcriptional activation in the nucleus. Here, we believe, one should be prepared for a long waiting game. It will take a great systematic effort in "conventional" Enzymology and Molecular Biology, before comparative hypotheses emerge and respective experiments can be designed concerning, for instance, age-related functional changes of the core of the Ras / Raf / MEK / ERK, or the G-protein-mediated pathways, and others. This infrastructure of information seems to be a rate-limiting step in understanding the ageing process, even at the level of cell proliferation.

Finally, one should bear in mind that ageing—from the level of molecular events within the cell, to higher orders of complexity at tissue or systemic level—remains a multiparameter process. In all likelihood, no knowledge of one single defect in growth factor binding, or secretion of a sole protein, or a change in gene expression, etc., could provide the answer to the ageing problem. Homeostasis is maintained by the organism in a concerted manner and its breakdown during ageing has to be addressed accordingly.

6. ACKNOWLEDGMENTS

We thank our colleagues Harris Pratsinis and Irene Zervolea for their collaboration within projects referred to in this article and for our stimulating discussions during the preparation of this manuscript.

7. REFERENCES

1. Deuel, T.F. and Kawahara, R.S. (1991). Growth factors and wound healing: Platelet-derived growth factor as a model cytokine. Annu. Rev. Med. 42, 567–584.
2. Hayflick, L. and Moorhead, P.S. (1961). The serial cultivation of human diploid fibroblasts. Exp. Cell Res. 25, 585–621.
3. Hayflick, L. (1965). The limited *in vitro* lifetime of human diploid strains. Exp. Cell Res. 37, 614–636.
4. Vojta, P.J. and Barrett, J.C. (1995). Genetic analysis of cellular senescence. Biochim. Biophys. Acta 1242, 29–41.
5. Cristofalo, V.J. and Pignolo, R.J. (1993). Replicative senescence of human fibroblast-like cells in culture. Physiol. Rev. 73, 617–638.
6. Macieira-Coelho, A. (1988). Relationship between behavior of cells in culture and the physiopathology of the organism. In: Interdisciplinary topics in gerontology. Biology of normal proliferating cells *in vitro*. Relevance for *in vivo* aging, edited by H.P. von Hahn, Karger, Basel, vol. 23., p. 6–21.
7. Dice, J.F. (1993). Cellular and molecular mechanisms of aging. Physiol. Rev. 73, 149–159.
8. Cristofalo, V.J., Phillips, P.D., Sorger, T. and Gerhard, G. (1989). Alterations in the responsiveness of senescent cells to growth factors. J. Gerontol. 44, 55–62.
9. Rattan, S.I.S. and Derventzi, A. (1991). Altered cellular responsiveness during ageing. BioEssays 13, 601–606.
10. Phillips, P.D., Kaji, K. and Cristofalo, V.J. (1984). Progressive loss of the proliferative response of senescing WI-38 cells to platelet-derived growth factor, epidermal growth factor, insulin, transferrin, and dexamethasone. J. Gerontol. 39, 11–17.
11. Bauer, E.A., Silverman, N., Busiek, D.F., Kronberger, A. and Deuel D.F. (1986). Diminished response of Werner's syndrome fibroblasts to growth factors PDGF and EGF. Science 234, 1240–1243.
12. Kletsas, D. and Stathakos, D. (1992). Quiescence and proliferative response of normal human embryonic fibroblasts in homologous environment. Effect of aging. Cell Biol. Int. Rep. 16, 103–113.
13. Harley, C.B., Goldstein, S., Posner, B.I. and Guyda, H. (1981). Decreased sensitivity of old and progeric human fibroblasts to a preparation of factors with insulinlike activity. J. Clin. Invest. 68, 988–994.
14. Hosokawa, M., Phillips, P.D. and Cristofalo, V.J. (1986).The effect of dexamethasone to epidermal growth factor binding and stimulation of proliferation in young and senescent WI38 cells. Exp. Cell Res. 164, 408–414.
15. Stathakos, D., Psarras, S. and Kletsas, D. (1993). Stimulation of human embryonic lung fibroblasts by TGF-β and PDGF acting in synergism. The role of cell density. Cell Biol. Int. 7, 55–63.
16. Heldin, C.-H. (1995). Dimerization of cell surface receptors in signal transduction. Cell 80, 213–223.
17. Hunter, T. (1995). Protein kinases and phosphatases: The Yin and Yang of protein phosphorylation and signaling. Cell 80, 225–236.
18. Marshall, C.J. (1995). Specificity of receptor tyrosine kinase signaling: Transient versus sustained extracellular signal-regulated kinase activation. Cell 80, 179–185.
19. Divecha, N. and Irvine, R.F. (1995). Phospholipid signaling. Cell 80, 269–278.
20. Paulsson, Y., Bywater, M., Pfeifer-Ohlsson, S., Ohlsson, R., Nilsson, S., Heldin, C.-H., Westermark, B. and Betsholtz, C. (1986). Growth factors induce early pre-replicative changes in senescent human fibroblasts. EMBO J. 5, 2157–2162.
21. Gerhard, G.S., Phillips, P.D. and Cristofalo, V.J. (1991). EGF- and PDGF-stimulated phosphorylation in young and senescent WI-38 cells. Exp. Cell Res. 193, 87–92.
22. Psarras, S., Kletsas, D. and Stathakos, D. (1994). Restoration of down-regulated PDGF receptors by TGF-β in human embryonic fibroblasts. Enhanced response during cellular in vitro aging. FEBS Lett. 339, 84–88.
23. Phillips, P.D., Kuhle, E. and Cristofalo, V.J. (1983). [^{125}I]EGF binding ability is stable throughout the replicative life-span of WI-38 cells. J. Cell. Physiol. 114, 311–316.

24. Phillips, P.D., Pignolo, R.J. and Cristofalo, V.J. (1987). Insulin-like growth factor-I: specific binding to high and low affinity sites and mitogenic action throughout the life span of WI-38 cells. J. Cell. Physiol. 133, 135–143.

25. Reenstra, W.R., Yaar, M. and Gilchrest, B.A. (1993). Effect of donor age on epidermal growth factor processing in man. Exp. Cell Res. 209, 118–122.

26. Carpenter, G. and Cohen, S. (1976). ^{125}I-labeled human epidermal growth factor. Binding, internalization, and degradation in human fibroblasts. J. Cell Biol. 71, 159–171.

27. Heldin, C.-H., Wasteson, A. and Westermark, B. (1982). Interaction of platelet-derived growth factor with its fibroblast receptor. Demonstration of ligand degradation and receptor modulation. J. Biol. Chem. 257, 4216–4221.

28. Kletsas, D., Stathakos, D., Sorrentino, V. and Philipson, L. (1995). The growth inhibitory block of TGF-β is located close to the G1/S border in the cell cycle. Exp. Cell Res. 217, 477–483.

29. Venable, M.E., Blobe, G.C. and Obeid, L.M. (1994). Identification of a defect in the phospholipase D/diacylglycerol pathway in cellular senescence. J. Biol. Chem. 269, 26040–26044.

30. Exton, J. H. (1994). Phosphatidylcholine breakdown and signal transduction. Biochim. Biophys. Acta 1212, 26–42.

31. De Tata, V., Ptasznik, A. and Cristofalo, V.J. (1993). Effect of the tumor promoter phorbol 12-myristate 13-acetate (PMA) on proliferation of young and senescent WI-38 human diploid fibroblasts. Exp. Cell Res. 205, 261–269.

32. Stein G.H. and Dulic, V. (1995). Origins of G1 arrest in senescent human fibroblasts. BioEssays 17, 537–543.

33. Sherwood, S.W., Rush, D., Ellsworth, J.L. and Schimke, R.T.(1988). Defining cellular senescence in IMR-90 cells: A flow cytometric analysis. Proc. Natl. Acad. Sci. USA 85, 9086–9090.

34. Seshadri, T. and Campisi, J. (1990). Repression of c-fos transcription and an altered genetic program in senescent human fibroblasts. Science 247, 205–209.

35. Riabowol, K., Schiff, J. and Gilman M.Z. (1992). Trancription factor AP-1 activity is required for initiation of DNA synthesis and is lost during cellular aging. Proc. Natl. Acad. Sci. USA 89, 157–161.

36. Atadja, P.W., Stringer, K.F. and Riabowol, K. (1994). Loss of serum response element-binding activity and hyperphosphorylation of serum response factor during cellular aging. Mol. Cell. Biol. 119, 4991–4999.

37. Rose, D.W., McCabe, G., Feramisco, J.R. and Adler, M. (1992). Expression of c-fos and AP-1 activity in senescent human fibroblasts is not sufficient for DNA synthesis. J. Cell Biol. 119, 1405–1411.

38. Schneider, C., King, R.M. and Philipson, L. (1988). Genes specifically expressed at growth arrest of mammalian cells. Cell 54, 787–793.

39. Ciccarelli, C., Philipson, L. and Sorrentino, V. (1990). Regulation of expression of growth arrest-specific genes in mouse fibroblasts. Mol. Cell. Biol. 10, 1525–1529.

40. Cowled, P.A., Ciccarelli, C., Coccia, E., Philipson, L. and Sorrentino, V. (1994). Expression of growth arrest-specific (gas) genes in senescent murine cells. Exp. Cell Res. 211, 197–202.

41. Benezra, R., Davis, R.L., Lockshon, D., Turner, D.L. and Weintraub, H. (1990). The protein Id: A negative regulator of helix-loop-helix DNA binding proteins. Cell 61, 49–59.

42. Peverali, F.A., Ramqvist, T., Saffrich, R., Pepperkok, R., Barone, M.V. and Philipson, L. (1994). Regulation of G1 progression by E2A and Id helix-loop-helix proteins. EMBO J. 13, 4291–4301.

43. Hara, E., Yamaguchi, T., Nojima, H., Ide, T., Campisi, J., Okayama, H. and Oda, K. (1994). Id-related genes encoding helix-loop-helix proteins are required for G1 progression and are repressed in senescent human fibroblasts. J. Biol. Chem. 269, 2139–2145.

44. Stein, G.H., Drullinger, L.F., Robetorye, R.S., Pereira-Smith, O.M. and Smith, J.R. (1991). Senescent cells fail to express cdc2, cycA and cycB in response to mitogen stimulation. Proc. Natl. Acad. Sci. USA 88, 11012–11016.

45. Pang, J.H. and Chen, K.Y. (1994). Global change of gene expression at late G1/S boundary may occur in human IMR-90 diploid fibroblasts during senescence. J. Cell. Physiol. 160, 531–538.

46. Stein, G.H., Beeson, M. and Gordon, L. (1990). Failure to phosphorylate the retinoblastoma gene product in senescent human fibroblasts. Science 249, 666–669.

47. Dimri, G.P., Hara, E. and Campisi, J. (1994). Regulation of two E2F-related genes in presenescent and senescent human fibroblasts. J. Biol. Chem. 269, 16180–16186.

48. Scher, C.J. (1994). G1 phase progression: Cycling on cue. Cell 79, 551–555.

49. Dulic, V., Drullinger, L.F., Lees, E., Reed, D.I. and Stein, G.H. (1993). Altered regulation of G1 cyclins in senescent human diploid fibroblasts: Accumulation of inactive cyclin E-Cdk2 and cyclin D1-Cdk2 complexes. Proc. Natl. Acad. Sci. USA 90, 11034–11038.

50. Noda, A., Ning, Y., Venable, S.F., Pereira-Smith, O.M. and Smith, J.R. (1994). Cloning of senescent cell-derived inhibitors of DNA synthesis using an expression screen. Exp. Cell Res. 211, 90–98.

51. Harper, J.W., Adami, G.R., Wei, N., Keyomarsi, K. and Elledge, S.J. (1993). The p21 Cdk-interacting pro-
    tein Cip-1 is a potent inhibitor of G1 cyclin-dependent kinases. Cell 75, 805–816.
52. El-Deiry, W.S., Tokino, T., Velculescu, V.E., Levy, D.B., Parsons, R., Trent, J.M., Lin, D., Mercer, E., Kin-
    zler, K.W. and Vogelstein, B. (1993). WAF1, a potential mediator of p53 tumor suppression. Cell 75,
    817–825.
53. Serrano, M. Hannon, G.J. and Beach, D. (1993). A new regulatory motif in cell-cycle control causing spe-
    cific inhibition of cyclin D/Cdk4. Nature 366, 704–707.
54. Evan, G.I. (1994). Old cells never die, they just apoptose. Trends in Cell Biol. 4, 191–192.
55. Lindenboim, L., Diamond, R., Rothenberg, E. and Stein, R. (1995). Apoptosis induced by serum depriva-
    tion of PC12 cells is not preceded by growth arrest and can occur at each phase of the cell cycle. Cancer
    Res. 55, 1242–1247.
56. Wang, E. (1995). Senescent human fibroblasts resist programmed cell death, and failure to suppress bcl2 is
    involved. Cancer Res. 55, 2284–2292.
57. Harrington, E.A., Bennett, M.R., Fanidi, A. and Evan, G.I. (1994). c-Myc-induced apoptosis in fibroblasts
    is inhibited by specific cytokines. EMBO J. 13, 3286–3295.
58. Barres, B.A., Hart, I.K., Coles, H.S.R., Burne, J.F., Voyodic, J.T., Richardson, W.D. and Raff, M.C. (1992).
    Cell death and control of cell survival in the oligodendrocyte lineage. Cell 70, 31–46.
59. Sell, C., Baserga, R. and Rubin, R. (1993). Insulin-like growth factor I (IGF-I) and the IGF-I receptor pre-
    vent etoposide-induced apoptosis. Cancer Res. 55, 303–306.
60. D'Mello, S.R., Galli, C., Ciotti, T. and Calissano, P. (1993). Induction of apoptosis in cerebrellar granule
    neurons by low potassium: Inhibition of death by insulin-like growth factor I and cAMP. Proc. Natl. Acad.
    Sci. USA 90, 10989–10993.
61. Baserga, R. (1995). The insulin-like growth factor I receptor: A key to tumor growth? Cancer Res. 55,
    249–252.
62. Yoshida, T., Satoh, M., Nakagaido, Y., Kuno, H. and Takeuchi, M. (1993). Cytokines affecting survival and
    differentiation of an astrocyte progenitor cell line. Dev. Brain. Res. 76, 147–150.
63. Fuks, Z., Persaud, R.S., Alfieri, A., McLoughlin, M., Ehleiter, D., Schwartz, J.L., Seddon, A.P., Cordon-
    Cardo, C. and Haimovitz-Friedman, A. (1994). Basic fibroblast growth factor protects endothelial cells
    against radiation-induced programmed cell death in vitro and in vivo. Cancer Res. 54, 2582–2590.
64. Araki, S., Shimada, Y., Kaji, K. and Hayashi, H. (1990). Apoptosis of vascular endothelial cells by fi-
    broblast growth factor deprivation. Biochem. Biophys. Res. Commun. 168, 1194–1200.
65. Coles, H.S.R., Burne, J.F. and Raff, M.C. (1993). Large-scale normal cell death in the developing rat kid-
    ney and its reduction by epidermal growth factor. Development 118, 777–784.
66. Sensibar, J.A., Sutkowski, D.M., Raffo, A., Buttyan, R., Griswold, M.D., Sylvester, S.R., Kozlowski, J.M.
    and Lee, C. (1995). Prevention of cell death induced by tumor necrosis factor α in LNCaP cells by overex-
    pression of sulfated glycoprotein-2 (clusterin). Cancer Res. 55, 2431–2437.
67. Nishihara, T., Okahashi, N. and Ueda, N. (1993). Activin A induces apoptotic cell death. Biochem. Bio-
    phys. Res. Commun. 197, 985–991.
68. Armstrong, D.K., Kaufmann, S.H., Ottaviano, Y.L., Furuya, Y., Buckley, J.A., Isaacs, J.T. and Davidson,
    N.E. (1994). Epidermal growth factor-mediated apoptosis of MDA-MB-468 human breast cancer cells.
    Cancer Res. 54, 5280–5283.
69. Yanagihara, K. and Tsumuraya, M. (1992). Transforming growth factor β1 induces apoptotic cell death in
    cultured human gastric carcinoma cells. Cancer Res. 52, 4042–4045.
70. Fukuda, K., Kojiro, M. and Chiu, J.-F. (1993). Induction of apoptosis by transforming growth factor-beta 1
    in the rat hepatoma cell line McA-RH7777: A possible association with tissue transglutaminase expression.
    Hepatology 18, 945–953.
71. Harvilesky, L.J., Hurteau, J.A., Whitaker, R.S., Elbendary, A., Wu, S., Rodriguez, G.C., Bast, R.C. Jr. and
    Berchuck, A. (1995). Regulation of apoptosis in normal and malignant ovarian epithelial cells by trans-
    forming growth factor-β. Cancer Res. 55, 944–948.
72. Jurgensmeier, J.M., Schmitt, C.P., Viesel, E., Hofler, P. and Bauer, G. (1994). Transforming growth factor-
    β-treated fibroblasts eliminate transformed fibroblasts by induction of apoptosis. Cancer Res. 54, 393–398.
73. Desmouliere, A., Redard, M., Darby, I. and Gabbiani, G. (1995). Apoptosis mediates the decrease in cellu-
    larity during the transition between granulation tissue and scar. Am. J. Pathol. 146, 56-66.
74. van Brunt, J. and Klausner, A. (1988). Growth factors speed wound healing. Biotechnology 6, 25–30.
75. Bennett, N.T. and Schultz, G.S. (1993). Growth factors and wound healing: Part II. Role in normal and
    chronic wound healing. Am. J. Surgery 166, 74–81.
76. Uitto, J. (1970). A method for studying collagen biosynthesis in human skin biopsies in vitro. Biochim.
    Biophys. Acta 291, 438–445.

77. Daly, C.H. and Odland, G.F. (1979). Age-related changes in the mechanical properties of human skin. J. Invest. Dermatol. 73, 84–87.

78. Montagna, W. and Carliste, K. (1979). Structural changes in aging human skin. J. Invest. Dermatol. 73, 47–53.

79. Holt, D.R., Kirk, S.J., Regan, M.C., Hurson, M., Lindblad, W.J. and Barbul, A. (1992). Effect of age on wound healing in healthy human beings. Surgery 112, 293–298.

80. Weeks, P.M. and Nath, R.K. (1993). Fetal wound repair: a new direction. Plast. Reconstr. Surgery 91, 922–924.

81. Frantz, F.W., Bettinger, D.A., Haynes, J.H., Johnson, D.E., Harvey, K.M., Dalton, H.P., Yager, D.R., Diegelmann, R.F. and Cohen I.K. (1993). Biology of fetal repair: The presence of bacteria in fetal wounds induces an adult-like healing response. J. Pediatr. Surg. 28, 428–434.

82. Pierce, G.F., Mustoe, T.A., Lingebach, J., Masakowski, V.R., Griffin, G.L., Senior, R.M. and Deuel, T.F. (1989). Platelet-derived growth factor and transforming growth factor-beta enhance tissue repair activities by unique mechanisms. J. Cell Biol. 109, 429–440.

9

IMMUNOSENESCENCE

Paradoxes and New Perspectives Emerging from the Study of Healthy Centenarians

Claudio Franceschi,[1] Daniela Monti,[2] Daniela Barbieri,[2] Stefano Salvioli,[2]
Emanuela Grassilli,[2] Miriam Capri,[2] Leonarda Troiano,[2] Franco Tropea,[2]
Marcello Guido,[2] Paolo Salomoni,[2] Francesca Benatti,[2] Sabrina Macchioni,[2]
Paolo Sansoni,[3] Francesco Fagnoni,[3] Roberto Paganelli,[4]
Gianpaolo Bagnara,[5] Roberto Gerli,[6] Giovanna De Benedictis.[7]
Giovannella Baggio,[8] and Andrea Cossarizza[2]

[1] INRCA, Department of Gerontological Research, Ancona, Italy
[2] Dipartimento di Scienze Biomediche, Sezione di Patologia Generale, Univ. di
    Modena, via Campi 287, 41100 Modena, Italy,
[3] Istituto di Clinica Medica Generale e Terapia Medica, Univ. di Parma, via
    Gramsci 14, 43100 Parma
[4] Dipartimento di Medicina Clinica, Cattedra di Allergologia e Immunologia
    Clinica, Univ. "La Sapienza," viale dell'Universit 37, 00185 Roma
[5] Istituto di Istologia, Univ. di Bologna, via Belmeloro 8, 40122 Bologna
[6] Istituto di Medicina Interna e Scienze Oncologiche, Univ. di Perugia,
    Policlinico Monteluce, 06100 Perugia
[7] Dipartimento di Biologia Cellulare, Univ. degli Studi della Calabria, 87030
    Arcavacata di Rende (CS)
[8] Istituto di Medicina Interna, Univ. di Padova, via Giustiniani 2, 35128 Padova

1. AGEING AND LONGEVITY: A THEORETICAL APPROACH

Ageing has been studied extensively[1,2]. On the contrary, longevity, and particularly human longevity, has been neglected[3]. Hundreds of theories are available on ageing, indicating that scientists are still far from understanding the biological and cultural basis of this process. To this long list of theories, we have added a new one, based on the consideration that the maintenance of soma integrity is the consequence of a continuous activity of a limited number of cellular defence mechanisms[4-6]. We have hypothesised that DNA repair, enzymic and non-enzymic antioxidants, production of heat shock and stress proteins, and activity of poly(ADP-ribose)polymerase form a network of interconnected cellular defence mechanisms, whose global efficiency has been set during evolution at different levels in different species and in different individuals of the same species. We have also speculated that apoptosis is a fundamental biological process which can join the list of cellular defence mechanisms, being an ancestral process used to eliminate damaged, mutated, viral-infected or transformed cells [7,8.] On the whole, the above-mentioned cellular defence mechanisms can be considered as the basic molecular and cellular anti-ageing systems.

However, it can be argued that ageing is not simply a cellular mechanism but that it represents the failure of more integrated systems whose purpose is to preserve body integrity. The nervous, the endocrine and the immune systems are all devoted to the maintenance of body homeostasis. Moreover, we and others have argued that these three systems, which evolved together, have to be studied as a whole, and that the more appropriate approach is to consider immunoneuroendocrine cells and organs as part of a unique system devoted to cope with all kinds of internal and external damaging agents [9]. In previous papers we have suggested that the immunoneuroendocrine system relies upon the above-mentioned molecular and cellular defence mechanisms for its continuous activity [5]. This point of view represents a tentative to combine molecular and cellular with systemic theories of ageing. It can be predicted that the optimal functioning of this immunoneuroendocrine system is of major importance for survival, ageing and longevity. Accordingly, as far as human longevity is concerned, we hypothesised that people who survived in good conditions for long periods, close to the maximum life span of our species, should be equipped with an optimal immunoneuroendocrine system [5]. For this reason, a research project was started some years ago on the biological basis of human longevity, in a selected group of healthy centenarians [6]. We will report here some of the data collected in the last few years on the immune system of centenarians. These data are the first part of a broader investigation in which immune and neuroendocrine parameters will be analysed in order to understand some of the molecular and cellular basis of human longevity. A genetic search for longevity assurance genes is also in progress.

2. AN OLD TENET: IMMUNOSENESCENCE EQUALS IMMUNODETERIORATION

The ageing of the immune system has been thoroughly studied in a variety of models and, particularly, in rodents and humans. The result of this intense investigation can be summarised with the equation immunosenescence = immunodeterioration [10]. Increased sensitivity to infectious diseases and cancer, decreased antibody production to non-self antigens and increased levels of autoantibodies, defective NK activity and decreased T lymphocyte proliferation have been considered a paradigm of a defective immune responsiveness with age. Moreover, it was also assumed that most of this age-related immune deterioration has to be ascribed to profound and early changes of thymus, whose involution starts immediately after puberty [11]. As a consequence, another popular tenet in immunogerontology is that the cellular, or T lymphocyte branch, suffers because of age more than the humoral, or B lymphocyte, branch. This scenario is mainly based on data obtained in rodents. We will try to demonstrate that in humans the deterioration of the immune system with age is not as dramatic as that reported in experimental animals. Moreover, we will review data suggesting that the time is mature to challenge the above-mentioned tenet.

2.1. Immunogerontological Bias

We and others have been interested in human immunosenescence. Some years ago, Dutch immunologists suggested that strict biochemical and clinical inclusion and exclusion criteria, known as SENIEUR Protocol, should be adopted in studies of immunosenescence [12,13]. This proposal was aimed at avoiding a classical bias in gerontological studies, i.e. the confusion between ageing and age-related diseases. Indeed, an entirely new scenario emerged when the immune system of carefully selected, healthy, elderly people was investigated. Lessons have been derived from this approach. In humans, physiological ageing of the immune system is probably not as altered as previously thought [14,15]. This hypothesis is a great challenge owing to the importance of the immune system for immunological pathologies but also for other diseases not traditionally included among immune diseases, in which immune responses can play an important or even crucial role (e.g. atherosclerosis, dementia, cancer).

In comparison with the enormous literature on immunosenescence in rodents and in elderly humans, only scanty and anecdotal data are available on people older than 80–85 years. This is strange for two main reasons. First of all because human lifespan is considerably longer, i.e. 110–120 years, and second because the number of old citizens is dramatically increasing in all countries, and particularly in those that are economically (and immunologically!) developed [16,17]. Thus, the last three/four decades of human life have been left unexplored immunologically. Moreover, the fact that most studies simply compare immune parameters from "young" and "old" subjects does not allow investigators to fully understand the biological significance of the changes as they occur over time. Human ageing is a slow process, and it is difficult to choose appropriate and reliable criteria for assessing it. Finally, the literature on immunosenescence is confused since, in many reports, data referring to lymphocyte subsets are presented as percentages but not as absolute numbers. These inconsistencies can create the illusion that nothing changes with age (e.g. the percentage of CD3+ cells) whereas upon closer examination, a significant decrease in these cells does occur [18,19].

2.2. Are Centenarians Exceptional Individuals?

For all these reasons (1. selection and healthy status of the subjects; 2. necessity to explore the last decades of human life; 3. importance of successful ageing to appreciate physiological immunosenescence) we began to investigate the immune system of centenarians. As we have studied healthy subjects, we will refer to centenarians as people who are not only older than 100 years, but also in good mental (e.g. practising pharmacist) and physical condition (able to chop firewood!). Generally, centenarians are considered to be a rare curiosity. However, this is not the case, as their number is increasing dramatically, and, according to recent predictions, those surviving longer than 95 or 100 years will soon represent a consistent group [16,17,20].

According to the most recent demographic data, an unprecedented proliferation of centenarians can be foreseen. According to J. Vaupel and B. Jeune [21], the number of people celebrating their 100th birthday multiplied several fold from 1875 to 1950, and doubled each decade since 1950 in developed countries. Thus, over the course of human existence the chance of enduring from birth to age 100 may have risen from one in twenty million to two in a hundred, a 400,000-fold increase. This development is largely attributable to reductions in death rate at advanced ages [22], i.e. from 80 to 100. These demographic data, together with those on the immune system of centenarians, represent a formidable challenge for all those biologists and demographers who think that there are biological barriers to longer life expectancy, and are in favour of those gerontologists who are sceptical on the existence of an upper limit to life expectancy. Indeed, if the present pace of improvement in old-age survival persists, it will be as likely for a child today to reach age 100 as it was for a child born eight decades ago to reach age 80 [22]. Thus, it has been speculated that a white female born in 1980 has 50% chance to become centenarian. The proliferation of oldest old and centenarians underlines the importance of the study of immunosenescence. In the next future, a great proportion of the population in developed countries will consist of people over 60 years. From a biomedical and social point of view, it is extremely urgent to have clear ideas about the positive or negative sign of immunosenescence, as ageing is a phenomenon in which most people will be involved. Immunologists and biomedical scientists cannot ignore this unprecedented phenomenon.

In Italy, preliminary data based on a nation-wide investigation coordinated by Professor Luciano Motta (University of Catania) and one of us (C.F.) indicate that the centenarians are nowadays about 4,000[23]. About one third of them are in relatively good mental and physical conditions, and can be considered healthy centenarians. We think that the SENIEUR Protocol, proposed by Ligthart et al. for elderly people[12,13,] should be revisited to be applied to people over 100 years. We are elaborating such a protocol, to be adapted to healthy centenarians, within the framework of MOLGERON, i.e. the Concerted Action Programme on Molecular Gerontology of the EU BioMed Research Programme.

Healthy centenarians are the best example of successful ageing, namely people who have escaped major age-related diseases and reached the extreme limit of human life in good clinical conditions[24]. In most cases, histories of these exceptional individuals reveal them to be free of cancer, dementia, diabetes, cardiovascular diseases, cataract. Moreover, as discussed above, in order to reach such an advanced age, centenarians should be equipped with well preserved and efficient immuno- and defence mechanisms and optimal combinations of an appropriate lifestyle and genetic background[25,26]. Thus, the study of centenarians, and particularly that of healthy centenarians is not only of broad biological and medical interest, but can help in identifying genes that prevent the above-mentioned age-related diseases.

2.3. Humoral Immunity and the First Paradox of Successful Immunosenescence

As far as humoral immunity is concerned, the tenet that ageing equals deterioration is based on the observation that there is an increased frequency with age of pathological processes involving B cells and antibody production, such as B chronic lymphocytic leukemia (B-CLL), presence of autoantibodies or monoclonal gammopathies, amyloidogenesis. Moreover, the decreased antibody response may also result in a propensity for infectious diseases, particularly pneumonia, or recurrent infections, as well as poor responses to vaccinations against the causative agents, resulting on the whole in an increased morbidity and mortality in elderly subjects. Most of the data on this topic have been collected before the revolution of immunosenescence caused by the use of SENIEUR Protocol in immunogerontological studies. For this reason, we begun the analysis of humoral immunity in healthy elderly people, including centenarians. First of all, we faced an unexpected paradox concerning humoral immunity with age, i.e. an increase of immunoglobulin (Ig) serum level and a concomitant decrease in peripheral blood B lymphocytes[19,27,28].

Most studies in the past two decades have addressed the assay of Ig classes and subclasses in sera from human aged subjects, in attempts to establish the normal range for laboratory purposes, as well as to investigate the physiology of this classical parameter of humoral immunity with ageing. By analysing 87 sera of healthy subjects belonging to several age groups carefully selected according to the established criteria of the SENIEUR

Protocol, and including a group of healthy centenarians, we found that both IgG and IgA serum levels significantly rise in correlation with age, whereas IgM levels remain unchanged[27]. Moreover, among IgG subclasses, we observed that IgG1, IgG2 and IgG3 showed a significant increase, whereas IgG4 did not [27]. An increased in vitro Ig production by B cells from aged people has been previously reported[29]. It is interesting to remind that IgG1 and IgG3 are mainly involved in the humoral responses to viral and bacterial antigens, IgG2 -together with IgM- are responsible for responses to polysaccharides (mainly outer wall antigens of capsulated bacteria), and IgG4, with IgE, are increased in response to parasite antigens, as well as being the memory isotype in conditions of chronic high dose exposure [30]. This increase of IgG and IgA antibodies may afford greater protection against viral and bacterial infections in healthy old aged people and centenarians. It is also noteworthy that very few IgG subclass defects were found in elderly subjects, with the exception of IgG4 deficiency[27,31].

Decreased numbers of lymphocytes are found in selected healthy elderly subjects[19,29]. Despite small changes in percentages of lymphocyte subsets, all are significantly decreased as absolute numbers[19]. B lymphocytes in centenarians were reported to be increased, although not significantly. In contrast, we observed a striking decrease of CD19+ B cells both in the 70–83 years age group, and in centenarians [27,28]. The percentage of these cells also decreased, from 13.5% in young controls, to 9.4% and 3.2% in the two groups of aged people. Further, we found that the CD19+ cells coexpressing the CD5 molecule also decreased with age[27,28]. These cells represent a distinct subset of B lymphocytes able to produce polyreactive autoantibodies, and originate chronic lymphocytic leukemia of the B cell type. This finding confirmed that we did not select subjects with unrecognized B-CLL, which is common in elderly subjects.

The scenario regarding the changes in humoral immunity which occur in elderly subjects with age is even more complex if we consider the problem of autoimmunity, and, in particular, that of autoantibodies. It has been reported that the frequency of subjects with detectable serum levels of organ-specific or non-organ-specific autoantibodies increases with age.

This tenet was challenged by our observation that organ-specific autoantibodies are practically absent in the plasma of healthy centenarians [28]. We subsequently showed that this is not a peculiar characteristic of centenarians, as the absence of autoantibodies also occurs in healthy old people[28]. In contrast, unselected, elderly people presented an age-related increase in these autoantibodies[28,32]. Non organ-specific autoantibodies (anti-dsDNA, anti-histones, rheumatoid factor, anti-cardiolipin) follow a different trend, increasing also in healthy aged donors, including centenarians (Gerli R. et al. manuscript in preparation).

These data on humoral immunity of elderly subjects raise several questions, and different possibilities can be envisaged to explain these age-related changes. In particular, it is possible to hypothesize that increased number of B lymphocytes and plasma cells occurs in organs other than peripheral blood, or that the lifespan of B lymphocytes and plasma cells in germinal centres is increased in aged people. An increased production of Ig per cell has to be ruled out. According to first and the second hypotheses, alterations with age of B cell homing and propensity to apoptosis can be predicted. Indeed, recent data from our laboratory suggest that the membrane expression of certain molecules involved in homing processes and of certain cell adhesion molecules changes with age (Cossarizza A. et al., manuscript in preparation). A different propensity to apoptosis has been observed in peripheral blood lymphocytes (PBL) from centenarians after exposure to damaging viral or chemical agents (Monti D. et al., manuscript in preparation).

A particular attention deserves the opposite behaviour of non organ- and organ-specific autoantibodies with age. The presence of high level of non organ-specific autoantibodies is thus apparently compatible with the capability to reach the extreme limit of human life. On the contrary, the absence of organ-specific autoantibodies is a marker of successful ageing and longevity, as we have observed a lack of such antibodies either in centenarians or in healthy aged subjects carefully selected according to the SENIEUR Protocol[28,32,33]. The hypothesis can be put forward that the cellular and molecular regulatory mechanisms controlling the production of these two types of autoantibodies are quite distinct, particularly in old people and centenarians, likely depending on their health status.

It remains to be established how these findings regarding autoantibodies and immunosenescence can be reconciled with the profound changes in the production of a variety of cytokines, which occur in aged people and centenarians, as described below. In centenarians, the predisposition to autoimmunity, due either to genetic factors or to aberrant self-recognition[34,35], can be different in comparison to that of young people. Moreover, non organ-specific autoantibodies could have a beneficial role, in contrast to an old tenet of immunosenescence, claiming that the presence of autoantibodies is a sign of the progressive blurring with age of the capability to distinguish between self and not self.

2.4 Cellular Immunity and the Second Paradox of Successful Immunosenescence

Unexpectedly, age-related changes in the T cell compartment were much less dramatic than what we would have predicted according to the data in rodents and in unselected aged subjects. Moreover, it is noteworthy that some parameters were found decreased, while others increased.

In particular, we found that the absolute number of CD3+, CD4+ and CD8+ T cells decreased with age[19,36], while activated peripheral T cells (HLA-DR+) were markedly augmented. In both cases centenarians did not escape this destiny. Another interesting trend has been recorded as far as percentage and absolute number of "virgin/unprimed" (CD45RA+) and "memory/activated" T cells (CD45R0+) was concerned: these parameters did not change significantly after the fourth decade of life[37], either among CD4+ or CD8+ T cells.

CD4+ and CD8+ peripheral blood T lymphocytes show mutually exclusive expression of CD45RA or CD45R0, two isoforms of the common leukocyte antigen that seem to recognise so-called virgin/unprimed and memory/activated T cells. The expression of these isoforms has been studied by three colour cytofluorimetric analysis on CD4+ or CD8+ peripheral blood CD3+ cells from 22 healthy centenarians, analysed in a context of 202 healthy donors 0–110 years old[38]. An age-related unbalance of virgin and memory cells was found between CD4+ and CD8+ subsets. As expected, at birth 95–99% of the CD3+ lymphocytes expressed CD45RA isoform. A rapid increase of CD45R0+ cells was observed in the first 2–3 decades of life, this phenomenon being much more pronounced on CD4+ cells. Subsequently, the increase of the memory compartment was much less rapid, so that in centenarians a consistent reservoir of CD45RA+ among CD4+ cells was still present (about 20%). In these exceptional individuals the percentage of CD45RA+ cells among CD8+ T lymphocytes was even higher (about 50%), and only slightly lower than that of young donors (about 55–60%). Thus, the main changes occurred at different rate in CD4+ and in CD8+ T cells, at an age comprised between 0 and 30 years, when the thymus is still functionally active. Interestingly, no difference in the usage of CD45 isoforms was observed within T cells bearing 4 different Vβ-T cell receptor.

Thus, the following questions arise: where do these T lymphocytes come from?, and why are sharp changes in expression of CD45 isoforms mainly observed in the first two-three decades of life? These modifications seem to mirror the involutive changes of the thymus, that take place immediately after puberty[39], i.e. at least 80 years ago. Assuming that a lifespan of several decades is highly improbable for most memory T cells, we need to understand origin and continuous renewal of virgin and memory T cell compartments when the thymus has likely undergone a profound involution, as in centenarians. In any case, it is difficult to explain why a consistent number of T cells shows a virgin phenotype, waiting for possible new antigens, even in far advanced age. Notwithstanding the fact that reversions from one isoform to the other have been described[40-43], and that the presence of antigens is required for memory cells to survive[44-46], the possibility exists that the thymus becomes progressively less important as the T lymphocyte producer, and its role is taken over by other "peripheral" lymphoid organ(s). We must assume therefore that negative and positive selection are occurring successfully in this (these) organ(s), as demonstrated by a lack of autoimmune responses in healthy centenarians. Thus, thymic remnants or substituting organ(s) are probably able to produce and select high numbers of T cells every day, until the extreme limit of human life.

In any case, tiny amounts of thymic tissues, such as those likely remaining in centenarians, should be capable of producing a number of CD4+ and CD8+ virgin T cells similar to those found in young people[37,38,47]. In this case, the redundancy of thymic tissue should be quite high. This is certainly possible, but still unproved, and deserves further attention. Moreover, preliminary data (Dellabona P. et al., manuscript in preparation) indicate that changes do occur in T cell repertoire of centenarians, but they are not due to holes in the repertoire. In this case, too, a tiny amount of thymic tissue should be able to produce a fairly complete T cell repertoire likely devoid of autoimmune T cells, as expected from the lack of overt autoimmune diseases and organ-specific autoantibodies in centenarians. Finally, in order to indirectly answer this difficult question, we approached this problem in another way. Indeed, we have evidence that most immune parameters are quite similar in centenarians and in people thymectomized for many years, due to several reasons, among which the most represented was Myasthenia Gravis (Gerli R. et al., submitted). Our data indicate that in adult humans, the immune function of thymus can be fully, or in part, vicariated by other, unknown organs. As we cannot exclude, at present, that a similar phenomenon could also occur in old people and centenarians, we think that the above quoted paradox is still waiting for an answer.

2.5 Cell Growth in Centenarians: A Paradoxical, Well Preserved Proliferative Activity Despite the Hayflick Limit

Ageing is characterised by a variety of alterations which occur in most organs and cell types. Loss of proliferative vigour is considered a marker of the ageing process and is related to the Hayflick phenomenon, i.e. the limited number of replications that normal cells can undergo. It has been demonstrated that there is a correlation between proliferative capability and maximum life span in different species, and an inverse correlation between proliferative capability and donor age (see articles by Derventzi et al., and Stathakos et al, in this book). Thus, a decreased capability to proliferate may be considered a characteristic of cellular senescence.

It has been suggested that a crucial change in the immune system is the reduced capability of its component cells to proliferate, a problem related to clonal expansion after exposure to antigenic stimuli[48]. We found that T lymphocytes from healthy centenarians were fully capable of proliferating, and that the only difference vis a vis young people (20–30 years old) is a delay in peak responsiveness[25].

We recently addressed this problem studying lymphocyte proliferation in 39 healthy centenarians (100–106 years), 33 middle aged (50–68 years) and 40 young (19–36 years) subjects. In centenarians, cell proliferation induced by some stimuli (human recombinant interleukin-2, phytohemagglutinin, autologous and allogeneic mixed lymphocyte reaction) was reduced in comparison to young subjects. However, PBL from the three groups of donors proliferated comparably in response to other stimuli (phorbol esters, pokeweed mitogen, anti-CD3 monoclonal antibodies). The pattern of cell growth was not modified when purified T cells were used. Costimulation of purified T cells via CD28 showed that cells from centenarians were less responsive than those from younger donors. FACS analysis showed that the percentage of CD28+ cells progressively decreased with age, being significantly reduced in centenarians. Our data suggest that, with age, T cell proliferation is not uniformly affected as previously thought, and that a complex remodelling occurs where some activation pathways are defective while others are well preserved until the last decades of life.

Recent data on the proliferative capability of fibroblasts from centenarians are in full agreement with the notion that they experience no major change in proliferative capability (Tesco G. et al., manuscript in preparation). An inverse relationship between replicative capability and donor age has been reported in human fibroblast strains. In order to test the proliferative vigour of cells from people who reached the extreme limit of human life, primary cultures of fibroblasts from forearm skin of 8 centenarians and 7 young donors (7–55 years old) were set up and compared. Fibroblasts from centenarians stopped growing after 27 ± 2 (mean $\pm$ SEM) cumulative population doublings (CPD), as predicted by the formula proposed by Schneider et al. (1981) [CPDmax = 44.6 - (0.18 x Donor age)]. When these fibroblasts were studied during in vitro ageing, we confirmed that cells undergo a progressive decline of [^{3}H]-thymidine incorporation, but this phenomenon was quite similar in fibroblasts from centenarians and young donors. Moreover, no difference in the responsiveness to several growth factors [basic-fibroblast growth factor, epidermal growth factor, insulin growth factor I, insulin growth factor II, 20% fetal bovine serum (FBS)] was observed between fibroblasts from centenarians and young donors, studied at the same population doubling level (PDL). The expression of c-fos and c-jun proto-oncogenes and the activation of AP-1 transcription factor in response to FBS stimulation following serum starvation was also indistinguishable between fibroblasts from centenarians and young donors studied at the same PDL.

On the whole, these data reveal an unpredicted well conserved capability of cells from the oldest old to proliferate, and are in line with the emerging concept that the process by which normal cells regulate proliferation is not correlated with the process which regulate senescence or, at least, ageing at an organismic level. Moreover, our data on fibroblasts, together with those concerning lymphocytes from centenarians, cast some doubt on the old tenet that a profound impairment of proliferative vigour is a marker of ageing.

3. AGEING, LONGEVITY AND APOPTOSIS

In a recent paper we argued that an intriguing relationship exists among cellular senescence, tumor growth and longevity[8]. All these phenomena are deeply related to programmed cell death or apoptosis[49]. A possible scenario is the following: cells may be equipped with genes which actively promote cellular senescence thus controlling cell death to avoid transformation[7,50]. This situation is balanced by other genes responsible for survival and viability. Circumstantial evidence suggests that cellular senescence may be considered a peculiar type of cell differentiation whose biological function is to counteract uncontrolled cell proliferation. From this point of view, cellular senescence can be considered one of the most important mechanisms in avoiding the continuous onset of tumors. The most effective evolutive way for a cell to control neoplastic growth is likely to set up genes which promote apoptotic cell death[51-59]. Moreover, we were able to show that an intriguing relationship at the molecular level between cell proliferation and apoptosis[60-65]. Interestingly, and unexpectedly, we have recently found that lymphocytes become progressively resistant to apoptosis with increasing age of the donor, and centenarians follow this trend (Monti D. et al., manuscript in preparation). At present, it is difficult to reconcile this findings with the above-mentioned hypothesis based on the consideration that apoptosis is the main mechanism to get rid of mutated and potentially transformed cells. In any case, resistance to apoptosis could contribute to cellular longevity, and, possibly, to organismic longevity.

Recently, we argued that mitochondria and intracellular ATP level could play a crucial role in apoptosis[66]. These considerations can be relevant for the ageing process, taking into account that alterations of mitochondria are supposed to play a major role in the ageing process[67]. The cytofluorimetric method we recently set up to study mitochondrial membrane potential in intact cells or in isolated mitochondria, which allows the analysis of mitochondrial functional heterogeneity, can help in clarifying the role of mitochondria in the ageing process or during apoptosis[68-74]. Preliminary data suggest that, unexpectedly, in lymphocytes from centenarians these organelles are quite resistant to several damaging agents, likely involving an oxidative stress (Cossarizza A. et al., manuscript in preparation). If confirmed in cells from other tissues and organs, these data challenge the idea that functional alterations of mitochondria occur with age, as a simple consequence of the well known accumulation with age of mitochondrial DNA damages.

4. INTERCELLULAR COMMUNICATION: THE IMBALANCE OF CYTOKINE NETWORK

In recent years it became clear that the immune orchestra depends on a subtle and well tuned network of humoral mediators, collectively called cytokines, that are responsible for differentiation, proliferation and survival of lymphoid cells. They include interleukins, colony stimulating factors, interferons, and others, such as tumor necrosis factors (TNF). These molecules, most of which have been characterised and cloned, constitute a complex network, and act by the interaction with, and binding to, specific membrane receptors, which must be considered as an integral part of the cytokine network. There are cytokines, such as interleukin-2 (IL-2), which have a particular importance for the proliferation and differentiation of T, B and NK cells. IL-2 and IL-10 lead to increased production of IgM, IgG and IgA, whereas IL-4 and IL-13 induce IgE and IgG4 synthesis[75,76].

Other cytokines, such as IL-1, IL-6 and TNF-α are considered pro-inflammatory agents, and play an important role not only in the immune responses but also in inflammation. IL-6 also amplifies Ig synthesis by committed B cells. It has been reported that the production and utilisation of one of the most important cytokines, IL-2, declines with age[77]. However, we have shown that the altered production and utilisation of this cytokine by cells from aged donors were rescued by the exposure of cells to low frequency pulsed electromagnetic fields, suggesting that the above-mentioned alterations are not irreversible, and can be positively modulated[78].

What about other cytokines, whose production and utilisation has not been critically analysed during human immunosenescence ? We measured the in vitro production of IL-6, TNF-α, IFN-γ and IL-1β by peripheral mononuclear cells from selected healthy young (mean age 26.8 years) and aged (mean age 80.2 years) subjects[79]. Significant increases of IL-6, TNF-α and IL-1β levels were found in mitogen-stimulated cultures from aged donors, occurring at 24 to 48 hours after stimulation. No significant differences were observed for IFN-γ production. Proliferative capability of cells stimulated with PHA was not impaired in aged subjects. Since the amounts of all cytokines studied were similar in unstimulated cultures from young and aged subjects, and also serum levels of TNF-α did not differ, these data indicate that the cellular machinery for the production of these cytokines is well preserved in ageing, and also that cells from old people are able to upregulate their production in response to appropriate stimuli. The increases of cytokine synthesis were not dependent on changes of monocyte numbers, nor were related to the significant rise of CD45R0+, and the concomitant decrease of CD45RA+, occurring among both CD4+ and CD8+ lymphocytes from aged subjects.

This study has been recently extended to cells from centenarians, and we found that centenarians follow the trend, being the levels of IL-1, IL-6 and TNF-α very high in the supernatant of stimulated cultures. However, in these donors we observed a great individual variability. In 3 groups of healthy subjects (young: 25–35 years old; old, 65–75 years old; centenarians), we have also studied the production of other cytokines. The results show that there is an age-dependent decrease of the production of IL-3 and leukemia inhibitory factor (LIF), as well as a tendency to an age-dependent decrease in the production of granulocyte-macrophage colony stimulating factor (GM-CSF) in PBMC cultures.

In these donors, we could measure plasma levels of IL-1β, IL-1 receptor antagonist (IL-1RA), TNF-α, TNF soluble receptor type I (sTNF-RI), sTNF-RII, IL-6 and IL-6 soluble receptor (sIL-6R) and stem cell factor (SCF). The main results were that: (i) IL-1 plasma level were not detectable in all groups; (ii) in healthy centenarians, IL-1RA, IL-6, sIL-6R and TNF-α were increased; (iii) sTNF-RI, sTNF-RII and SCF increased in an age-dependent manner. Interestingly, in the peripheral blood of centenarians a number of granulocytic-macrophagic, erythroid and multipotent hemopoietic progenitors similar to that of young and aged people was found, as assessed in cultures added with IL-3, GM-CSF and erythropoietin (Bagnara G. et al., manuscript in preparation). Adding SCF significantly increases colony formation in all subjects. Thus, centenarians still maintain a normal pool of functionally effective hemopoietic progenitors in the presence of a remodelled network of hemopoietic cytokines, some of which (SCF) are increased, while others (IL-3, GM-CSF) are decreased with age.

The increased production of proinflammatory cytokines by stimulated mononuclear cells of healthy aged subjects may be relevant to several aspects of age-associated pathological events, including atherosclerosis, osteoporosis, fibrosis and dementia. However, the fact that the same phenomenon is observed in healthy centenarians, individuals who escaped the above-mentioned pathologies, indicates that the age-related increase in these cytokines can have a different biological meaning in subjects of different ages (young and old subjects versus centenarians).

On the whole, our data suggest that cytokine network undergo profound and complex changes with age. Indeed, the production and/or utilisation of some cytokines decreases with age, while the production of other cytokines increases. This field is far from being clear, as data on possible changes of other cytokines are lacking, especially in humans. Moreover, no data are available on possible changes in cytokine receptors on the plasma membrane of target cells (number per cell, affinity, etc.), and on possible changes in intracellular transduction signal pathways. Such data are urgently needed, considering that cytokines and their receptors are involved in the mechanisms responsible for many age-associated pathologies (atherosclerosis, dementia, autoimmune diseases, etc.).

5. INNATE IMMUNITY: THE FIRST TO COME, THE LAST TO GO

T and B cells are classical examples of adaptive immunity, being clonally distributed and capable of specifically reacting with single epitopes of a given antigen. However, for millions of years lower creatures had to survive in environments full of pathogens, such as viruses and bacteria, despite the absence of an immune system functioning at a clonal level of recognition. Mechanisms have been used, and evolved, in order to overcome such a problem: they are collectively called innate immunity, and comprehend responses such as chemotaxis, phagocytosis, natural cytotoxicity, among others. Evolutive studies from different laboratories, including ours, suggest that these mechanisms are fully capable of preserving body integrity even in the absence of adaptive immunity[80–94]. However, for save of economy, they did not disappear with the onset of adaptive immunity. Indeed, they are still present and effective in more evoluted animals such as mammals, where they are intermixed to, and collaborate with, clonally distributed T and B cells, and still represent a first line of defence against a variety of pathogens.

On the basis of these considerations, it can be predicted that the most sophisticated immune responses are also the most fragile, and prone to age-related alterations. On the contrary, ancestral innate immune responses should be more resistant to age-related changes, being simpler, more economical and conserved throughout evolution. Accordingly, innate natural immunity was studied in healthy elderly subjects and centenarians. In particular, we focused on natural killer (NK) cell activity and chemotaxis. NK cells and activity during ageing have been studied extensively by several groups. Different results (decrease, increase, no change) have been reported, probably because of a poor selection and insufficient inclusion and exclusion criteria[15,19]. In large groups of healthy centenarians, middle aged (40–50 years old) and young (20–30 years old) subjects a detailed cytofluorimetric analysis allowed us to demonstrate an age-related increase in cells with high NK activity (CD16+,CD57-)[19]. On the contrary, cells with intermediate (CD16+,CD57+) or low (CD57+,CD16-) NK activity showed only minor modifications. In centenarians, an increase in this high activity NK subset is mirrored by well preserved cytotoxicity, as measured by both NK and redirected killing assays. In Down's syndrome, an example of precocious ageing in humans[39,95–106], an expansion of NK cells occurs, suggesting that this is peculiar to immunosenescence[104,107]. However, in this syndrome, NK cells were, functionally, highly inefficient[107]. Recent data suggest that a persistently low NK activity is a predictor of impending morbidity[108]. Conversely, it can be speculated that well preserved NK activity can help in becoming a centenarian. The age-related increase of cells bearing NK markers, and of non-MHC-restricted T lymphocytes[19], could be interpreted as a compensatory mechanism to cope with decreases in T cells. Recent preliminary data on chemotaxis suggest that the capability of peripheral blood mononuclear cells to respond to chemotactic stimuli is well preserved in centenarians (Genedani S. et al., manuscript in preparation).

On the whole, the above-mentioned data indicate that, as predicted, innate immunity does not undergo a significant deterioration with age. This is probably one of the reasons to explain why healthy aged subjects such as healthy centenarians are apparently fully capable of coping with infectious agents, and have no increased frequency of infectious diseases. Elderly subjects who show an increased susceptibility to infections (influenza, tuberculosis, etc.) are probably those in whom pathological changes in immune system occurred. The hypothesis can be put forward that physiological ageing per se does not represent a major risk factor for most infectious pathologies.

6. CONCLUSIONS

Taking into account the complex (positive and negative) changes which occur in the immune system with age, we prefer to use the words reshaping and retuning, instead of "alteration", "deterioration", or "decline", to describe the complexity of immunosenescence[47,109–111]. It is our opinion that these pejorative descriptions do not grasp its substance, and that terms such as continuous remodelling are clearly more appropriate to describe a situation where some immune parameters increase, others decrease, whereas still others remain unchanged.

In other words, we think that the body undergoes a continuous adaptation as a consequence of the continuous exposure to low levels of internal and external damaging agents, such as oxygen free radicals, glucose and other reducing sugars, radiations, among others. This is a dynamic point of view, that considers centenarians as the end-product of very effective cellular defence mechanisms selected throughout phylogenesis and ontogenesis.

Investigations on centenarians, the best example of successful ageing, can clarify the trend and direction of the immunosenescence, and go far beyond immunology in a strict sense, since the most important age-related pathologies (e.g. atherosclerosis, dementia, cancer) do have an immunological component.

7. EUROPEAN PERSPECTIVES: THE MOLECULAR GENETICS OF LONGEVITY, AND NATIONAL WAYS TO BECOME CENTENARIAN

It is reasonable to assume that most of the above-mentioned data on aged subjects and particularly on centenarians have a strong genetic component. Indeed, ageing and longevity are likely genetically controlled, even if in a paradoxical way[112], i.e. through the action of anti-ageing genes. We are addressing this problem by studying multiallelic system to identify significant genetic associations with human longevity. Indeed, longevity can be regarded as a multifactorial trait resulting from the interaction between environmental factors and a set of epistatic alleles having pleiotropic age-dependent effects. A possible approach to the identification of these longevity-alleles is the comparative analysis of polymorphic markers at candidate loci between a long lived group and a control group[113]. It is to be expected that, if a certain gene affects lifespan, the allele pool at that locus would be different between the long lived and the control group. So far significant longevity-alleles associations have been found at HLA, ACE, APO E, APO C loci[26,113–117]. We are using a slightly different approach taking into account that the information content of a genetic marker is a function of the number and the relative frequencies of different alleles found at a locus in a population[118]. DNA markers resulting from a variable number of tandem repeats (VNTR) are highly informative multiallelic systems by which human genetic studies have, in recent years, received formidable impulse[118,119].

The above polymorphisms can be analysed by polymerase chain reaction (PCR) amplification of the variable region, thus allowing unequivocal allele detection. The analysis of VNTR markers, within or near genes candidate in longevity, could reveal possible difference in the allele pool at a particular locus between a long lived and a control group. We predict that this research on the genetic basis of longevity will be particularly fruitful if focused not only on candidate genes per se, but particularly on their regulatory flanking regions (promoters). Indeed, several circumstantial evidences suggest that changes in gene expression can be crucial for ageing and longevity. Moreover, we predict that this approach should be more sensitive in comparison with the analysis of single diallelic polymorphisms such as that previously performed by other groups[113].

With this approach, we are currently analysing the following genes and the relative promoters: APO B and renin, chosen because healthy centenarians we analysed were remarkably free of cardiovascular disease and hypertension. APO B has been previously studied with negative results[113]; superoxide dismutase Cu/Zn, chosen because it is the key enzyme of antioxidant defence system which plays a crucial role in ageing[120], and, according to our previous studies, is particularly efficient in healthy centenarians[7,8]; bcl-2, chosen because is the key enzyme in cell proliferation and cell death which, according to studies from several groups, including ours[7,8,121-123] changes with age; thyroid peroxidase, the key enzyme in thyroid metabolism, a target molecule for autoantibodies, which increase with age but are remarkably lacking in centenarians[28,32,33,124]; TNF-α one of the most important proinflammatory cytokines, whose production, according to our studies, is deregulated with age[79], which is also involved in the production of oxyradicals and in the subsequent damage of mitochondria, a key organelle in the ageing process[67]; and mitochondrial DNA, for the above-mentioned reasons.

Preliminary data show that for most of these genes there are significant associations with longevity, even if differences between centenarians from the North and South of Italy occur. Thus, as expected, genetic and environment likely play a role in longevity. It can be predicted that genetic differences among centenarians from different European countries will emerge, and this type of studies are urgently needed to unravel the role of the different genetic background and cultural environment of the various European countries.

Studies on supercentenarians (over 110 years old) are also needed to test in these extremely selected people the most important theories of ageing. As their number is at present very low, a strong effort to co-ordinate the European Countries which have reliable demographic records is starting.

ACKNOWLEDGMENTS

This work has been partially supported by grants from M.U.R.S.T. (40% and 60%) and C.N.R. (Progetto Finalizzato "Invecchiamento").

REFERENCES

1. Finch, E.C. Longevity, Senescence and the Genome; The University of Chicago Press; Chicago; Vol. 1990.
2. Rose, M.R. Evolutionary Biology of Aging; Oxford University Press; New York; Vol. 1991.
3. Franceschi, C. and Fabris, N. (1993) Human longevity: the gender difference (Editorial). Aging, Clinical and Experimental Research, 5, 333–336.
4. Franceschi, C. (1989) Cell proliferation, cell death and aging. Aging, Clinical and Experimental Research, 1, 3–15.

5. Kirkwood, T.B.L. and Franceschi, C. (1992) Is ageing as complex as it would appear? New perspectives in gerontological research. Ann. N.Y. Acad. Sci., 663, 412–417 .

6. Franceschi, C., Monti, D., Scarfî, M.R., Zeni, O., Temperani, P., Emilia, G., Sansoni, P., Lioi, M.B., Troiano, L., Agnesini, C., Salvioli, S. and Cossarizza, A. (1992) Genomic instability and aging. Studies in centenarians (successful aging) and in Down's syndrome patients (accelerated aging). Ann. N.Y. Acad. Sci., 663, 4–16.

7. Monti, D., Troiano, L., Tropea, F., Grassilli, E., Cossarizza, A., Barozzi, D., Pelloni, M.C., Tamassia, M.G. and Franceschi, C. (1992) Apoptosis - Programmed cell death: a role in the aging process? Am. J. Clin. Nutr., 55, 1208s–1214s.

8. Monti, D., Grassilli, E., Troiano, L., Cossarizza, A., Salvioli, S., Barbieri, D., Agnesini, C., Bettuzzi, S., Ingletti, M.C., Corti, A. and Franceschi, C. (1992) Senescence, immortalization and apoptosis: an intriguing relationship. Ann. N.Y. Acad. Sci., 673, 70–82.

9. Franceschi, C. and Ottaviani, E., Invertebrate/vertebrate neuroendocrine and immune systems: commonality of mechanisms and signal molecules In Advances in Comparative and Environmental Physiology; (Cooper, E. L., Ed.); Springer-Verlag: Berlin, 1996; Vol. 24; pp 213–244.

10. Makinodan, T. and Kay, M.M.B. (1980) Age influence on the immune system. Advances in Immunology, 29, 287–330.

11. Miller, J.F.A.P. and Osoba, D. (1967) Current concepts of the immunological function of the thymus. Physiol. Rev., 47, 437–520.

12. Ligthart, G.J., Corberand, J.X., Fournier, C., Garanaud, P., Hijmans, W., Kennes, B., M,ller-Hermelink, H.K. and Steinmann, G.G. (1984) Admission criteria for immunogerontological studies in man: the Senieur Protocol. Mech. Ageing Dev., 28, 47–55.

13. Ligthart, G.J., Corberand, J.X., Geertzen, H.G.M., Meinders, A.E., Knook, D.L. and Hijmans, W. (1990) Necessity of the assessment of health status in human immunogerontological studies: evaluation of the SENIEUR Protocol. Mech. Ageing Dev., 55, 89–97.

14. Ligthart, G.J., Schuit, H.R.E. and Hijmans, W. (1985) Subpopulations of mononuclear cells in ageing: expansion of the null cell compartment and decreased in the number of T and B cells in human blood. Immunology, 55, 15–21.

15. Ligthart, G.J., Schuit, H.R.E. and Hijmans, W. (1989) Natural killer function is not diminished in the healthy aged and is proportional to the number of NK cells in the peripheral blood. Immunology, 68, 396–402.

16. Olshansky, S.J., Carnes, B.A. and Cassel, C. (1990) In search of Methuselah: estimating the upper limits to human longevity. Science, 250, 634–640.

17. Barinaga, M. (1991) How long is the human life-span? Science, 254, 936–938.

18. Sansoni, P., Brianti, V., Fagnoni, F., Snelli, G., Marcato, A., Passeri, G., Monti, D., Cossarizza, A. and Franceschi, C. (1992) NK cell activity and T-lymphocyte proliferation in healthy centenarians. Ann. N.Y. Acad. Sci., 663, 505–507.

19. Sansoni, P., Cossarizza, A., Brianti, V., Fagnoni, F., Snelli, G., Monti, D., Marcato, A., Passeri, G., Ortolani, C., Forti, E., Fagiolo, U., Passeri, M. and Franceschi, C. (1993) Lymphocyte subsets and natural killer cell activity in healthy old people and centenarians. Blood, 80, 2767–2773.

20. Jeune, B. Morbus centenarius or sanitas longaevorum? A critical review of centenarian studies; Odense University Press; Odense; Vol. 15 1994.

21. Vaupel, J.W. and Lundström, H. The future of mortality at older ages in developed countries; Odense University Press; Odense; Vol. 3 1993.

22. Kannisto, V., Lauristel, J., Thatcher, A.R. and Vaupel, J.W. Reduction of mortality at advanced ages; Odense University Press; Odense; Vol. 4 1993.

23. Motta, L., Receputo, G., Franceschi, C. I centenari in Italia: aspetti biologici e clinico-epidemiologici, Congresso della Societè Italiana di Medicina Interna, Roma, 1995.

24. Rowe, J.W. and Kahn, R.L. (1987) Human aging: usual and successful. Science, 237, 143–149.

25. Franceschi, C., Monti, D., Cossarizza, A., Fagnoni, F., Passeri, G. and Sansoni, P. (1991) Aging, longevity and cancer: studies in Down's syndrome and in centenarians. Ann. N.Y. Acad. Sci., 621, 428–440.

26. Morellini, M., Trabace, S., Lulli, P., Cappellacci, S., Brioli, G., Orr·, D., Pennesi, G., Monti, D., Cossarizza, A., Lambert-Gardini, S., Sansoni, P. and Franceschi, C. (1992) HLA antigens and aging. Ann. N.Y. Acad. Sci., 663, 499–500.

27. Paganelli, R., Quinti, I., Fagiolo, U., Cossarizza, A., Ortolani, C., Guerra, E., Sansoni, P., Pucillo, L.P., Cozzi, E., Bertollo, L., Monti, D. and Franceschi, C. (1992) Changes in circulating B cells and immunoglobulin classes and subclasses in a healthy aged population. Clin. Exp. Immunol., 90, 351–354.

28. Mariotti, S., Sansoni, P., Barbesino, G., Caturegli, P., Monti, D., Cossarizza, A., Giacomelli, T., Passeri, G., Fagiolo, U., Pinchera, A. and Franceschi, C. (1992) Thyroid and other organ-specific autoantibodies in healthy centenarians. Lancet, 339, 1506–1508.

29. Lehtonen, L., Eskola, J., Vainio, O. and Lehtonen, A. (1990) Changes in lymphocyte subsets and immune competence in very advanced age. J. Gerontol. Med. Sci., 45, 108–112.

30. Aalberse, R.C., van der Gaag, R. and van Leeuwen, J. (1983) Serological aspects of IgG4 antibodies. I. Prolonged immunization results in an IgG4-restricted response. J. Immunol., 130, 722–726.

31. Paganelli, R., Scala, E., Quinti, I. and Ansotegui, I.J. (1994) Humoral immunity in aging. Aging, Clinical and Experimental Research, 6, 143–150.

32. Mariotti, S., Barbesino, G., Caturegli, P., Bartalena, L., Sansoni, P., Fagnoni, F., Monti, D., Fagiolo, U., Franceschi, C. and Pinchera, A. (1993) Complex alteration of thyroid function in healthy centenarians. J. Clin. Endocr. Metab., 177, 1130–1134.

33. Mariotti, S., Franceschi, C., Cossarizza, A. and Pinchera, A. (1995) The aging thyroid. Endocr. Rev., 16, 686–715.

34. Theofilopoulos, A.N. (1995) The basis of autoimmunity: part I. Mechanisms of aberrant self-recognition. Immunol. Today, 16, 90–98.

35. Theofilopoulos, A.N. (1995) The basis of autoimmunity: part II. Genetic predisposition. Immunol. Today, 16, 150–159.

36. Cossarizza, A., Kahan, M., Ortolani, C., Franceschi, C. and Londei, M. (1991) Preferential expression of Vβ6.7 domain on human peripheral CD4+ T cells. Implications for positive selection of T cells in man. Eur. J. Immunol., 21, 1571–1574.

37. Cossarizza, A., Ortolani, C., Paganelli, R., Monti, D., Barbieri, D., Sansoni, P., Fagiolo. U., Forti, E., Londei, M. and Franceschi, C. (1992) Age-related imbalance of virgin (CD45RA+) and memory (CD45R0+) cells between CD4+ and CD8+ T lymphocytes in humans: a study from newborns to centenarians. J. Immunol. Res., 4, 118–126.

38. Cossarizza, A., Ortolani, C., Paganelli, R., Barbieri, D., Monti, D., Fagiolo, U., Castellani, G., Bersani, F., Sansoni, P., Londei, M. and Franceschi, C. (1996) CD45 isoforms expression on CD4+ and CD8+ T cells throughout life, from newborns to centenarians: implications for T cell memory. Mech. Ageing Dev., 86, 173–195.

39. Fabris, N., Mocchegiani, E., Amadio, L., Zannotti, M., Licastro, F. and Franceschi, C. (1984) Thymic hormone deficiency in normal ageing and in Down's syndrome: is there a primary failure of the thymus? Lancet, i, 983–986.

40. Brod, S.A., Rudd, C.E., Purvee, M. and Hafler, D.A. (1989) Lymphokine regulation of CD45R expression on human T cell clones. J. Exp. Med., 170, 2147–2152.

41. Bell, E.B. and Sparshott, S.M. (1990) Interconversion of CD45R subsets of CD4 T cells in vivo. Nature, 348, 163–166.

42. Rothstein, D.M., Yamada, A., Schlossman, S.F. and Morimoto, C. (1991) Cyclic regulation of CD45 isoform expression in a long term human CD4+CD45RA+ T cell line. J. Immunol., 146, 1175–1183.

43. Michie, C., McLean, A., Alcock, C. and Beverley, P.C.L. (1992) Lifespan of human lymphocyte subsets defined by CD45 isoforms. Nature, 360, 264–265.

44. Gray, D. and Skarvall, H. (1988) B-cell memory is short-lived in the absence of antigen. Nature, 336, 70–73.

45. Gray, D. and Matzinger, P. (1991) T cell memory is short-lived in the absence of antigen. J. Exp. Med., 174, 969–974.

46. Gray, D. (1993) Immunological memory. Annu. Rev. Immunol., 11, 49–77.

47. Cossarizza, A., Barbieri, D. and Londei, M. (1995) T cell repertoire usage in humans, from newborns to centenarians. Int. Rev. Immunol., 12, 41–55.

48. Murasko, D.M., Weiner, P. and Kaye, D. (1987) Decline in mitogen-induced proliferation of lymphocytes with increasing age. Clin. Exp. Immunol., 70, 440–448.

49. Monti, D., Troiano, L., Grassilli, E., Agnesini, C., Tropea, F., Barbieri, D., Capri, M., Salvioli, S., Ronchetti, I., Bellomo, G., Cossarizza, A. and Franceschi, C. (1992) Cell proliferation and cell death in immunosenescence. Ann. N.Y. Acad. Sci., 663, 250–261.

50. Cohen, J.J. (1991) Programmed cell death in the immune system. Advances in Immunology, 50, 55–85.

51. Lennon, S.V., Martin, S.J. and Cotter, T.G. (1990) Induction of apoptosis (programmed cell death) in tumour cell lines by widely diverging stimuli. Biochem. Soc. Transact., 18, 343–345.

52. Sentman, C.L., Shutter, J.R., Hockenbery, D., Kanagawa, O. and Korsmeyer, S.J. (1991) bcl-2 inhibits multiple forms of apoptosis but not negative selection in thymus. Cell, 67, 879–888.

53. Strasser, A., Harris, A.W. and Cory, S. (1991) bcl-2 transgene inhibits T cell death and perturbs thymic self-censorship. Cell, 67, 889–899.

54. Sen, S. and D'Incalci, M. (1992) Apoptosis. Biochemical events and relevance to cancer chemotherapy. FEBS Lett., 307, 122–127.

55. Gratiot-Deans, J., Ding, L., Turka, L.A. and Nuñez, G. (1993) bcl-2 proto-oncogene expression during human T cell development. J. Immunol., 151, 83–91.

56. Lee, S., Christakos, S. and Small, B.M. (1993) Apoptosis and signal transduction: clues to a molecular mechanism. Curr. Opin. Cell Biol., 5, 286–291.

57. Fisher, D.E. (1994) Apoptosis in cancer therapy: crossing and threshold. Cell, 78, 539–542.

58. NuÒez, G., Merino, R., Grillot, D. and Gonzales-Garcia, M. (1994) Bcl-2 and Bcl-x: regulatory switches for lymphoid death and survival. Immunol. Today, 15, 582–588.

59. Rebollo, A., Gomez, J., Martinez De Aragon, A., Lastres, P., Silva, A. and Perez-Sala, D. (1995) Apoptosis induced by IL-2 withdrawal is associated with an intracellular acidification. Exp. Cell Res., 218, 581–585.

60. Bettuzzi, S., Troiano, L., Davalli, P., Tropea, F., Ingletti, M.C., Grassilli, E., Monti, D., Corti, A. and Franceschi, C. (1991) In vivo accumulation of sulfated glycoprotein-2 (SGP-2) mRNA in rat thymocytes upon dexamethasone-induced cell death. Biochem. Biophys. Res. Commun., 175, 810–815.

61. Grassilli, E., Bettuzzi, S., Monti, D.,·Ingletti, M.C., Franceschi, C. and Corti, A. (1991) Studies on the relationship between cell proliferation and cell death: opposite patterns of SGP-2 and ornithine decarboxilase mRNA accumulation in PHA-stimulated human lymphocytes. Biochem. Biophys. Res. Commun., 180, 59–63.

62. Grassilli, E., Bettuzzi, S., Troiano, L., Ingletti, M.C., Monti, D., Corti, A. and Franceschi, C. (1992) SGP-2, apoptosis and aging. Ann. N.Y. Acad. Sci., 663, 471–474.

63. Grassilli, E., Carcereri de Prati, A., Monti, D., Troiano, L., Menegazzi, M., Barbieri, D., Franceschi, C. and Suzuki, H. (1992) Studies on the relationship between cell proliferation and cell death. II. Early gene expression during concanavalin A-induced proliferation or dexamethasone-induced apoptosis of rat thymocytes. Biochem. Biophys. Res. Commun., 188, 1261–1266.

64. Sikora, E., Grassilli, E., Radziszewska, E., Bellesia, E., Barbieri, D. and Franceschi, C. (1993) Transcription factors DNA-binding activity in rat thymocytes undergoing apoptosis after heat shock or dexamethasone treatment. Biochem. Biophys. Res. Commun., 197, 709–715 .

65. Sikora, E., Grassilli, E., Bellesia, E., Troiano, L. and Franceschi, C. (1993) Studies on the relationship between cell proliferation and cell death. III. AP-1 DNA-binding activity during concanavalin A-induced proliferation or dexamethasone-induced apoptosis of rat thymocytes. Biochem. Biophys. Res. Commun., 192, 386–391.

66. Richter, C., Schweizer, M., Cossarizza, A. and Franceschi, C. (1996) Control of apoptosis by the cellular ATP level. FEBS Lett., 378, 107–110.

67. Linnane, A.W. (1992) Mitochondria and aging: the universality of bioenergetic disease. Aging, Clinical and Experimental Research, 4, 267–271.

68. Cossarizza, A., Baccarani Contri, M., Kalashnikova, G. and Franceschi, C. (1993) A new method for the cytofluorimetric analysis of mitochondrial membrane potential using the J-aggregate forming lipophilic cation 5,5',6,6'-tetrachloro-1,1',3,3'-tetraethylbenzimidazolcarbocyanine iodide (JC-1). Biochem. Biophys. Res. Commun., 197, 40–45.

69. Cossarizza, A., Kalashnikova, G., Grassilli, E., Chiappelli, F., Salvioli, S., Capri, M., Barbieri, D., Troiano, L., Monti, D. and Franceschi, C. (1994) Mitochondrial modifications during rat thymocyte apoptosis: a study at the single cell level. Exp. Cell Res., 214, 323–330.

70. Cossarizza, A., Cooper, E.L., Quaglino, D., Salvioli, S., Kalachnikova, G. and Franceschi, C. (1995) Mito chondrial mass and membrane potential in coelomocytes from the earthworm Eisenia foetida: studies with fluorescent probes in single intact cells. Biochem. Biophys. Res. Commun., 214, 503–510.

71. Cossarizza, A., Franceschi, C., Monti, D., Salvioli, S., Bellesia, E., Rivabene, R., Biondo, L., G., R., Tinari, A. and Malorni, W. (1995) Protective effect of N-acetylcysteine in Tumor Necrosis Factor a-induced apoptosis in U937 cells: the role of mitochondria. Exp. Cell Res., 220, 232–240.

72. Cossarizza, A., Salvioli, S., Franceschini, M.G., Kalashnikova, G., Barbieri, D., Monti, D., Grassilli, E., Tropea, F., Troiano, L. and Franceschi, C. (1995) Mitochondria and apoptosis: a cytofluorimetric approach. Fund. Clin. Immunol., 3, 67–68.

73. Tropea, F., Troiano, L., Monti, D., Lovato, E., Malorni, W., Rainaldi, G., Mattana, P., Viscomi, P., Portolani, M., Cermelli, C., Cossarizza, A. and Franceschi, C. (1995) Sendai virus and herpes virus type I induce apoptosis in human peripheral blood mononuclear cells. Exp. Cell Res., 218, 63–70.

74. Cossarizza, A., Ceccarelli, D. and Masini, A. (1996) Functional heterogeneity of isolated mitochondrial population revealed by cytofluorimetric analysis at the single organelle level. Exp. Cell Res., 222, 84–94.

75. Sutton, B.J. and Gould, H.J. (1993) The human IgE network. Nature, 366, 421–428.

76. Defrance, T., Carayon, P., Billian, G., Guillemot, J.-C., Minty, A., Caput, D. and Ferrara, P. (1994) Interleukin-13 is a B cell stimulating factor. J. Exp. Med., 179, 135–143.

77. Gillis, S., Kozak, R., Durante, M. and Weksler, M. (1981) Immunological studies of aging. Decreased production of and response to T cell growth factor by lymphocytes from aged humans. J. Clin. Invest., 67, 937–942.

78. Cossarizza, A., Monti, D., Cantini, M., Paganelli, R., Montagnani, G., Cadossi, R., Bersani, F. and Franceschi, C. (1989) Extremely low frequency pulsed electromagnetic fields increase interleukin-2 (IL-2) utilization and IL-2 receptor expression in lymphocytes from old subjects. FEBS Lett., 248, 141–144.

79. Fagiolo, U., Cossarizza, A., Scala, E., Fanales-Belasio, E., Ortolani, C., Cozzi, E., Monti, D., Franceschi, C. and Paganelli, R. (1993) Increased cytokine production in mononuclear cells of healthy elderly people. Eur. J. Immunol., 23, 2375–2378.

80. Ottaviani, E., Petraglia, F., Montagnani, G., Cossarizza, A., Monti, D. and Franceschi, C. (1990) Presence of ACTH- and β-endorphin-immunoreactive molecules in the freshwater snail Planorbarius corneus (L.) (Gastropoda, Pulmonata) and their possible role in phagocytosis. Regul. Pept., 27, 1–9.

81. Ottaviani, E., Petraglia, F., Genedani, S., Bernardi, M., Bertolini, A., Cossarizza, A., Monti, D. and Franceschi, C. (1990) Phagocytosis and ACTH- and β-endorphin-like molecules in invertebrate (molluscan) and in vertebrate (human) cells: possible significance for the evolution of the immunoneuroendocrine system. Ann. N.Y. Acad. Sci., 594, 454–457.

82. Franceschi, C., Cossarizza, A., Ortolani, C., Monti, D. and Ottaviani, E. (1991) Natural cytotoxicity in a mollusc: an unorthodox comparative approach. Advances in Neuroimmunology, 1, 99–113.

83. Franceschi, C., Cossarizza, A., Monti, D. and Ottaviani, E. (1991) Cytotoxicity and immunocyte markers in cells from the freshwater snail Planorbarius corneus (L.) (Gastropoda, Pulmonata): implications for the evolution of NK cells. Eur. J. Immunol., 21, 489–493.

84. Ottaviani, E., Cossarizza, A., Ortolani, C., Monti, D. and Franceschi, C. (1991) ACTH-like molecules in gastropod molluscs: a possible role in ancestral immune response and stress. Proc. Royal Soc. Ser. B, 245, 215–218.

85. Ottaviani, E., Caselgrandi, E., Bondi, M., Cossarizza, A., Monti, D. and Franceschi, C. (1991) The "immuno-mobile" brain: evolutionary evidence. Advances in Neuroimmunology, 1, 27–39.

86. Ottaviani, E., Franchini, A., Cossarizza, A. and Franceschi, C. (1992) ACTH-like molecules in lymphocytes. A study in different vertebrate classes. Neuropept., 23, 215–219.

87. Ottaviani, E., Caselgrandi, E., Petraglia, F. and Franceschi, C. (1992) Stress response in the freshwater snail Planorbarius corneus (L.) (Gastropoda, Pulmonata): interaction between CRF, ACTH, and biogenic amines. Gen. Compar. Endocrinol., 87, 354–360.

88. Ottaviani, E., Caselgrandi, E., Fontanili, P. and Franceschi, C. (1992) Evolution, immune responses and stress: studies on molluscan cells. Acta Biologica Hungarica, 43, 293–298.

89. Ottaviani, E., Franchini, A. and Franceschi, C. (1993) Presence of several cytokine-like molecules in molluscan hemocytes. Biochem. Biophys. Res. Commun., 195, 984–988.

90. Ottaviani, E., Caselgrandi, E., Franchini, A. and Franceschi, C. (1993) CRF provokes the release of norepinephrine by hemocytes of Viviparus ater (Gastropoda, Prosobranchia): further evidence in favour of the evolutionary hypothesis of the "mobile immune-brain". Biochem. Biophys. Res. Commun., 193, 446–452.

91. Ottaviani, E. and Franceschi, C. (1993) Phylogeny of neuroendocrine-immune orchestra. Trends Compar. Biochem. Physiol., 1, 109–117.

92. Ottaviani, E., Franchini, A., Caselgrandi, E., Cossarizza, A. and Franceschi, C. (1994) Relationship between CRF and IL-2: evolutionary evidence. FEBS Lett., 351, 19–21.

93. Franceschi, C., Paganelli, R., Fagiolo, U. and Ottaviani, E. (1994) Cytokines, aging and evolution: the problem of promiscuity. Int. J. Immunopathol. Pharmacol., 7, 227–233.

94. Cooper, E.L., Cossarizza, A., Suzuki, M.M., Salvioli, S., Capri, M., Quaglino, D. and Franceschi, C. (1995) Autogeneic but not allogeneic earthworm effector coelomocytes kill the mammalian tumor cell target K562. Cell. Immunol., 166, 113–122.

95. Franceschi, C., Licastro, F., Paolucci, P., Masi, M., Cavicchi, S. and Zannotti, M. (1978) T and B lymphocyte subpopulations in Down's syndrome. Study on not-institutionalized subjects. J. Ment. Defic. Res., 22, 179–181.

96. Franceschi, C., Licastro, F., Chiricolo, M., Fantini, M.P., Paolucci, P., Fabris, N., Zannotti, M. and Masi, M. (1981) Failure of autologous mixed lymphocyte cultures and increased replicative capability after gamma irradiation in not-institutionalized subjects with Down's syndrome. Hum. Genet., 2 (suppl.), 238–239.

97. Franceschi, C., Licastro, F., Chiricolo, M., Bonetti, F., M., Z., Fabris, N., Mocchegiani, E., Fantini, M.P., Paolucci, P. and Masi, M. (1981) Deficiency of autologous mixed lymphocyte reactions and serum thymic factor level in Down's syndrome. J. Immunol., 126, 2161–2164.

98. Licastro, F., Chiricolo, M., Tabacchi, P.L., Barboni, F., Zannotti, M. and Franceschi, C. (1983) Enhancing effect of lithium and potassium ions on lectin-induced lymphocyte proliferation in aging and Down's syndrome subjects. Cell. Immunol., 75, 111–121.

99. Chiricolo, M., Minelli, L., Licastro, F., Tabacchi, P.L., Zannotti, M. and Franceschi, C. (1984) Alteration of the capping phenomenon on lymphocytes from aged and Down's syndrome subjects. Gerontology, 30, 145–152.

100. Franceschi, C., Chiricolo, M., Licastro, F., Zannotti, M., Masi, M., Mocchegiani, E. and Fabris, N. (1988) Oral zinc supplementation in Down's syndrome. Restoration of thymic endocrine activity and of some immune defects. J. Ment. Defic. Res., 32, 169–181.

101. Cossarizza, A., Monti, D., Montagnani, G., Dagna-Bricarelli, F., Forabosco, A. and Franceschi, C. (1989) Fetal thymic differentiation in Down's syndrome. Thymus, 14, 163–170.

102. Cossarizza, A., Monti, D., Dagna-Bricarelli, F., Montagnani, G. and Franceschi, C. (1990) LAK activity is inducible in blood mononuclear cells from human fetus. Immunol. Lett., 24, 137–140.

103. Scarfì, M.R., Cossarizza, A., Monti, D., Bersani, F., Zannotti, M., Lioi, M.B. and Franceschi, C. (1990) Age-related increase of mitomycin-C-induced micronuclei in lymphocytes from Down's syndrome subjects. Mutat. Res., 237, 217–222.

104. Cossarizza, A., Monti, D., Montagnani, G., Ortolani, C., Masi, M., Zannotti, M. and Franceschi, C. (1990) Precocious aging of the immune system in Down's syndrome: alterations of B-lymphocytes, T-lymphocyte subsets and of cells with NK markers. Am. J. Med. Genet., 7 (suppl.), 213–218.

105. Licastro, F., Chiricolo, M., Mocchegiani, E., Fabris, N., Cossarizza, A., Masi, M., Arena, G., Zannotti, M., Monti, D., Beltrandi, E., Mancini, R., Casadei-Maldini, M. and Franceschi, C. (1990) Impairment of T cells subpopulations and thymic endocrine function in children with Down's syndrome: a possible pathogenetic role for zinc deficiency. J. Immunol. Res., 2, 95–100.

106. Cossarizza, A., Monti, D., Bersani, F., Scarfì, M.R., Zannotti, M., Cadossi, R. and Franceschi, C. (1991) Extremely low frequency pulsed electromagnetic fields increase lymphocyte proliferation in Down's syndrome. Aging, Clinical and Experimental Research, 3, 241–246.

107. Cossarizza, A., Ortolani, C., Forti, E., Montagnani, G., Paganelli, R., Zannotti, M., Marini, M., Monti, D. and Franceschi, C. (1991) Age-related expansion of functionally inefficient cells with markers of NK activity in Down's syndrome. Blood, 77, 1263–1270.

108. Levy, S.M., Herberman, R.B., Lee, J., Whiteside, T., Beadle, M., Heiden, L. and Simons, A. (1991) Persistently low natural killer cell activity, age, and environmental stress as predictors of infectious morbidity. Nat. Immun. Cell Growth Regul., 10, 289–307.

109. Franceschi, C. and Cossarizza, A. (1995) Introduction: the reshaping of immune system with age. Int. Rev. Immunol., 12, 1–3.

110. Franceschi, C., Monti, D., Barbieri, D., S., S., Negro, P., Capri, M., Guido, M., Azzi, R., P., S., Paganelli, R., Fagiolo, U., Baggio, G., Donazzan, S., Mariotti, S., D'Addato, S., Gaddi, A., Ortolani, C. and Cossarizza, A. (1995) Immunosenescence in humans: deterioration or remodelling? Int. Rev. Immunol., 12, 57–74.

111. Franceschi, C., Monti, D., Sansoni, P. and Cossarizza, A. (1995) The immunology of exceptional individuals: the lesson of centenarians. Immunol. Today, 16, 12–16.

112. Franceschi, C. and Monti, D. (1990) Il controllo genetico dell'invecchiamento e della longevità: peculiarite paradossi. Giornale di Gerontologia, 38, 127–134.

113. Schächter, F., Cohen, D. and Kirkwood, T. (1993) Prospects for the genetics of human longevity. Hum. Genet., 91, 519–526.

114. Takata, H., Suzuki, M., Ishii, T., Sekiguki, S. and Iri, H. (1987) Influence of MHC genes on human longevity among Okinawan-Japanese centenarians and nonagenarians. Lancet, ii, 824–826.

115. Schächter, F., Faure-Delanef, L., Guenot, F., Rouger, H., Frouguel, P., Lesueur-Ginot, L. and Cohen, D. (1994) Genetic association with human longevity at the APOE and ACE loci. Nature Genet., 6, 29–32.

116. Kervinen, K., Savolainen, M.J., Salokannel, J., Hynnine, A., Heikkinen, J., Ehnholm, C., Koistinen, M.J. and Kesaniemi, Y.A. (1994) Apolipoprotein E and B polymorphisms-longevity factors assessed in nonagenarians. Atherosclerosis, 105, 89–95.

117. Louhija, J., Miettinen, H.E., Kontula, K., Tikkanen, M.J., Miettinen, T.A. and Tilvis, R.S. (1994) Aging and genetic variations of plasma apolipoproteins. Relative loss of the apolipoprotein E4 phenotype in centenarians. Arterioscl. Thromb., 14, 1084–1089.

118. Harding, R.M. (1992) VNTRs in review. Evolut. Anthropol., 1, 62–71.

119. Weissenbach, J., Gyapay, G., Dib, C., Vignal, A., Morissette, J., Millaseau, P., Vaysseix, G. and Lathrop, M. (1992) A second-generation linkage map of the human genome. Nature, 359, 794–801.

120. Orr, W.C. and Sohal, R.S.S. (1994) Extension of lifespan by overexpression of superoxide dismutase and catalase in Drosophila melanogaster. Science, 263, 1128–1130.

121. Cristofalo, V.J. and Pignolo, R.J. (1993) Replicative senescence of human fibroblast-like cells in culture. Physiol. Rev., 73, 617–638.
122. Wang, E. (1995) Senescent human fibroblasts resist programmed cell death, and failure to suppress bcl2 is involves. Cancer Res., 55, 2284–2292.
123. Zhou, T., Edwards III, C.K. and Mountz, J.D. (1995) Prevention of age-related T cell apoptosis defect in CD2-fas-transgenic mice. J. Exp. Med., 182, 129–137.
124. Pinchera, A., Mariotti, S., Barbesino, G., Bechi, R., Sansoni, P., Fagiolo, U., Cossarizza, A. and Franceschi, C. (1995) Thyroid autoimmunity and aging. Horm. Res., 43, 64–68.

10

DROSOPHILA AS A MODEL SYSTEM FOR MOLECULAR GERONTOLOGY

Christine Brack,[*] Ruedi Ackermann,[†] Noriko Shikama,[‡] Elisabeth Thüring,[°] and Martin Labuhn

Biozentrum
Department of Cell Biology
University Basel
CH-4056 Basel, Switzerland

1. INTRODUCTION

The use of fruit flies in molecular ageing research will be the topic of this chapter. We like to point out the value of this animal model to study senescence mechanisms, including its advantages and limitations. In the first part we shall give a short survey on the reasons why gerontologists choose Drosophila as a model system and on the kind of answers we can expect to gain with this model. A brief introduction into Drosophila biology should serve for the understanding of experiments described later. Finally, we discuss our standpoint on viewing ageing as part of development.

In the second part we shall discuss a few examples of the ageing research done with this organism, addressing mainly some of the essential maintenance functions. This allows us to tackle single questions, although it is obvious that all these individual aspects are only part of the intricate network of interactions influencing homeostasis, as discussed in other chapters of this book. None of them should actually be examined separately, but this is done here for a matter of simplicity. This contribution should not be taken as an exhaustive review; the examples chosen here are questions to which our group has made some contributions. Excellent review papers on other aspects of gerontology can be consulted elsewhere[1,2,3].

In the last part we like to give an outlook on future perspectives using Drosophila in ageing research. We shall discuss some of the problems encountered with this model, and try to suggest possible solutions and future ways to go.

* Correspondence to: Dr. Christine Brack, Biozentrum Dept. Cell Biology, University Basel, Klingelbergstrasse 70, CH-4056 Basel.
† Present address: Zoologisches Institut, University Basel, CH-4051 Basel, Switzerland.
‡ Present address: Department of Biochemistry, University Glasgow, Glasgow G12 8QQ, Scotland.
° Present Address: Ciba-Geigy AG, 4057 Basel.

1.1. The Fruit fly as Model for Ageing Research

Drosophila melanogaster is a valuable model for research not only into early development but also into ageing. The fruit fly has several advantages over other model organisms. Because the culturing of large numbers of flies is very easy, and because Drosophila has a rather short generation time it is possible to select for life span variants[4,5]. Furthermore, both classical genetics and modern gene technology are very well developed and documented, and it is relatively easy to generate transgenic flies. Information on genes, mutations, map positions, markers, therefore, is easily accessible[6]. The methods to visualize gene activity in situ have been developed mainly in Drosophila[7].

It has been debated whether the fruit flies can serve as an appropriate model for ageing mechanisms in higher eukaryotes such as mammals, including man. However, recent successful research—mainly done with Drosophila and other invertebrates—has demonstrated that important molecular mechanisms for embryonic development and differentiation, for cell communication and signal transduction, etc., have been conserved through evolution. Thus it has become apparent that there is striking evolutionary conservation of molecular components and their interplay within organisms. The homeobox gene clusters and their role in early development[8] and the recently discovered conservation in eye development[9,10] may serve as just a few examples. The same is probably true for almost all essential maintenance functions like DNA replication and repair, defense mechanisms, transcription, translation, and protein degradation; most of the components have been conserved, at least functionally.

For these reasons we expect that ageing mechanisms discovered in fruit flies will be comparable to those operating in other organisms. There is one major limitation to the fly model: Drosophila is composed almost exclusively of postmitotic cells (with the exception of the reproductive organs and some intestinal cells). Therefore, it is suited only as model for ageing mechanisms in postmitotic cells and not for specific mechanisms concerning replicative senescence or cell replacement at the organismic level. Still, we think that the analysis of age-related changes in maintenance functions at the cellular and molecular level may give us important insights into ageing mechanisms common to mitotic and postmitotic cells.

On one hand, the insect model has been useful to study normal ageing, such as physiological age-related changes, biochemical changes, changes in gene expression, behavior, reproductive or physical fitness, and others. On the other hand, Drosophila has been a favorite model for experimentally influencing the ageing process, e.g., by altering the expression level of specific genes of interest and then examining their effect, mainly on life span. Two experimental approaches are used for this purpose: transgenic animals, and epigenetics (alteration of living conditions such as temperature, food, antioxidant, stress, etc.). As we mentioned in the beginning of this chapter, we have decided to discuss a few of the specific homeostatic maintenance functions and explain for each of them which experimental approach has been taken and what results were obtained.

Development and Ageing. When gerontologists look at the life cycle of a higher organism, they are often confronted with the question of defining the time point of when development ends and when ageing starts. Depending on the species and the specific function studied, this may vary considerably. In our view it is not necessary to make this distinction. Looking at the question with the eyes of a developmental biologist, ageing can be defined as the continuation of the development process, encompassing the latest phase in an organism's life cycle. This concept is very useful particularly also for the Drosophila model. At first glance, the definitions seem to be quite obvious for this

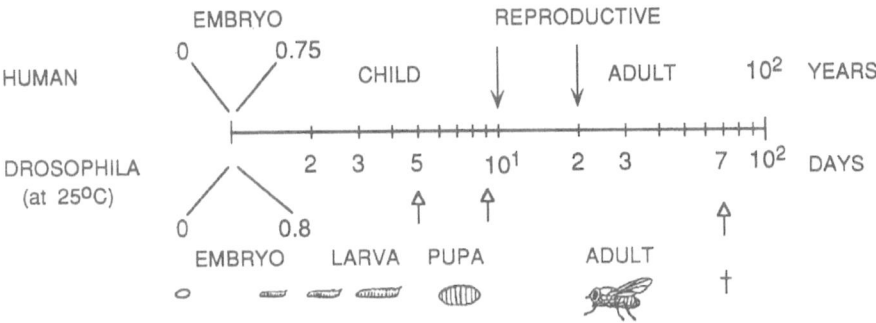

Figure 1. The life cycle of male Drosophila (in days) and man (in years) drawn to logarithmic scale.

holometabolous insect species with its well-defined larval and pupal stage, culminating in the metamorphosis to adult flies. However, it was shown that environmental impacts on juvenile stages, such as larval density, can definitely influence the senescence program, resulting in life span alterations[11]. Other examples, where environmental and/or dietary factors are considered to affect life spans are C. elegans [12] or the honeybee[13]. In both cases, external changes applied at the larval stages induce an alternative differentiation program that results in a new phenotype with increased life span (the dauer larvae of C. elegans, or the honeybee queen). Obviously, development, differentiation, and ageing processes cannot be separated here either and it is questionable, whether these organisms should really be used as examples for ageing in the classical sense. For this reason, it will be important not to consider only age-related changes occurring in adult Drosophila, but to include also the other developmental stages in a global analysis of development-ageing. Thus we have, for example, included embryos, larvae and pupae in many of our own gene expression analyses.

In most organisms early development is almost exclusively governed by genetic processes: temporal and spatial control of gene expression determines pattern formation and cell differentiation. This phase of the life cycle comprises only ~10% of an organism's maximal life span, e.g., 8–9 days in the life of a fly with a maximal life span of 60–70 days, or 10–15 years in the life of humans with a maximal life span of 100–120 years (illustrated in Figure 1). The adult reproductive phase and the senescence phase, on the other hand, represent ~90% of the life span. During this long period, additional components, like environmental factors and effects from metabolic by-products (e.g., oxygen radical induced damage) gain a much higher importance, particularly in postmitotic organisms like Drosophila. Mechanisms controlling the ageing process, therefore, have to include genetic as well as epigenetic factors. It is these interplays between genes and environment that complicate straightforward analysis of the ageing processes.

1.2. Drosophila Biology

The life cycle of the fruit fly D. melanogaster consists of an embryo stage, three larval stages, a pupal stage, and the imago (adult reproductive fly). In Figure 1 the life of a fly is schematically represented in a logarithmic scale and compared to human life span. It is important to point out that the time scale presented here for the fly is variable, since early development as well as adult life span of Drosophila are to a certain extent temperature dependent (reviewed in[14]).

At 25°C the development from a fertilized egg to an adult Drosophila takes about 11 days. Within 24 hours after egg fertilization the embryo develops into a first instar larva. Early during embryonic development, at the cellular blastoderm stage, the cells determined to form adult tissues are segregated. After the first instar larva hatches from the egg these cells are organized into imaginal discs. The larva undergoes two molts at about 48 and 72 hours after egg deposition, to give rise to the second and third instar larva, respectively. The third instar larva grows till pupariation occurs at 120 hours. From the time when the first instar larva hatches to the time of pupariation the larval cells do not divide, although the larva undergoes a considerable increase in size. This size increase is almost exclusively due to cell growth. In contrast, the cells of the imaginal discs divide during the larval stages.

It takes further 120 hours from pupariation to the eclosion of the adult fly. During metamorphosis most larval cells, except the cells of the Malpighian tubules and some nerve cells, undergo autolysis. The adult fly is reconstructed from the imaginal discs, some of which cease cell division at pupariation, others during the early stage of metamorphosis. The imaginal disc cells undergo growth and morphogenetic movements until they sculpture the adult body. During the larval stages and mainly during metamorphosis important tissue remodeling, growth and differentiation processes take place. This is why our considerations discussed above become important. The whole life cycle is one continuous process; one cannot make any clear distinction between changes related to development or metamorphosis processes or related to ageing.

2. GENES AND LIFE SPAN: EXPERIMENTAL APPROACHES

One of the fundamental questions amenable to experimental analysis for which short-lived animals like the fruit fly are particularly suitable, is: can one define genes that influence life span? Although various parameters such as geotactic motility[15], the ability to fly[16], or resistance to paraquat[17] have been proposed to serve as biomarkers for ageing in flies, the most widely used assay to test genetic and/or epigenetic influences on senescence is to measure the life span. Longevity is thus a measure for optimal survival and maintenance functions at the organismic level. Figure 2 shows an example of a survival curve of a wild type Drosophila strain at 25°C. From such data the mean and the maximum life span can be calculated, and experimentally induced variations in life span can be

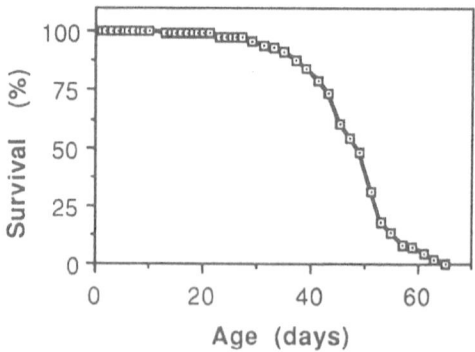

Figure 2. Life span of D. melanogaster (Oregon R) males at 25°C. The survival curve is averaged from ten independent experiments corresponding to a total number of > 3000 animals.

statistically evaluated. The following approaches have been used to determine the genetic components involved:

- identification of genes affecting life span by subjecting animals to various biological selection pressures[4,5] and selection of long lived strains.
- analysis of age-related changes in gene expression that correlate with life span.
- modulation of candidate genes, e.g., genes involved in specific homeostatic maintenance functions (proteins involved in translation and ROS detoxification). This can be achieved by making transgenic animals and testing, whether changing the copy number and/or the expression level of such genes affects life span.
- experimentally induce alterations in life span by environmental factors (temperature, diet, etc.) and examine the influence on gene expression patterns. This approach may allow to analyze the interplay between epigenetic and genetic factors.
- in combination with the use of the enhancer trap and reporter gene techniques unique to Drosophila, screen for temporal and spatial patterns of gene expression in whole animals[18]. This approach has allowed to detect marker genes that scale with life span under conditions of experimentally or genetically altered life span[19,20].

Examples of these approaches will be discussed in more detail in the following sections.

2.1. Selection for Altered Lifespan

Several laboratories have used various reproducible selection schemes to derive genetically selected long lived and short lived Drosophila lines with the hope to identify genes that influence life span. The strains were directly selected for delayed fecundity, which indirectly resulted in selection for extended longevity (ELP = extended longevity phenotype)[4,5,21,22]. Other laboratories selected for stress resistance on the first hand[23]. The success of the various selection experiments has been interpreted as a conclusive demonstration for the presence of genes influencing life span and ageing.

However, it turned out that subsequent identification of responsible genes is difficult. Obviously these selection schemes do not operate at the level of single genes, but select for pleiotropic traits. Biochemical characterization by 2D gel analysis of proteins has shown differences in protein patterns between long and short lived strains[24]. For other strains, genetic analyses have allowed to localize recessive genes responsible for the ELP on the 3rd chromosome[25]. In most of these selected strains, however, the ELP is expressed only as long as the selection pressure is maintained, which suggests that although longevity is genetically determined, it can be environmentally modulated. Thus a high larval density at a critical period (larval age 60–120 hrs) is required to express adult ELP[26]. This led Arking to propose that senescence determining genes are active early in development in the presenescence period[27]. It seems that the selection of Drosophila for extended life span mainly works through an elevated efficiency of antioxidant defense system genes (see below).

2.2. Age-Related Changes in Gene Transcription

As mentioned in the introduction, regulation of gene expression plays a major role in development. The selection experiments strongly suggest that there are genes that can influence life span. Therefore, one can speculate that regulation of gene expression may also be involved in controlling the senescence phase of the adult life. The fly has been used by many investigators to study age-related changes in gene regulation. Various approaches are employed.

The most straightforward approach is to measure RNA levels in ageing flies. Of course, this does not allow to draw any conclusions on the activity of the final gene product, but it gives a first evaluation on gene activity at the transcriptional level. It is possible to examine changes in total RNA, in classes of molecules, or in individual transcripts of candidate genes, depending on the availability of gene probes. Because flies are small, these analyses are usually done with RNAs extracted from whole animal homogenates and give information at the level of a population and not from individual flies or tissues. Alternatively, RNA is extracted from separated body parts (heads, thoraces, abdomen).

A second approach has recently been proposed that allows the analysis of gene expression patterns in intact organs (e.g., the antennae) of individual flies[19,20]. It is based on the enhancer trap and reporter gene technology developed with transgenic animals[18]. In the following, some examples shall be described.

2. 2.1. Changes in Steady State RNA Levels. To quantitate age-related changes in gene expression at the RNA or at the protein levels, the common procedure is to extract total RNA and protein from flies of different age groups and to compare the abundance of a particular transcript or protein relative to equivalent amounts of total RNA or protein, or to a marker transcript (e.g., EF-1α, actin mRNA). When we measured absolute amounts of RNA and protein per fly equivalent, we found that the total soluble protein amount per fly does not change considerably during ageing. In contrast, the yields of total RNA in old flies were always ~60% lower than in young flies. This drastic decrease in total RNA can be observed irrespective of the method of RNA extraction used[28], which would mean that it is not an experimental artifact, but that indeed the amount of total RNA per fly decreases with increasing age. Figure 3 shows changes in total RNA extracted from whole flies, in comparison to changes in RNA extracted from head and thorax (= exclusively postmitotic tissues). When talking about changes in gene expression, it is of advantage, therefore, to compare always RNA amounts per fly equivalent (or protein), or to give relative amounts.

The question then arises of which RNA species are subject to age-related changes. We have examined steady state levels of transcripts from RNA polymerase I, II, and III genes; first, in total RNA extracted from whole flies (Figure 4A), and then in RNA extracted from separated body parts (head, thorax, and abdomen, Figure 4B). RNAs were

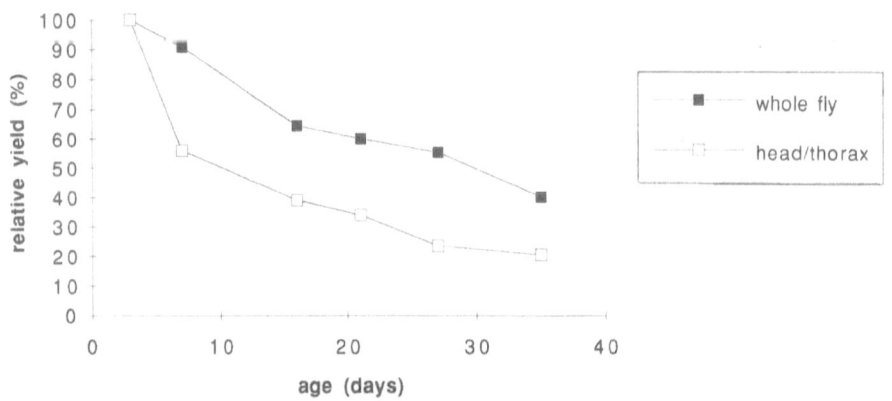

Figure 3. Changes in total RNA. The RNA concentration was determined by OD 260 absorption. Numbers are given in relative values compared to 2–3 day old flies (100%).

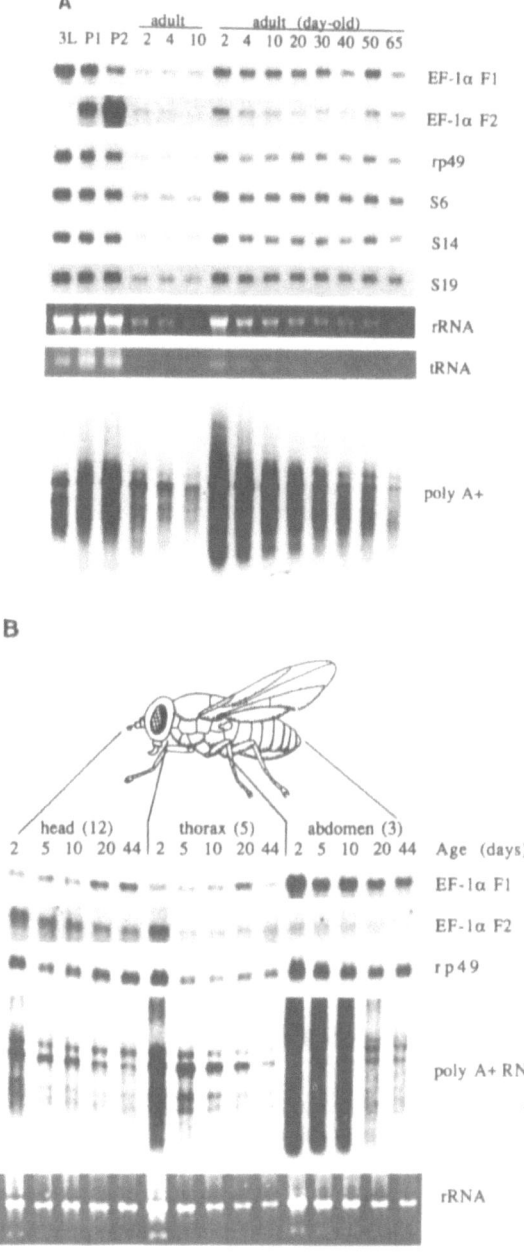

Figure 4. Changes in gene transcription during development and ageing.. A) Northern blot showing changes in rRNAs, tRNAs, total polyA[+] RNA, and specific mRNAs coding for EF-1α F1 and F2. and four ribosomal proteins. Total RNA was extracted from whole animals; the first six lanes, third instar larvae (3L). early pupae (P1), late pupae (P2), and 2–10 day old adult flies, contain RNA extracted from aliquots equivalent to one animal. In the following eight lanes, adult flies of 2–65 days, total RNA equivalent to two animals was loaded. rRNA and tRNAs were detected by EthBr staining; all mRNAs were detected by hybridization with specific probes. B) Northern blot showing changes in RNAs exteracted from head, thorax, and abdomen of male flies.

Table 1. Age-related changes in RNA steady state levels in male *D. melanogaster*. Quantitation of steady state RNA levels in extracts from whole male flies

	Age (days)	
	8–10 d	40 d
	relative amounts (% of 2 day old)	
Total RNA (OD260)	60–80	35–45
RNA Pol II transcripts: poly A$^+$ mRNA	40	18
RNA Pol I transcripts: rRNA	38	20
RNA Pol III transcripts: tRNA	55–60	8–10
Specific genes		
ribosomal proteins		
S6, S14, S19, rp49	60–80	50–60
EF-1α F1	75	40
cytoplasmic actins:		
Act 42A	110	120
Act 5C	60–70	60
muscle actins:		
Act 87E	70	60
Act 57A	35	30
Act 88F, 79B	5–15	0.5–6

The data are expressed in relative amounts, the values of two day old flies are set at 100%. The first column shows the changes occurring during the maturation phase (8-10 days). The second column displays the age-related decrease occurring in the senescent period.

size fractionated on agarose gels, stained with Ethidium Bromide to visualize ribosomal RNAs and tRNAs, and then blotted to Nylon membranes for hybridization with specific gene probes.

Ribosomal RNAs (rRNAs) make up for > 90% of total cellular RNA and it was expected that large changes might affect mainly these RNA species. As shown on the gels, the rRNA amount per fly is decreasing by approximately 80% during the adult life (Figures 4A,5). In the abdomen, which contains the largest quantities of rRNA, the age-related decrease is continuous (and probably reflects the age-related decline in germline tissue), whereas in head and thorax the main drop occurs in the first 8 days of adult life (Figure 4B).

tRNAs, the second abundant RNA species visible on Ethidium Bromide stained gels, also decline continuously in whole fly extracts. In the somatic postmitotic tissues (head and thorax) they behave like the rRNA (Figure 4B). These results suggest that RNA polymerase I and III products, the two most abundant RNA species, diminish in old flies (Table 1, Figure 5). It is important to point out that the concentration of both RNA species is very high in third instar larvae and pupae; compared to 1–2 day old adult flies, these developmental stages contain at least twice as much RNA.

What about RNA polymerase II products? Hybridization of Northern blots to a labeled oligo dT probe allows an approximate evaluation of the quantities of poly A$^+$ mRNAs. Again, we detect a general decrease: poly A$^+$ mRNAs are most abundant in the pupal stage and decrease rapidly and continuously in whole-fly samples of adult males, almost in parallel to the rRNAs (Fig. 4A). In thorax and abdomen, poly A$^+$ RNA decreases constantly, while after the initial decrease of the first five days, the steady-state amount of poly A$^+$ RNA remains almost unchanged in the head of adult males (Figure 4B). The age-related decline in total mRNA suggests that RNA polymerase II-mediated transcription

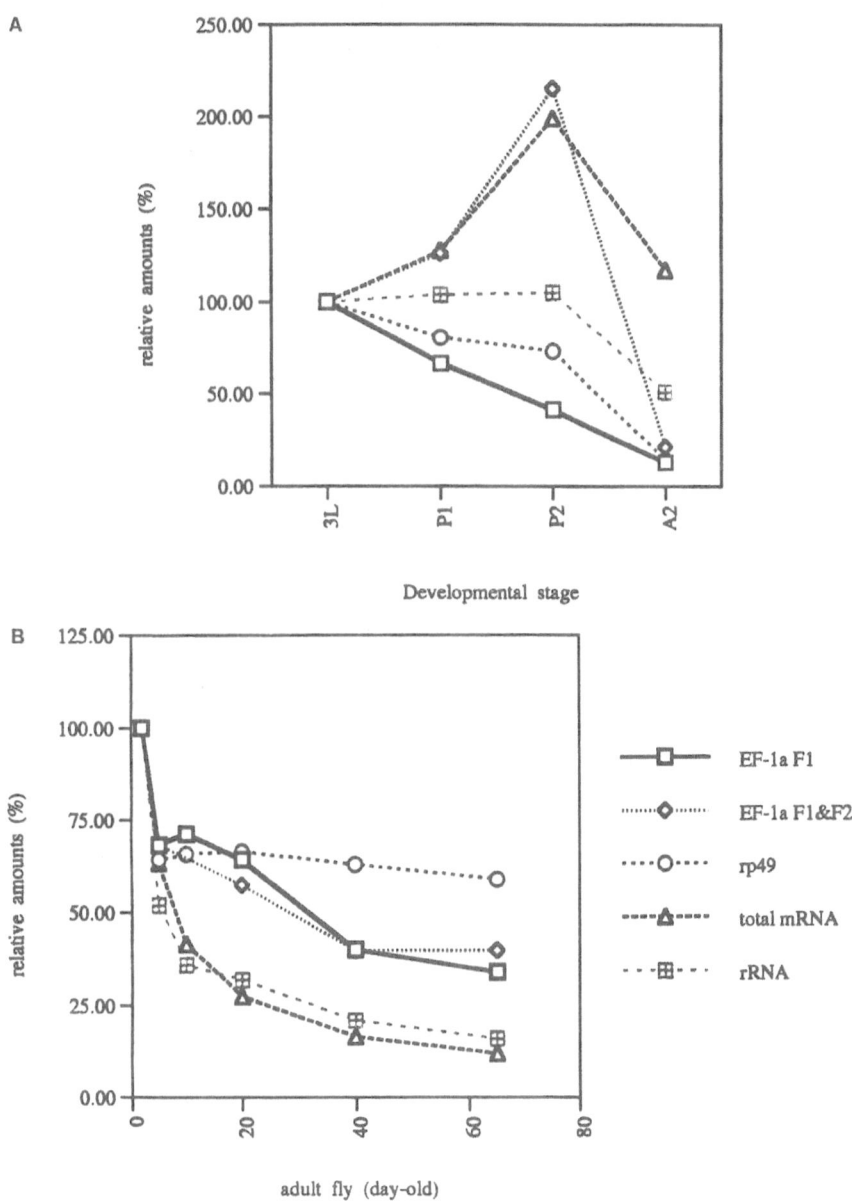

Figure 5. Changes in gene transcription during development and ageing. Quantitation of RNAs from Northern blots. The data are averaged from 4–5 independent experiments; they are expressed in relative amounts, setting either the values of the third instar larvae (A) or of the 2 day old flies as 100% (B).

slows down with age. Thus it appears that all RNA species are affected by ageing. This is consistent with a common regulatory mechanism affecting transcription mediated by all three classes of RNA polymerases.

An increasing number of specific RNA polymerase II genes are being analyzed by hybridization of Northern blots to specific gene probes. Of particular interest are candidate genes coding for proteins or enzymes involved in homeostatic maintenance functions.

Thus we have examined transcript levels of several genes of the protein synthesis machinery[28]. The most extensively studied genes, peptide synthesis elongation factor EF-1α and some antioxidant defense genes, will be presented in separate sections (2.3. and 2.4.). Other examples are presented here.

Ribosomal protein genes. Many investigators use as an internal standard one of the ribosomal protein transcripts, with the assumption that these housekeeping genes are constitutively expressed at the same level throughout development and ageing. We have quantitated steady state levels of four ribosomal protein mRNAs: rp49, S6, S14, and S19. They all change in a similar way, decreasing mainly in the first few days of adult life and then remaining approximately constant (Figure 4, Table 1). This suggests that in Drosophila the ribosomal protein genes are coordinately regulated at the transcriptional level, as was reported to be the case for senescent human fibroblasts[29]. When we determined ribosome activity, however, we found that active ribosomes decrease to much lower levels, almost parallel to rRNA[28]. These results indicate that in the fly transcription of ribosomal protein genes is not coupled to rRNA gene transcription and ribosome activity.

Heat shock genes are induced by a number of stresses, and thus have been the object of several analyses. Whereas hsp70 expression in response to heat shock was found to be reduced in old flies[30,31], a tissue-specific increase of hsp70 gene expression in response to oxidative stress was detected in muscles of old flies[32]. The small heat shock proteins hsp23 and hsp26 exhibit distinct spatial and temporal patterns of constitutive expression[33]. Ubiquitin, a protein involved in degradation of aberrant proteins, is another heat shock protein whose transcription level is affected by aging[34].

Actin gene expression. D. melanogaster contains six different actin genes, which are expressed in distinct temporal and spatial patterns during embryogenesis[35]. Two alleles code for cytoskeletal actins (Act 5C and Act 42A), whereas the four other genes code for larval specific and adult specific muscle actins[35]. The actin gene family, therefore, presents an interesting example of a multigene family consisting of both highly conserved housekeeping genes, and genes expressed exclusively in terminally differentiated muscle cells. Expression in adult life had not been studied in detail, although many investigators use cytoplasmic actin mRNAs as internal standard when comparing different RNA samples from ageing flies.

We have quantitated steady state levels of all six actin isoform mRNAs on Northern blots (Figure 6). Our results show that the two adult specific muscle genes coding for flight and jump muscle (Act 79B and Act 88F) are shut down almost completely after eclosion, whereas two other muscle actins (Act 57A and Act 87E) decrease to intermediate levels. The two cytoskeletal actins behave differently: although Act 5C is expressed in all tissues, and codes for a protein with a high turnover rate, its mRNA decreases to intermediate levels comparable to ribosomal protein genes. In contrast, the gene coding for the cytoskeletal actin Act 42A, which is expressed in a subset of cells only (e.g., nervous system, sensory organs, gonads) is abundantly transcribed in old flies, it is the only transcript that increases during senescence (Figure 6)[36].

Although the coding regions of the actin genes have been highly conserved during evolution, the noncoding sequences are quite diverged. Comparison of promoter regions among the differentially regulated actin genes on one hand, and with promoters of other housekeeping genes, on the other hand, might help to define regulatory sequences required for maintaining high expression levels of housekeeping genes.

These few examples have shown that, although RNA levels in general decrease after eclosion of adult flies, individual changes can be observed as soon as specific transcripts are analyzed. The information we can obtain by analyzing steady state levels of RNAs is

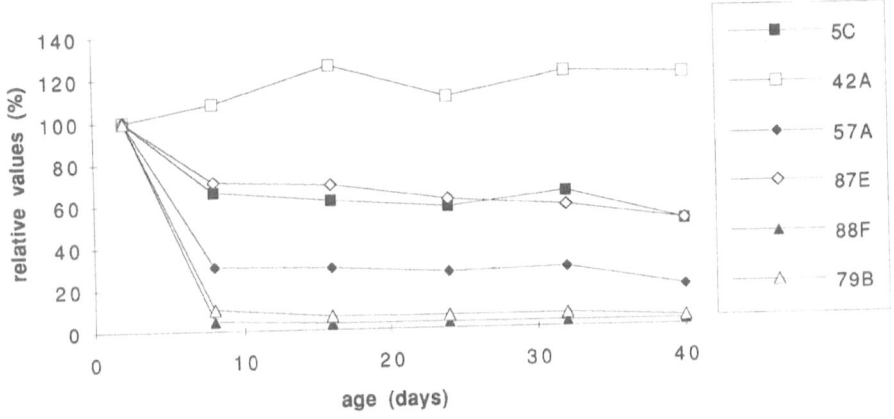

Figure 6. Changes in actin mRNAs during ageing. The numbers are averaged from five independent experiments: values from 2 day old flies were taken as 100%.

of course limited. It does not tell us anything about RNA stability, turnover, and translatability, nor about the protein products.

The results presented here also illustrate the problem addressed in the introduction. We observe that transcripts of most genes studied decrease rapidly in the first 5–10 days after eclosion (maturation stage), and then more slowly in the later "senescent" stage. This could reflect changes related to the pre-imago developmental program and might have nothing to do with senescence. A high level of gene expression, both transcription and protein synthesis, is required in the rapidly dividing cells during embryogenesis, for the rapid body growth at the larval stage, and for cell differentiation during metamorphosis and pupal stage. The postmitotic cells of the adult flies probably require lower gene activity and protein turnover. Arking proposed that senescence determining genes may be active in early adult life, i.e., in the presenescence period[21]. The important decline in all the RNAs studied within a few days after eclosion is still within the presenescence period. Does it involve any of the postulated senescence-determining genes?

The only way to elucidate whether changes in gene expression are causally related to a senescence program and thereby influence life span is to compare animals with different life span and see, whether altered expression patterns correlate with longevity.

2.2.2. Gene Expression Patterns and Life Span. The elegant work of Helfand et al.[19,20] presents the first direct evidence for a linkage between gene expression and life span. The authors applied the enhancer trap technique that was originally developed for gene analysis of early development in Drosophila embryos[18]. The enhancer trap consists of a stably integrated P-element carrying the bacterial β-galactosidase gene linked to a minimal promoter. Depending on the insertion site of the P-element in the genome, nearby enhancers may influence the expression of the β-gal reporter gene. The enhancer specific gene activity can easily be visualized in situ by staining intact flies for β-gal activity. This method provides, therefore, an indirect way of determining temporal and spatial activity of specific, even yet unknown, genes. Using this approach, Helfand et al.[19] found several genes expressed in the antenna whose pattern of expression scales with physiological life span, either when life span was altered by temperature shifts[19] or when the analysis was done with short-lived hyperactive mutants[20]. Both of these conditions cause changes in the metabolic activity of the animals. Besides demonstrating a correlation of gene activity with life span, these results provide further evidence for the rate of living hypothesis (see 2.4.).

2.3. Protein Synthesis and Ageing in Drosophila

In many animal species, as well as in cultured cells, age-related changes in the rate of protein synthesis have been observed[37-41]. Protein synthesis and turnover are key homeostatic functions that are particularly important also for postmitotic cells. Therefore, changes in the protein synthesis pathway, either at the transcriptional, at the translational, or at the post-translational level have been intensively studied, particularly in insects[37,39].

In their systematic studies on all steps involved in protein synthesis, carried out in D. melanogaster, Webster et al. have concluded that the age-related decrease in protein synthesis is due to the transcriptional down-regulation of the gene coding for peptide synthesis elongation factor EF-1α[41]. EF-1α is a component of the EF-1 complex that includes other subunits β, γ, and δ. EF-1α promotes binding of aminoacyl-tRNA to the A-site of the ribosome by hydrolyzing GTP, while EF-1$\beta\gamma\delta$ catalyzes GDP/GTP exchange on EF-1α[42,43]. Webster's finding stimulated a number of investigators to follow up the hypothesis that EF-1α may play a key role in the loss of protein synthesis activity during ageing.

2.3.1. EF-1α in Transgenic Animals. First, a series of experiments was done using the transgenic animal approach. Shepherd et al.[44] reported that Drosophila transformed with a P-element vector carrying an additional copy of the EF-1α gene under control of a heat shock gene (hsp70) promoter live longer than transgenic control flies when grown at elevated temperature. The authors concluded that the EF-1α gene may play a key role in longevity determination. These results were considered spectacular, because they predicted that a single gene might be involved in age-related down regulation of translation and in life span determination.

Stearns et al.[45] have extended these experiments to examine the relative contribution of EF-1α overexpression versus P-element position effect and genetic background on the life span and on other fitness parameters. By P-element jump-out from Shepherd's line they created new transgenic EF-1α and control lines with six different insert positions on the third chromosome, and finally crossed one of these lines to six different wild type strains. (The experimental design is illustrated in Figure 7). From their results the authors concluded that position of P-elements and genetic backgrounds have statistically relevant influences that affect life span and other biological fitness parameters more drastically than does the overexpression of the EF-1α transgene[45].

However, our detailed investigations of Shepherd's and Stearns' transgenic flies have shown that these transgenic EF-1α flies do not express more EF-1α mRNA or protein, nor do they have higher EF-1α activity than the control flies[46,47]. In fact, the transgene could not be induced at the experimental conditions used. Moreover, in the course of these experiments we found that in none of the transgenic lines tested does the EF-1α mRNA really disappear during senescence[46-48]. From all these experiments we have to conclude that, although there are clear differences in mean life span between the transgenic lines at the different experimental conditions tested, the extension of the life span does not result from overexpression of the EF-1α transgene, but rather from position effects (this point will be further discussed in the last section). We have tried in vain to generate transgenic animals overexpressing the EF-1α gene, using also other promoters and

A) GENETIC BACKGROUND

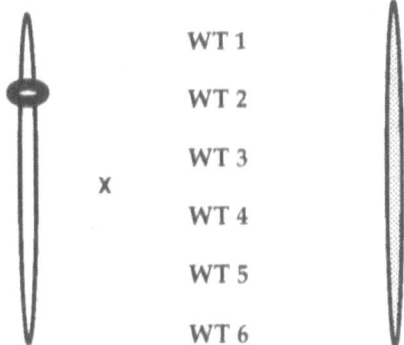

B) POSITION EFFECT

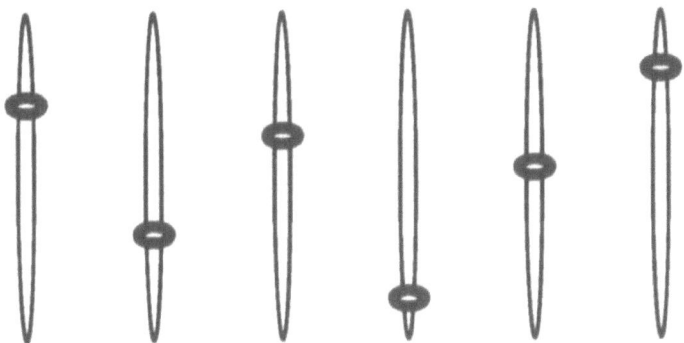

Figure 7. Experimental design for testing genetic background (A) and position effect (B) in EF-1α transgenic animals (after Keiser and Stearns[45]). A) One of the newly created homozygous EF-1α transgenic lines was crossed to six different isogenic wild type lines. B) Each of the newly created homozygous EF-1α transgenic lines, carrying the insert on different positions on the third chromosome was crossed to the same isogenic wild type line. In both experiments the life span was measured in the heterozygous F1.

other vectors[48,49]. In all cases, either transgenic EF-1α flies were not viable, or the transgene was not active. It seems that either these P-elements integrate into silent chromosome regions, or that EF-1α overexpression is lethal. This observation is nevertheless interesting and points to the importance of a tight regulation of this gene[50,51]. Recent reports demonstrated that deregulated expression of translation factor genes leads to drastic changes in cell growth, including transformation and tumorigenesis[52] (review also in[53]). Experiments done with other model organisms also point to the importance of a tight regulation of the EF-1α gene. Overexpression of EF-1α in yeast leads to a decrease in translational fidelity[54]. On the other hand, a high fidelity mutant in Podospora with an increased life span was found to contain a mutation in the EF-1α gene[55]. The regulation of the EF-1α gene must have adapted during evolution to give for each organism an optimal elongation rate and translational fidelity.

2.3.2. EF-1α Expression in Wild Type Animals. Because the transgenic animal approach still had not solved the question of whether EF-1α is really involved in the age-related decrease of protein synthesis, further studies were carried out on a wild type (Oregon R) strain of Drosophila. We have re-investigated age-related and development-related changes of EF-1α expression, at the level of mRNA, protein, and EF-1α activity[28]. D. melanogaster contains two genes coding for EF-1α, F1 and F2, which are developmentally regulated[56]. We have examined mRNA levels of both alleles. We have extended the RNA analysis to other genes involved in protein synthesis (see 2.2.1).

EF-1α transcription. Changes in the EF-1α F1 and F2 mRNA during development and ageing were analyzed using total RNA extracted from embryos, larvae, pupae, and adult male flies of different ages (Fig. 4A, 5). In whole animals, the EF-1α F1 mRNA decreases continuously from the third instar larval stage to old adult flies. When we look at individual body parts, however, it becomes apparent that this decrease in adult flies comes mainly from a loss of EF-1α F1 message in the abdomen. The steady-state levels of F1 mRNA remain almost unchanged in the head and the thorax, i.e., in postmitotic tissues (Fig. 4B).

The EF-1α F2 message is most heavily expressed in the pupal stage and reduced to low levels in adult flies. During ageing, however, the ratio of F2/F1 does not change at all, indicating that F2 expression might be equally important for adult senescent flies[28]. Interestingly, the F2 gene seems to be preferentially expressed in the head part of adult flies: the ratio of F2/F1 is much higher in the head than in the abdomen. This suggests that the EF-1α F2 gene might be expressed in a tissue-specific manner. Our recent in situ hybridization and immune localization experiments confirmed this: F2 mRNA and F2 protein can be detected exclusively in the central nervous system, whereas F1 mRNA and protein are ubiquitously expressed[57]. The F2 gene might be a Drosophila homologue of the recently described EF-1α variants found in mammals: both the S1 gene of rodents[58] and the EF-1α 2 allele in humans[59] are specifically expressed in terminally differentiated postreplicative tissues, such as brain, heart, and skeletal muscle.

EF-1α protein and EF-1α activity. Total soluble protein was extracted from equivalent fly numbers and the amount of EF-1α protein was quantitated on Western blots. We could show that EF-1α decreases mainly within the first few days of adult flies and then remains at a constant level[28]. The same is true also for other EF-1 components, EF-1β and EF-1γ, examined by Shikama[60]. The EF-1α binding activity, however, which was measured as the amount of [^{14}C]Phe-tRNA bound to ribosomes, drops to much lower values than the amount of EF-1α protein[46,47]. As seen in Fig. 8, old flies contain less than 20% of the active EF-1α protein compared to 1–2 day-old flies, whereas the protein amount de-

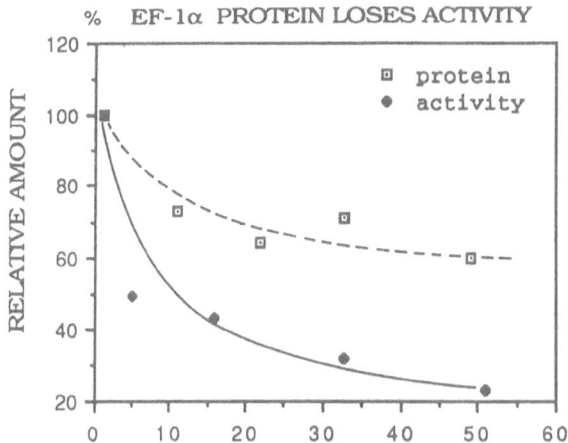

Figure 8. Age-related changes in EF-1α protein and EF-1α activity. The amount of EF-1α protein was determined on Western blots. Active EF-1α was quantitated by formation of the complex of EF-1α with ribosome and [14C]Phe-tRNA[28]. The values from one day old flies were taken as 100%.

creases only to 40%. We conclude that during senescence EF-1α protein loses catalytic activity, either by changes in post-translational modifications, or by oxidative damage. This loss in activity is a good example for the age-related accumulation of altered, inactive protein postulated by the free radical theory of ageing. Clearly, gene expression and information transfer have to be examined at all levels: mRNA, protein, as well as specific activity of a protein.

Assuming that the activity assay measures both F1 and F2 binding to ribosomes, we might explain the strong decrease in EF-1α activity observed soon after eclosion as being a result from the transcriptional down-regulation of the F2 gene, which takes place between the pupal stage and eclosion. The drastic decrease in total EF-1α message (F1 plus F2) starts after the late pupal stage and continues till the first five days after eclosion, which correlates with the loss of EF-1α activity. As discussed above, this may reflect changes related to massive differentiation and growth processes taking place during metamorphosis, rather than changes related to adult senescence.

In conclusion, all these recent results on EF-1α expression do not support the hypothesis proposed by Webster[41] and Shepherd[44]. Although EF-1α mRNAs decrease immediately after eclosion, they do not disappear completely in senescent flies, nor do they decrease faster than any other RNAs examined (see previous section). In contrast, we now know that the EF-1α promoter is still quite active in old flies; it has even been proposed as a ubiquitous promoter-enhancer for transgene expression in ageing flies[61]. In agreement with our observations are also results published by Dudas and Arking[62] who have measured steady state levels of EF-1α mRNAs in their Drosophila lines selected for different life spans; they did not detect any correlation of longevity with the EF-1α transcript levels.

2.4. Oxygen Radicals, Gene Expression, and Drosophila Ageing

2.4.1. Free Radicals and Drosophila Life Span. Life span of flies can be modified by environment, e.g., by changing temperature, oxidative stress or physical activity, i.e., parameters that influence the metabolic rate [14,63,64]. Thus it was shown that flies maintained at conditions of high physical activity (or temperature) have shorter life span than flies kept at low activity (low temperature)[65,66]. Alterations in life span, induced either by higher temperature or activity correlate with the accumulation of several types of altered macromolecules that are products of free radical reactions[14]. Most of these negative effects can be compensated by antioxidants. This inverse relationship between metabolic rate and life span lead Sohal to propose that oxidative stress plays an important role in the ageing process of flies and thus in life span determination[65].

Reactive oxygen species (ROS) are generated as metabolic by-products, mainly in oxidative phosphorylation in mitochondria[67,68] and have been implicated in the ageing process because they induce oxidative damage to macromolecules, which leads to loss of homeostatic functions. All organisms, from prokaryotes to higher eukaryotes have developed a complex network of defense and repair mechanisms to fight deleterious effects of ROS; e.g., antioxidant enzymes like superoxide dismutase (SOD), catalase, GSH-peroxidase (review[69]); DNA repair enzymes; proteases for the degradation of modified proteins; but also non enzymatic protective molecules, among which glutathione (GSH) plays the most important role. GSH is a powerful ROS and thiol scavenger and thereby the main redox buffer in all eukaryotic cells; it also plays a key role for accurate formation of disulfide bonds in proteins. The various defense mechanisms constitute by themselves a homeostatic system and many groups working on different biological model systems have shown that it is very important to maintain the balance between an optimal concentration of free radicals and the oxidant defenses.

Enzymatic as well as non-enzymatic defenses against free radicals and ROS decline with age[14,17,63,69–72]. Age-related changes in intracellular GSH levels and membrane bound thiol levels have been reported to occur in the housefly and in Drosophila[66,72]. Thus, decreases in GSH levels seem to be a common phenomenon of ageing cells and organisms (discussed in more detail in 2.4.4). The free radical theory of ageing predicts that the decline of the defense systems results in accumulation of damaged macromolecules like oxidized lipids, modified proteins[68,74,75], or DNA damage[67,76]. This is associated with a general decrease in transcription activity[77,78]; our results presented here (2.2.2), mitochondrial transcription[79], mitochondrial function[68,76], protein synthesis activity[41,772,80]. It is not surprising then, that antioxidant defense genes have been extensively studied both in normal ageing flies and in transgenic animals, at the levels of transcription and of enzyme activity.

Arking et al. have measured the mRNA concentration of several genes coding for antioxidant defense enzymes[27,82]. They could show that their genetically selected long lived flies had a statistically significant increase in the levels of SOD, catalase, and Xanthine dehydrogenase mRNA, but not Glutathione transferase (GST), at day five of the adult life. The relative enzyme activities of SOD, catalase, and GST also showed a significant increase that tracks with the ELP. Furthermore, all the ELP lines possess an elevated resistance to paraquat, an oxigen radical generator[27]. It is important to point out again, that all these traits are expressed only when the selected lines are kept under high larval density[26]. At low larval density, the enzyme levels are comparable to the ones of control strains.

2.4.2. Transgenic Animals and Antioxidant Defense Genes. Because of the postulated correlation between antioxidant enzyme expression and longevity, several transgenic animal experiments have been performed. They consisted mainly in testing the overexpression of genes coding for catalase and/or SOD. Results were controversial when the levels of either SOD alone[71,83,84] or catalase alone[70] were modified. For example, in one case, the SOD enzyme activity increased by 30–70% in flies containing two copies of the genomic SOD gene, but no effect was observed on life span and paraquat resistance[71].

Use of a tandem duplication of the chromosomal region containing the SOD gene allowed to study the dosage effect of SOD expression. Only a slight, dosage dependent effect on life span, enzyme activity, and paraquat resistance was observed[85]. The interpretation of the observed effect on life span (20% increase), however, is questionable, because experiments were not well controlled regarding the genetic background.

Because both enzymes, SOD and catalase, act in tandem to eliminate ROS, it is not surprising that overexpression of either enzyme alone did not result in considerable alterations of longevity. The most conclusive experiment consisted in making transgenic flies containing additional copies of both genes[86.] Although the same problems were encountered as discussed earlier—i.e., variation of life span and transgene expression depend on insert position—there was a positive correlation in some lines between longevity and antioxidant effect. Some of the double transgenic fly lines expressed higher enzyme levels and had an increased mean and maximum life span. In addition, these flies displayed higher geotactic motility, a higher rate of oxygen consumption and an increased metabolic rate in old age. The elevated enzyme levels also protected flies from oxidative damage: the carbonyl content (a measure of the amount of oxidized protein) was shown to be lower at all ages. In conclusion, these experiments present strong evidence that the coordinate overexpression of SOD and catalase slow down the ageing process.

2.4.3. Reactive Oxygen Species, Redox Changes, and Gene Expression. The experiments described above present indirect evidence that ROS influence the ageing process by inducing damage in macromolecules. A second role for ROS has recently been discovered: ROS are involved in the regulation of oxidative stress response genes, which are mediated by redox sensitive transcription factors (TF). It was proposed that ROS may serve as second messengers in signal transduction pathways in various stress induced responses in the cell, e.g., apoptotic signals, growth factor withdrawal, oxidative stress, UV or X-ray irradiation, H2O2, phorbol esters, TNF, and others.[87-91].

A common consequence of these various signals is that they result in immediate redox changes in target cells, which can be measured as changes in GSH. Since intracellular GSH levels modulate the redox state of cells they may also be involved in modulating the redox-dependent TFs[88-92]. At least two classes of such TFs have been well characterized: the AP1 family (fos/jun) and the rel family (NFκB). Several others have also been shown to respond to redox changes[93]. Most of these redox sensitive TFs are multifunctional, multimeric proteins (usually active as heterodimers), which allows for combinatorial effects in different cell types or at different times of development. Uncontrolled expression of them is oncogenic (e.g., fos, jun, rel, myb, hormone receptors, reviewed in[94]).

So far only members of redox regulated TFs involved in early development or immediate stress response have been characterized, but it is well possible that other yet unidentified functional partners are modulated in age-related processes. Drosophila homologues of the rel family are dorsal, a transcription factor involved in the dorso-ventral polarity determination, and dif, a protein also functionally related to the mammalian NFκB, inducing immune response genes (reviewed[95]). Drosophila homologues of the AP1 family have recently been cloned they are even functionally conserved in mammalian cells[96,97]. Among the many genes containing AP1 and/or NFκB binding sites in their promoter are housekeeping and maintenance genes, like EF-1α, SOD, or stress response genes (hypoxia induced TF, ferritin, metallothionein), all of which may be relevant not only for age-related pathologies but also for physiological ageing. It is well possible that such TFs are sensitive not only to immediate response, but also to gradual changes in the oxidative state of cells that must occur during ageing.

An interesting correlation between redox state (GSH levels) and gene activity was pointed out, again based on studies with flies. Sohal proposed that cellular differentiation may be correlated with changes in antioxidant defense and free radical levels: differentiation is always associated with low GSH concentrations and high SOD activity[14]. Similar results were reported by Collatz et al.[64] who uses Phormia flies as model: GSH decreases in late larval stage, is very low in the pupal stage, increases again at eclosion and then gradually declines with age.

The only way to test the causal relationship between redox changes and putative gene activity changes influencing ageing, is to experimentally alter the redox state and examine the influence on gene expression. The Drosophila model might be suitable to study the effect of antioxidants on redox regulated transcription in whole organisms. Experiments addressing this question are presented in the following section.

2.4.4. Antioxidant Treatment and Life Span. As we discussed above, the age-related decrease in all RNA species is consistent with a general mechanism that affects all three RNA polymerases. This could, e.g., involve common transcription factors or RNA polymerase subunits that are sensitive to redox changes. We proposed to address the following question: if general down-regulation of transcription is a result of direct or indirect changes in the oxidative state of ageing cells, it might be possible to influence the age-related decline in gene expression by maintaining the redox state at a constant level. This could be achieved by feeding flies with antioxidants that restore GSH levels. An interesting molecule and good candidate for this purpose is the antioxidant N-Acetylcysteine (NAC). It is known as radical scavenger and/or a precursor for cysteine and GSH[88,90,98,99]. Furthermore, it was shown that NAC inhibits the redox sensitive transcription factor NFκB[88,90,92]. In cell culture experiments, NAC was shown to protect neuronal cells from apoptotic cell death by increasing intracellular levels of GSH[99].

The following experiment was carried out to test the hypothesis that NAC might influence ageing. When adult male Drosophila are kept on food containing different concentrations of NAC, we observe a dose-dependent increase in longevity. The survival curves in Figure 9 show that both mean and maximum life span are increased by NAC treatment.

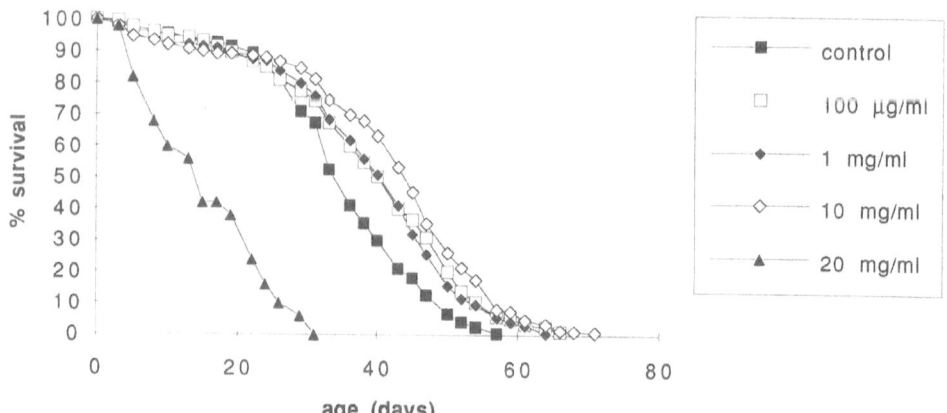

Figure 9. Life span of male flies treated with N-Acetylcysteine (NAC). The survival curves are averages of two experiments with a total number of 200 flies per NAC concentration. The life span increases in a dose dependent way, up to a concentration of 10 mg/ml NAC in the food. Higher doses are toxic.

At the lower doses, increasing NAC concentrations result in a dose dependent life extension. Higher NAC doses, however, are toxic: at 20 mg/ml NAC the flies live much shorter, and at 50 mg/ml they die within 2 days (not shown). The maximal statistically significant differences in mean life span between control flies and NAC treated flies was observed at 10 mg/ml NAC and corresponds to an increase of ~27%.

Further experiments are required to elucidate the molecular mechanisms responsible for NAC effect on longevity. First, the postulated correlation between transcription activity, GSH levels, and life span should be examined. Preliminary experiments indicate that NAC treated flies contain higher levels of total RNA and rRNA and are more resistant to paraquat (unpublished results[57]).

Interestingly, data obtained by other research groups working with a completely different experimental system have recently pointed to a possible correlation between NAC survival function and transcription activity[100]. Thus it was shown that NAC can promote the survival of neuronal cells by preventing apoptotic cell death induced either by growth factor deprival, oxidative stress, GSH loss or cysteine loss. This survival-promoting effect of NAC on neuronal PC12 cells was then shown to be transcription dependent, suggesting that NAC may act through redox sensitive transcription factors[100].

The effect of dietary supplements on insect life span (antioxidants and others) has been documented before. Centrophenoxine extended life span of Phormia manduca but Vitamin C had no effect[64]. Kinetin, a plant growth hormone delays ageing and prolongs the life span of the fruitfly Zaprionus pravittiger[101]. In the mosquito Aedes aegypti an age-related decrease in GSH could be corrected by feeding with the cysteine precursor magnesium thiazolidine-4-carboxylic acid[102]. This treatment resulted in an increase in cysteine and GSH levels and was accompanied by an increase in longevity[103]. The authors suggested that a decrease in cysteine might be responsible for the GSH decline and thus be an important determinant of the ageing process. All these findings that dietary supplements can alter the lifespan of insects give further evidence that the molecular mechanisms determining the ageing process are multifactorial, including both genetic and epigenetic factors and their mutual interactions.

3. FUTURE PERSPECTIVES

3.1. Transgenic Animals

One of the major advantages of the fruit fly as a model organism for research in developmental biology is the relative ease with which transgenic Drosophila can be generated. There is little doubt about the value of transgenic animals for investigating short term gene activity during early embryonic development. Here we like to point out some of the problems and disadvantages encountered when this technology is applied in ageing research. So far, the transgenic animal approach has been used mainly to study the effect of specific maintenance genes (EF-1α, SOD, and catalase) on life span and fitness. The examples discussed above also show where the problems lie.

Several important parameters have to be considered, like the choice of recipient strains, P-element vectors, regulatable promoter, insert position, genetic background, fly handling, many of which are difficult to control.

Recipient strains and P-element vectors: the vectors should contain a selectable genetic marker that allows for rapid screening of large fly numbers and easy identification of transgenic offspring. The most currently used system consists of using recipients flies carrying an eye colour mutation in the rosy or in the white gene. The P-element contains the wild type gene, besides the transgene, and allows for selection by eye colour. However, these mutations by themselves can have considerable effects on fitness and longevity of the fly. For example, small amounts of Xanthine dehydrogenase are sufficient for rescue of the eye coulour in the rosy mutation, whereas higher enzyme amounts may be required to restore full viability. As an alternative, we are currently investigating the possibility to use a P-element vector containing an antibiotic resistance gene (neomycin resistance) as selectable marker[104], which can be transformed into wild type flies. This requires feeding of larvae on neomycin containing food, a regime that may have other effects on life span still to be tested.

Promoters: to test the influence of a candidate transgene on ageing and life span, it is often desirable to modulate its expression levels. Overexpression can be achieved by increasing the copy number of a gene of interest, introducing additional copies of a transgene by classical genetic crosses. Alternatively, the expression level of a gene can be modulated by using inducible promoters. The heat shock gene promoter (hsp70 promoter) has been preferentially used; first, because it is ubiquitously expressed, and second, because heat induction should allow to compensate at least partially for the insert position. The disadvantage is that increasing the temperature shortens life span. Local overexpression can be achieved by use of tissue-specific promoters. A possible alternative for an inducible system to express a transgene could be the tetracycline dependent repressor/transactivator[105] or the yeast gal4 system[106]. Both would allow for a controlled transgene expression, at a desired time of development or ageing, either ubiquitously or tissue specifically. Because so far no one has succeeded to do homologous recombination in flies, the gene knock-out approach is not possible. However, it is conceivable that by expressing antisense RNA, specific genes may be repressed.

Genetic background and insert position: the experiments performed by Stearns' group have shown that both the insert position and the genetic background can have a tremendous effect on life span as well as other life history parameters. The effects on life span were in the same order of magnitude as the effects measured by Shepherd at al.[44] and Orr and Sohal[86], and were independent of the expression of the transgene. Because homologous recombination in Drosophila is not possible, one cannot create test and control lines that differ only in the expression of the transgene. The insert position thus remains one of the main problems: test and control transgenes will always be at different chromosomal loci. It would be worthwhile to consider the use of site specific recombination systems, like the inversion of the yeast FLP[107], or the phase variation in salmonella. The insert position as well as the genetic background would be identical for both constructs and the lines could be directly compared. To avoid problems with the genetic background, isogenic lines can be used, which allows for direct comparison of test and control lines. However, an obvious limitation is that isogenic lines are usually less fit than outbred lines. Other parameters to consider are the size of the inserts, as well as the handling of flies, both of which may also influence life span and other life history parameters.

It will be very difficult to control all of these parameters, each of which may influence the senescence program and life span. The ideal test and control lines to measure the effect of a single gene of interest on longevity might look as follows: the inserts of the controls should contain the same transgene with mutations in the reading frame resulting in RNA expression, but no protein. Test and control inserts should be at the same position in the chromosome, in an isogenic recipient line. This would allow to control for the genetic background, the insert positions and the size of the insert. At the moment it is almost impossible to take all these factors into account at the same time. Therefore, we think that transgenic Drosophila cannot be used yet for measuring single gene influences on life span. Elaborate experimental design could minimize other effects that cannot be attributed to the single gene of interest. Usually a large number of transgenic lines are used to circumvent some of the difficulties.

Another possible use for transgenic Drosophila in gerontology has been illustrated by the experiments of Helfand et al.[19,20]. The enhancer trap and reporter gene technology could be further extended to identify as yet unknown genes, e.g., mutations that postpone senescence. Enhancer trap lines (made with a P-element containing the neomycin resistance gene as selectable marker and carrying the insert at different locations) could be screened for alterations in longevity. If one or several lines show a long lived phenotype, the gene adjacent to the inserted P-element can be easily cloned by plasmid rescue with the bacterial vector sequences contained in the P-element[18]. At the same time, the bacterial β-galactosidase reporter gene gives information about the temporal and spatial expression of the gene involved in the long lived phenotype. The big disadvantage of this experiment would clearly be the laborious screens for extended life span; good biomarkers for senescence would help to reduce the amount of work.

Transposable elements of Drosophila may be used for another type of experiment involving identification and selection of longevity mutants. An active transposon like copia, which can integrate about 50 to 100 times into the fly genome, could be used to mutagenize flies and create a large population with random distributions of multiple transposon insertions. By selecting for late fecundity and/or stress resistance (indirect selection for longer life span) one might expect to map hotspots of insertion positions that correlate with longevity. The same approach could be used to map other life history loci on the genome. Compared with the enhancer trap P-element approach discussed above, cloning of the insertion sites is less straightforward, but still easier than cloning the putative longevity genes of the ELP lines identified in selection experiments.

3.2. Gene Expression Studies

The results presented in this chapter have shown the main advantages and disadvantages of the fly model for gene expression studies. There are two principal ways to examine age-related changes of gene activity; the more biochemical type of approach with extracted molecules (RNA or protein), and the "organismic" approach using whole flies. Both can be used to study activity and expression levels of known candidate genes or for the identification of new "gerontogenes".

The disadvantage of the first approach is that information gained by using fly extracts is limited and may even be misleading. When macromolecules are extracted from whole flies, e.g., massive changes in specific organs, like the decrease in germ cell production in the gonads, may obstruct more subtle changes in the somatic tissues. This is even more dramatic when mixed populations of both sexes are used. An example of such a misinterpretation that had far reaching consequences is the earlier paper on EF-1α expression[41]. We now think that the discrepancy between those results and our own data come mainly from the fact that Webster's group used mixed populations of males and females. Oocytes contain huge amounts of mRNA, are metabolically very active and synthesize large amounts of protein. The concentration of EF-1α mRNA and protein in young females is considerably higher than in males. For this reason one should certainly separate the sexes for some experiments and/or try to use separated body parts or isolated tissues.

Clearly, the most relevant results are obtainable only when gene expression patterns are compared in different types of animals: young and old ones, short lived and long-lived ones (e.g., selected ELP lines), wild type control and experimentally induced longevity (e.g., NAC treated flies). The ultimate goal of future experiments with the antioxidant treated flies will be to identify genes that are up- or down regulated by the NAC treatment and postulated redox changes. In combination with the enhancer trap technique, tissue specific differences induced by NAC treatment could be determined. Another powerful method to screen for such differences in gene expression is the differential RNA display (DDRT-PCR) method[108]. In addition, specific DDRT-PCR patterns might serve as fingerprints or biomarkers, if they correlate with life span. Subsequent cloning of the affected genes is now relatively easy. At the protein level, reproducible 2-D gel electrophoresis patterns could similarly serve as protein fingerprints. Appropriate methods for reproducible electrophoresis conditions and sequencing have now been developed[109].

3.3. Drosophila: A Good Model for Molecular Gerontology?

In the end we like to review briefly the main points that speak for or against using the fruit fly as a model organism for molecular gerontology. Positive aspects are: Drosophila has a small genome and cloning of interesting genes is easier than in mammals. It is foreseeable also, that the fruit fly will be one of the next organisms whose genome will be entirely sequenced. The versatile genetic methods, as well as some of the most advanced molecular genetic technologies are available. In situ visualization of gene products (RNA or protein) are easier to carry out on this small organism than in mammals. It is certainly the combination of all the Drosophila specific technology (genetics, transgenic animals, enhancer trap lines, etc.) with modern molecular biology methods that make the fruit fly an interesting and valuable model organism.

On the negative side we have to mention: only some aspects of ageing can be studied. Drosophila cannot serve as a model for replicative senescence. Tissue-specific phenomena are difficult to analyze, because of the small size of the animal, although the enhancer trap technology has allowed to see differential gene expression in individual organs[19, 20]. For the same reason, longitudinal studies on single animals are not possible. The fly is certainly not a good model for many human-specific age-related diseases. On the other hand, general mechanisms influencing the ageing process common to all species, could be more easily studied in the small animal. Questions concerning the influence of environmental factors and their interactions with gene expression can be approached. Thus Drosophila might be a suitable organism to find out what molecular mechanisms are at the basis of the beneficial effect dietary restriction has on all aspects of ageing. The different cellular responses to ROS reviewed by Remacle et al.[87] could be individually tackled.

In conclusion, we think that the positive aspects prevail. Keeping well in mind all the advantages and disadvantages inherent to the model system, to continue research into molecular aspects of ageing with Drosophila will be a challenge.

REFERENCES

1. Finch, C. E. (1990) Longevity, Senescence, and the Genome (Univ. Chicago Press)
2. Holliday, R. (1995) Understanding Aging. Cambridge University Press
3. Rattan, S. I. S. (1995) Ageing—a biological perspective. Molec. Aspects Med. 16, 439–508
4. Rose, M. R. (1984) Laboratory evolution of postponed senescence in Drosophila melanogaster. Evolution 38, 1004–1010
5. Luckinbill, L. S.; Arking, R.; Clare, M. J.; Cirocco, W. C. & Buck, S. A. (1984) Selection for delayed senescence in Drosophila melanogaster. Evolution 38, 996–1003

6. Ashburner, M. (1989) Drosophila. Cold Spring Harbor Lab. Press

7. Hafen, E. & Levine, M. (1986) The localization of RNAs in Drosophila tissue sections by in situ hybridization. in: Drosophila—a practical approach. Ed. D. B. Roberts, IRL Press Oxford, pp. 139–174

8. Gehring, W. J. (1987) The homeobox: structural and evolutionary aspects. Molecular Approaches to Developmental Biology, 115–129

9. Quiring, R.; Walldorf, U.; Kloter, U. & Gehring, W. J. (1994) Homology of the eyeless gene of Drosophila to the small eye gene in mice and aniridia in humans. Science 265, 785–789

10. Halder, G.; Callaerts, P. & Gehring, W. J. (1994) Induction of ectopic eyes by targeted expression of the eyeless gene in Drosophila. Science 267, 1788–1792

11. Buck, S.; Nicholson, M.; Dudas, S.; Wells, R.; Force, A.; Baker III, G. T. & Arking, R. (1993) Larval regulation of adult longevity in a genetically-selected long-lived strain of Drosophila. Heredity 71, 23–32

12. Zuckerman, B. M. (1980) Nematodes as Biological Models. Academic Press, New York

13. Maurizio, A. (1961) Lebensdauer und Altern bei der Honigbiene (Apis mellifica l.). Gerontologia (Basel) 5, 110–128

14. Sohal, R. S. & Allen, R. G. (1985) Relationship between metabolic rate, free radicals, differentiation and aging: a unified theory. in: Molecular Biology of Aging, Eds. A. Woodhead & A. D. Blackett, Plenum, pp 1995, 75–104

15. Leffelaar, D. & Grigliatti, T. (1984) Age-dependent behavior loss in adult Drosophila melanogaster. Dev. Genet. 4, 211–22

16. Williams, C. M.; Barness, L. A. & Sawyer, W. H. (1943) The utilization of glycogen by flies during flight and some aspects of the physiological aging of Drosophila. Biol. Bull. 84, 263–272

17. Arking, R.; Buck, S.; Berrios, A.; Dwyer, S. & Baker III, G. T. (1991) Elevated paraquat resistance can be used as a bioassay for longevity in a genetically based long-lived strain of Drosophila. Dev. Genet. 12, 362–370

18. Bellen, H. J.; O'Cane, C. J.; Wilson, C.; Grossniklaus, U.; Pearson, R. K. & Gehring, W. J. (1989) P-element-mediated enhancer detection: a versatile method to study development in Drosophila. Genes & Dev. 3, 1288–1300

19. Helfand, S. L., Blake, K. J., Rogina, B., Stracks, M. D., Centurion, A., Naprta, B. (1995) Temporal patterns of gene expression in the antenna of adult Drosophila melanogaster. Genetics 140, 549–555

20. Rogina, B., Helfand, S. L. (1995) Regulation of gene expression is linked to life span in adult Drosophila. Genetics 141, 1043–1048

21. Arking, R. (1987) Successful selection for increases longevity in Drosophila: Analysis of the survival data and presentation of a hypothesis on the genetic regulation of longevity. Exp. Gerontol. 22, 199–220

22. Arking, R. & Wells, R. A. (1990) Genetic alteration of normal aging processes is responsible for extended longevity in Drosophila. Dev. Genet. 11, 141–148

23. Hoffmann, A. A. & Parsons, P. A. (1988) Selection for increased desiccation resistance in Drosophila melanogaster : Additive genetic control and correlated responses for other stresses. Genetics 122, 837–845

24. Fleming, J. E.; Quattrocki, E.; Latter, G.; Miquel, J.; Marcuson, R.; Zuckerkandl, E. & Bensch, K. G. (1986) Age-dependent changes in proteins of Drosophila melanogaster. Science 231, 1157–1159

25. Buck, S.; Wells, R. A.; Dudas, S. P.; Baker III, G. T. & Arking, R. (1993) Chromosomal location and regulation of the longevity determinant genes in a selected strain of Drosophila melanogaster. Heredity 71, 11–22

26. Buck, S.; Nicholson, M.; Dudas, S.; Wells, R.; Force, A.; Baker III, G.T. & Arking, R. (1993) Larval regulation of adult longevity in a genetically-selected long-lived strain of Drosophila. Heredity 71, 23–32

27. Arking, R. (1995) Antioxidant genes and other mechanisms involved in the extended longevity of Drosophila. in: Oxidative stress and Aging. Eds. R. G. Cutler, L. Packer, J. Bertram & A. Mori. Birkhäuser Basel, pp. 123–139

28. Shikama, N. & Brack Ch. (1996) Changes in the expression of genes involved in protein synthesis during Drosophila aging. Gerontology 42, 123–136

29. Seshadri, T.; Uzman, J. A.; Oshima, J. & Campisi, J. (1993) Identification of a transcript that is downregulated in senescent human fibroblast. Cloning, sequencing analysis, and regulation of the human L7 ribosomal protein gene. J. Biol. Chem. 268, 18474–18480

30. Niedzwiecki, A. & Fleming, J. E. (1990) Changes in protein turnover after heat shock are related to accumulation of abnormal proteins in aging Drosophila melanogaster. Mech. Aging Dev. 52, 295–304

31. Niedzwiecki, A.; Kongpachith, A. M. & Fleming, J. E. (1991) Aging affects expression of 70kDa heat shock proteins in Drosophila. J. Biol. Chem. 266, 9332–9338

32. Wheeler, J. C.; Bieschke, E. T. & Tower, J. (1995) Muscle-specific expression of Drosophila hsp70 in response to aging and oxidative stress. Proc. Natl. Acad. Sci. USA 92, 10408–10412

33. Marin, R.; Valet, J. P. & Tanguay, R.M. (1993) hsp23 and hsp26 exhibit distinct spatial and temporal patterns of constitutive expression in Drosophila adults. Dev. Genetics 14, 69–77

34. Niedzwiecki, A. & Fleming, J. E. (1993) Heat shock induces changes in the expression and binding of ubiquitin in senescent Drosophila melanogaster. Dev. Genetics 14, 78–86

35. Fyrberg, E. A.; Mahaffey, J. W.; Bond, B. J. & Davidson, N. (1983) Transcripts of the six Drosophila actin genes accumulate in a stage- and tissue-specific manner. Cell 33, 115–123

36. Labuhn, M. & Brack, Ch. (1996) Age-related changes in the mRNA expression of actin iroforms in Drosophila melanogaster. Gerontology 42 (in press)

37. Levenbook, L. (1986) Protein synthesis in relation to insect aging: An overview. In Insect Aging: Strategies and Mechanism, Eds. K.-G. Collatz & R. S. Sohal, Springer-Verlag, Heidelberg, pp. 200–206

38. Webster, G. C. (1988) Protein synthesis. In: Drosophila as a Model Organism for Ageing Studies. Eds. F. A. Lints and M. H. Soliman, Blackie Press, Glasgow and London, pp. 119–128

39. Rattan, S. I. S. (1991) Protein synthesis and the components of protein synthetic machinery during cellular ageing. Mutat. Res. 256, 115–125

40. Lints, F. A. and Bourgois, M. (1985) Aging and lifespan in insects with special regard to Drosophila: review 1982–1984. Rev. Biol. Res. Aging 2, 61–84

41. Webster, G. C. (1986) Effect of aging on the components of the protein synthesis system. in: Insect Aging: Strategies and Mechanism, Eds. K.-G. Collatz & R. S. Sohal, Springer-Verlag, Heidelberg, pp. 207–216

42. Riis, B.; Rattan, S .I.; Clark, B. F. C. & Merrick, W. C. (1990) Eukaryotic protein elongation factors. Trends Biochem. Sci. 15, 420–424

43. Van Damme, H. T. F.; Karssies, R.; Timmers, C. J.; Janssen, G. M. C. & Möller, W. (1990) Elongation factor 1β of Artemia: localization of functional sites and homology to elongation factor 1δ. Biochim. Biophys. Acta 1050, 241–247

44. Shepherd, J. W. C.; Walldorf, U.; Hug, P. and Gehring, W. J. (1989) Fruit flies with additional expression of the elongation factor EF-1α live longer. Proc. Natl. Acad. Sci. USA 86, 7520–7521

45. Stearns, S. C. & Kaiser, M. (1994) The effects of enhanced expression of elongation factor EF-1α on lifespan in Drosophila melanogaster. IV. A summary of three experiments. Genetica 91, 167–182

46. Brack, Ch.; Shikama, N.; Ackermann, R. & Wichser, U. (1993) The role of protein synthesis elongation factor EF-1α in aging of Drosophila. in: Recent advances in aging science, Eds. E. Beregi, I.A. Gergely, K. Rajczi, Monduzzi Bologna, pp. 67–73

47. Shikama, N.; Ackermann, R. & Brack, Ch. (1994) Protein synthesis elongation factor EF-1α expression and longevity in Drosophila melanogaster. Proc. Natl. Acad. Sci. USA 91, 4199–4203

48. Ackermann, R. (1996) Transcriptional regulation of the EF-1α gene in Drosophila melanogaster during aging. PhD Thesis, University Basel

49. Ackermann, R. & Donati, F. (unpublished results)

50. Thomas, G. & Thomas, G. (1988) Translational control of mRNA expression during the early mitogenic response in Swiss mouse 3T3 cells: identification of specific proteins. J. Cell Biol. 103, 2137–2144

51. Jefferies, H .B. J.; Thomas, G. & Thomas, G. (1994) Elongation factor-1α mRNA is selectively translated following mitogenic stimulation. J. Biol. Chem. 269, 4367–4372

52. Tatsuka, M.; Mitsui, H.; Wada, M.; Nagata, A.; Nojima, H. & Okayama, H. (1992) Elongation factor-1α gene determines susceptibility to transformation. Nature 359, 333–336

53. Sonenberg, N. (1993) Translation factors as effectors of cell growth and tumorigenesis. Curr Opin Cell Biol 5, 955–960

54. Song, J. M.; Picologlou, S.; Grant, C. M.; Firoozan, M.; Tuite, M. F. & Liebman, S. (1989) Elongation factor EF-1α gene dosage alters translational fidelity in Saccharomyces cerevisiae.. Molec. Cell. Biol. 9, 4571–4575

55. Silar, P. & Picard, M. (1994) Increased longevity of EF-1α high-fidelity mutants in Podospora anserina. J. Mol. Biol. 235, 231–236

56. Hovemann, B.; Richter, S.; Walldorf, U. & Cziepluch, C. (1988) Two genes encode related cytoplasmic elongation factors 1α (EF-1α) in Drosophila melanogaster with continuous and stage specific expression. Nucl. Acids Res. 16, 3175–3194

57. Brack, Ch. & Thüring, E. (unpublished results)

58. Lee, S.; Francoeur, A. M.; Liu, S. & Wang, E. (1992) Tissue-specific expression in mammalian brain, heart, and muscle of S1, a member of the elongation factor-1α gene family. J. Biol. Chem. 267, 24064–24068

59. Knudsen, S. M.; Frydenberg, J.; Clark, B. F. C. & Leffers, H. (1993) Tissue-dependent variation in the expression of elongation factor-1a isoforms: isolation and characterization of a cDNA encoding a novel variant of human elongation facor 1α. Eur. J. Biochem. 215, 549–554

60. Shikama, N. (1994) Expression of the peptide synthesis elongation factor EF-1α and ageing in Drosophila melanogaster. PhD Thesis, University Basel

61. Ackermann, R. & Brack. Ch. (1996) A strong ubiquitous promoter-enhancer for development and ageing of Drosophila melanogaster. Nucl. Acids Res. 24, 2452–2453

62. Dudas, S. P. & Arking, R. (1994) The expression of the EF-1α genes of Drosophila melanogaster is not associated with the extended longevity phenotype in a selected lon-lived strain. Exp. Gerontol. 29, 645–657

63. Sohal, R. S.; Arnold, L. & Orr, W. C. (1990) Effect of age on superoxide dismutase. catalase, glutathione reductase, inorganic peroxides, TBA-reactive material, GSH/GSSH, NADPH/NADP and NADH/NAD in Drosophila melanogaster. Mech. Aging Dev. 56, 223–235

64. Collatz, K. G.; Jaekel, K.; Haebe, M. & Flury, T. (1993) Insects as models for aging mechanism studies. in: Recent advances in aging science. Eds. E. Beregi. I. Gergely, K. Rajczi. Monduzzi. Bologna 1993. pp. 75–82

65. Sohal, R. S. (1986) The rate of living theory: a contemporary interpretation. in: insect Aging: Strategies and Mechanism, Eds. K.-G. Collatz & R. S. Sohal, Springer-Verlag, Heidelberg, pp. 23-44

66. Sohal, R. S.; Agarval, S.; Dubey, A. & Orr, W. C. (1993) Protein oxidative damage is associated with life expectancy of houseflies. Proc. Natl. Acad. Sci. USA 90, 7255–7259

67. Ames, B. N.; Shigenaga, M. K. & Hagen, T.M. (1993) Oxidants, antioxidants, and the degenerative diseases of aging. Proc. Natl. Acad. Sci. USA 90, 7915–7922

68. Miquel, J. (1992) An update on the mitochondrial-DNA mutation hypothesis of cell aging. Mutation Res. 275, 209–216

69. Pacifici, R. E. & Davies, K. J. A. (1991) Protein, lipid and DNA repair systems in oxidative stress: the free radical theory of aging revisited. Gerontology 37, 166–180

70. Orr, W. C. & Sohal, R. S. (1992) The effects of catalase gene overexpression on life span and resistance to oxidative stress in transgenic Drosophila melanogaster. Arch. Biochem. Biophys. 297, 35–41

71. Orr, W. C. & Sohal, R. S. (1993) Effects of Cu-Zn superoxide dismutase overexpression on life span and resistance to oxidative stress in transgenic Drosophila melanogaster. Arch. Biochem. Biophys. 301, 34-40

72. Rattan. S. I. S. & Derventzi, A. (1991) Altered cellular response during aging. BioEssays 13, 601–606

73. Agarwal, S. & Sohal, R. S. (1994) Aging and protein oxidative damage. Mech. Aging Dev. 75, 11–19

74. Stadtmann, E. R. (1992) Protein oxidation and aging. Science 257, 1220–1224

75. Rosenberger, R. F. (1991) Senescence and the accumulation of abnormal proteins. Mutation Res. 256, 255–262

76. Shigenaga. M. K.; Hagen, T. M. & Ames, B. N. (1994) Oxidative damage and mitochondrial decay in aging. Proc. Natl. Acad. Sci. USA1 91, 10771–10778

77. Richardson, A.; Sparks, M. B.; Staecker, J. L.; Hardwick, J. P. and Liu, D. S. H. (1982) The transcription of various types of ribonucleic acid by hepatocytes isolated from rats of various ages. J. Gerontol. 37. 666–672

78. Richardson, A.; Rutherford, M. S.; Birchenall-Sparks, M. C.; Roberts, M. S.; Wu. W. T. & Cheung, H. T. (1985) Levels of specific messenger RNA as a function of age. Molecular Biology of Aging: Gene Stability and Gene Expression, Ed. R. Sohal, Raven Press, N.Y.

79. Kristal, B. S.; Park, B.-J. & Yu, B. P. (1994) Antioxidants reduce peroxyl-mediated inhibition of mitochondrial transcription. Free Radical Biol & Med. 16, 653–660

80. Richardson, A. (1981) The relationship between aging and protein synthesis. CRC Handbook of Biochemistry in Aging. CRC Press, Boca Raton, Ed. J. R. Florini, pp. 79–101

81. Rattan, S. I. S. (1991) Protein synthesis and the components of protein synthetic machinery during cellular aging. Mutation Res. 256, 115–125

82. Dudas, S. P. & Arking R. (1995) A coordinate upregulation of antioxidant gene activities is associated with delayed onset of senescence in a long-lived strain of Drosophila. J. Gerontol. 50A, 117–127

83. Seto, N. O.; Hayashi, S. & Tener, G. M. (1990) Overexpression of Cu-Zn superoxide dismutase in Drosophila does not affect life-span. Proc. Natl. Acad. Sci. USA 87, 4270–4274

84. Reveillaud, I.; Niedzwiecki, A; Bensch, K.G. & Fleming, J. F. (1991) Expression of bovine superoxide dismutase in Drosophila melanogaster augments resistance to oxidative stress. Mol. Cell. Biol. 11/2, 632–640

85. Staveley, B. E.; Phillips, J. P. & Hilliker, A. J. (1990) Phenotypic consequences of copper-zinc superoxide dismutase overexpression in Drosophila melanogaster. Genome 33, 867–872

86. Orr, W. C. & Sohal, R. S. (1994) Extension of lifespan by overexpression of superoxide dismutase and catalase in Drosophila melanogaster. Science 263, 1128–1130

87. Remacle, J.; Raes, M.; Toussaint, O.; Renard, P. & Rao, G. (1995) Low levels of reactive oxigen species as modulators of cell function. Mutation Res. 316, 103–122

88. Staal, F. J . T.; Anderson, M. T.; Staal, G. E. J.; Herzenberg L. A.; Gitler, C. & Herzenberg, L. A. (1994) Redox regulation of signal transduction: tyrosine phosphorylation and calcium flux. Proc. Natl. Acad. Sci. USA 91, 3619–3622

89. Xanthoudakis, S. & Curran, T. (1994) Analysis of c-fos and c-jun redox-dependent DNA binding activity. Meth. Enzymol. 234, 163–174

90. Schreck, R. & Baeuerle, P. A. (1994) Assessing oxigen radicals as mediators in activation of inducible eukaryotic transcription factor NFκB. Meth. Enzymol. 234, 151–163

91. Devary, Y.; Gottlieb, R. A.; Smeal, T. & Karin, M. (1992) The mammalian ultraviolet response is triggered by activation of src tyrosine kinase. Cell 71, 1081–1091

92. Schreck, R.; Rieber, P. & Baeuerle, P. (1991) Reactive oxygen intermediates as apparently widely used messengers in the activation of NFkB transcription factor and HIV-1. EMBO J. 10, 2247–2258

93. Xanthoudakis, S.; Miao, G.; Wang, F.; Pan, Y. C. & Curran, T. (1992) Redox activation of fos-jun DNA binding activity is mediated by a DNA repair enzyme. EMBO J. 11, 3323–3335

94. McCormick, A. & Campisi, J. (1991) Cellular aging and senescence. Current Opinion Cell Biol. 3, 230–234

95. Hultmark, D. (1994) Ancient relationships. Insect immunity. Science 367, 116–117

96. Zhang, K.; Chaillet, J. R.; Perkins, L. A.; Halzonetis, T. D. & Perrimon, N. (1990) Drosophila homolog of the mammalian jun oncogene is expressed during embryonic development and activates transcription in mammalian cells. Proc. Natl. Acad. Sci. USA 87, 6281–6285

97. Perkins, K. K.; Admon, A.; Patel, N. & Yijan, R. (1990) The Drosophila Fos-related AP-1 protein is a developmentally regulated transcription factor. Genes Dev. 4, 822–834

98. Moldeus, P. and Cotgreave, I. A. (1994) N-Acetylcysteine. Meth. Enzymol. 234, 482–492

99. Ferrari, G.; Yan, C. Y. I. & Greene, L. A. (1995) N-acetylcysteine (D- and L-stereoisomers) prevents apoptotic death of neuronal cells. J. Neurosci. 15, 2857–2866

100. Yan, C. Y. I.; Ferrari, G. & Greene, L. A. (1995) N-acetylcysteine-promoted survival of PC12 cells is glutathione-independent but transcription-dependent. J. Biol. Chem. 270, 26827–26832

101. Sharma, S. P.; Kaur, P. & Rattan, S. I. S. (1995) Plant growth hormone kinetin delays ageing, prolongs the lifespan and slows down development of the fruitfly Zaprionus pravittiger. Biochim. Biophys. Res. Comm. 216, 1067–1071

102. Richie, J. P. Jr. & Lang, C. A. (1988) A decrease in cysteine levels causes the glutathione deficiency of aging in the mosquito. Proc. Soc. Exp. Biol. Med. 187, 235–240

103. Richie, J. P. Jr.; Mills, B. J. & Lang, C. A. (1987) Correction of a glutathione deficiency in the aging mosquito increases its longevity. Proc. Soc. Exp. Biol. Med. 184, 113–117

104. Thummel, C. S.; Boulet, A. M. & Lipshitz, H. D. (1988) Vectors for Drosophila P-element-mediated transformation and tissue culture transfection. Gene 74, 445–456

105. Gossen, M. & Bujard, H. (1992) Tight control of gene expression in mammalian cells by tetracyclin responsive promoters. Proc. Natl. Acad. Sci. USA 89, 5547–5551

106. Brand, H. B. & Perrimon N. (1993) Targeted gene expression as a means of altering cell fates and generating dominant phenotypes. Development 118, 401–415

107. Golic, K. G. & Lindquist, S. (1989) The FLP recombinase of yeast catalyzes site-specific recombination in the Drosophila genome. Cell 59, 499–509

108. Liang, P. & Pardee, A. B. (1992) Differential display of eukaryotic messenger RNA by means of the polymerase chain reaction. Science 257, 967–971

109. Tsugita, A.; Kamo, M.; Kawakami, T. & Ohki, Y. (1996) Two-dimensional electrophoresis of plant proteins and standardization of gel patterns. Electrophoresis 17, 855–865

11

SKIN AGEING

The Relevance of Antioxidants

Paolo U. Giacomoni[1] and Patrizia D'Alessio[2]

[1] R.A.D. L'Oréal
Chevilly-Larue, France
[2] C.H. U. Necker
Paris, France

1. INTRODUCTION

Ageing can be defined as the time-related impairment of organ functionality or as a general decline in organic functions as well as a decrease in adaptiveness to change, and to restore disrupted homeostasis [1]. This definition, though satisfactory for the layman interested and, we dare say, preoccupied by his own ageing, is non satisfactory, at least to the biochemist, in so far as it does not allow to make quantitative measurements which can be easily generalised to individuals of other species. On the other hand, such a possibility is allowed, for instance, by a more reductionistic definition, which can be gathered by considering that ageing is the time-related accumulation of molecular modifications in an organ or in an organism[2]. Such modifications can be mutations or deletions in DNA, carbonylation of proteins in the extra cellular matrix and so on.

This second definition has some advantages and might appear as having a few drawbacks. The main advantage is that it allows one to study the phenomenon of ageing from a molecular point of view, down to cell cultures or to reconstructed organs, since the biochemical analysis allows quantitative determination of molecular changes introduced by any kind of treatment. Being reductionistic, this definition might nevertheless appear to miss the aspect of the ageing phenomenon as something related to the entire organism. For instance, using the above definition might induce one to undertake sophisticated experiments and calculations in order to learn about the age of an individual, whereas it is generally accepted that, after a glance, it is easy to guess the age of an individual with an error of about ten percent. We have indeed acquired a capability to perform instant image analysis and do recognise those elements which make that an individual appear more or less "old", independently of reductionistic definitions.

Yet, the contradiction is only apparent: it is not because we have learned to perform instant image analysis that the differences between "young" and "old" so recognised do

Molecular Gerontology, edited by Rattan and Toussaint
Plenum Press, New York, 1996

not rest on molecular modifications which sum up to morphological differences. Modifications in the connective tissues (bones and skin) more than in other tissues or in cells allow this image analysis. As a matter of fact, with age the volume of the bones is reduced, the skin changes several physical properties, cartilage continuously develop and an old individual is characterised by wrinkles in skin, shrinking and voûting of the backbone, larger nose and ears, which are immediately recognised. It is therefore reasonable to surmise that the results of instant image analysis after visual inspections is the consequence of deeper, molecular phenomena.

We are interested in the ageing of the skin. We confine our analysis to ageing as reductionistically defined above. We are therefore compelled to identify the agents able to increase the number of molecular damages in the skin. These agents can be external to the human body, such as UV radiation from the sun or the forces exerted on the dermis by the gravitational field of the earth. Other agents can be internal, in so far they are present within the human body. Yet some of them originate externally, such as food and cigarette smoke, while others are endogenously synthesised and controlled, such as hormones.

Aging provoking agents can thus be external-only, external-metabolised or internal-only. Gravitational and tractional forces, UV radiation, physical traumas and foreign intruders can be considered as major external-only inducers of ageing. Nutrition and smoking habits are examples of external-metabolised agents, and hormonal status plays a role as an internal-only inducer of ageing.

We share the opinion that one of the major contributors to the ageing phenomenon is the so-called oxidative stress, which is the consequence of the imbalance between the production of free radicals and the individual capacity of anti-oxidant defence. We can define stress as the sudden variation of the value of at least one environmental parameter. Biological responses to stress may follow similar pathways, independently of the physico-chemical nature of the stress (UV, infrared, H2O2, heavy metals etc). We shall describe below how different types of stress exert their action by inducing oxidative stress within the organism. Free radicals, such as superoxide, hydroxyl radical, nitric oxide, hydrogen peroxide and singlet oxygen, are formed in vivo[3]. Nitric oxide reacts with superoxide to form peroxynitrite, a membrane damaging agent, potential mediator of oxidant-induced cellular injury[4]. Oxidative stress may arise because of exhaustion of the anti-oxidant pool (such as ascorbate, glutathione and α-tocopherol) or because of increased formation of reactive oxygen species (ROS)[5] generated, for instance, by UV irradiation and by the associated tissue reaction. We shall see below that oxidative stress may trigger, be part of, and amplify, the inflammatory reaction.

"Spontaneous" oxidative stress can be generated in mitochondria which produce ROS such as superoxide anion in fairly high amount during oxidative phosphorylation. Superoxide anion has a long mean free path and, approaching the acidic surrounding of mitochondrial DNA, can become protonated ($pK_a \approx 4$). It can thus force the electrostatic barrier around DNA, react in the close vicinity of it and introduce breaks in the sugar-phosphate backbone. Since mitochondrial DNA repair is not the most efficient enzymatic machinery ever developed, damages in mitochondrial DNA are likely not to be repaired. Transmitted from generation to generation as post replicative mutations or deletions, they provide an excellent marker of cellular ageing[6]. Within a cell, damages to mitochondrial DNA are likely to be distributed at random, e.g. according to the law of Poisson. This law can also be assumed to describe the distribution of cells in an organ, having a given amount of damaged mitochondrial DNA. It could thus be expected that above a threshold of damages, or of damage-carrying cells, the functionality of that organ be lost or seriously impaired. This leads to the apparently paradoxical conclusion that spontaneous cell death may induce organ ageing. Incidentally, this conceptual result fills the gap between the two definitions of ageing given at the beginning of this chapter, because the accumulation of molecular modifications results in the impairment of organ functionality.

We shall point out below that all the inducers of skin ageing quoted above are either able to provoke oxidative stress, or to induce responses similar to those induced by oxidative stresses, such as the synthesis of ICAM-1 (Inter Cellular Adhesion Molecule-1). We think that upon interaction with external-only, external-metabolised or internal-only agents, some cells die or are otherwise damaged. These cells will secrete molecular signals, cytokines or other "death-signals" which shall stimulate resident mast cells to secrete histamine, thus invoking macrophages and neutrophils from the capillaries into the dermis. These cells shall accomplish diapedesis (i.e. sneak out of the vessel) by binding to endothelial adhesion molecules and by secreting hydrogen peroxide onto endothelial cells. Once in the dermis, they shall fulfil their physiological function via a non-targeted release of ROS. The resulting damage of the surrounding extracellular matrix. and also of some neighbouring cells, will perpetuate the inflammatory reaction. The damages in the extracellular matrix will be monitored by nearby fibroblasts which shall synthesise and excrete new fibers, without the capability to get them to be oriented. The net result of this catabolic process is a dermis devoid of its initial elastic properties. This is why it is important, in the field of the biology of skin ageing, to learn about the physiological role of the major physical agents able to induce oxidative stress.

1.1. Outline of Our Previous Work

One of us (P.U.G.) has been working for a decade on the molecular effects of UV radiation on animal epidermis, cultured cells and purified DNA. He noted that the response of epidermal cells to UV radiation closely resembles the response elicited by oxidative stress. In particular UV radiation induces a dramatic drop in the level of nicotinamide adenosine dinucleotide[7] and the accumulation of hsp70 mRNA[8] which are also provoked by ionizing radiation and by treatments with hydrogen peroxide or heavy metals. The observation that the in vitro nicking of DNA by UV requires iron and oxygen[9] contributed to the conclusive demonstration that UV radiation is endowed with pro-oxidant properties. A consequence of this observation was that the morphological analysis of UV irradiated epithelial cells was undertaken on cultured epidermoid A 431 cells as well as on cultured normal human keratinocytes. Both cell types were observed to undergo zeiosis upon UV irradiation. Zeiosis consists of loss of the cell-to-cell or cell-to-substrate contact, rounding up of the cell and appearance of several surface blebs. The number and the size of the blebs are dependent of wavelength, dose and cell type[10,11,12]. Zeiosis is a phenomenon typically induced by superoxide anion and by other ROS. Cells can be rescued from UV induced damages and cell death by the addition of α-tocopherol thus confirming the pro-oxidant nature of UV radiation[11,12].

The pro-oxidant properties of UV radiation can be mimicked by treating cells or purified nucleic acids with hydrogen peroxide in the presence of iron and it is well known that relatively high concentrations of hydrogen peroxide can be generated within the skin during the inflammatory reaction, because of the action of neutrophils and macrophages. It was therefore reasonable to undertake the analysis of the parameters able to modulate the effects of oxidative stress on cells in culture and on purified nucleic acids. Thereby it was observed that histidine increases dramatically the cytotoxicity of hydrogen peroxide and that it modulates the oxidative nicking of DNA[13-17].

These results are of importance because they point out that ROS kill cells by introducing double-strand breaks in their DNA. Moreover they allow one to examine the potential damages induced in the cells by agents capable to generate oxidative stress.

2. GENERAL OUTLINE OF THE AGEING PROCESS, AS WE SEE IT

Let us consider a skin cell in the absence of external stress, resting in an organism fed with glucose, water and oxygen. It will metabolise glucose and oxygen in order to survive and in the process to reduce oxygen to water its mitochondria will necessarily generate superoxide. Some of these anions have a non negligible chance to interact with mitochondrial (mt) DNA, thus modifying (damaging) it. Since DNA repair is poor in the mitochondria, those damages will be transmitted to the cell of the next generation under the form of mutations or deletions. In the course of oxidative phosphorylation, the mitochondria of the daughter cell, too, will generate superoxide, damage their own DNA etc. Once the mutations and deletions in the mtDNA are in excess to what the cell can bear, cell metabolism is impaired, membrane potential drops, ATP is no longer synthesised and "death signals" are diffused, for instance under the form of metabolites of the arachidonic acid cascade, triggered by enzymes like the lipoxygenases. These signals will be detected by resident mast cells and will induce them to secrete molecular signals, among which one finds the cytokine TNF-α and the autacoid histamine. This autacoid induces endothelial cells, lining the vascular wall, to immediately release P-selectin for neutrophil adhesion[18] while TNF-α induces neo-synthesis of ICAM-1. Neutrophils will thus bind to these endothelial adhesion molecules and secrete hydrogen peroxide which will induce further synthesis of ICAM-1 by the endothelium. Other macrophages and neutrophils will bind to it, while secreting hydrogen peroxide which damages endothelial cells and alters monolayer confluence by inducing cell retraction (Figure 1). While all this happens, macrophage and neutrophils perform diapedesis and sneak out of the vessel to reach the cell to be scavenged.

We notice here that in the response to damages generated by endogenously produced superoxide within one cell, it is expected that endothelial cells suffer from damages generated by hydrogen peroxide, released by immune cells.

This is the first step of the physiological cascade of the oxidative stress and it occurs at the vascular site.

Macrophages and other immune cells are known to pass from the blood vessel to the tissue, destroy the intruder and be drained into the lymphatic circulation. When in the tissue, these immune cells will try and reach the "death signal" emitting cell which has to be removed. In order to do so, they induce the release of collagenases by neighbouring fibroblasts to digest the extracellular matrix[19]. Upon reaching the cell to be scavenged, source of the chemoattractant signal, the immune cells are supposed to engulf it and to destroy it. For doing so, they have first to digest it in order to produce small size debris, which they perform by releasing nitric oxide, hydrogen peroxide and singlet oxygen. These reactive molecules are not targeted, but can diffuse in the tissue, thus altering molecules of the surrounding extracellular matrix and, possibly, some neighbouring cells.

We notice here that, in response to damages induced by endogenously produced superoxide within one cell, molecules from the extracellular matrix can be digested or damaged by proteases and by reactive oxygen species, released by immune cells. This is the second step of the physiological cascade of the oxidative stress and it occurs at the perivascular site.

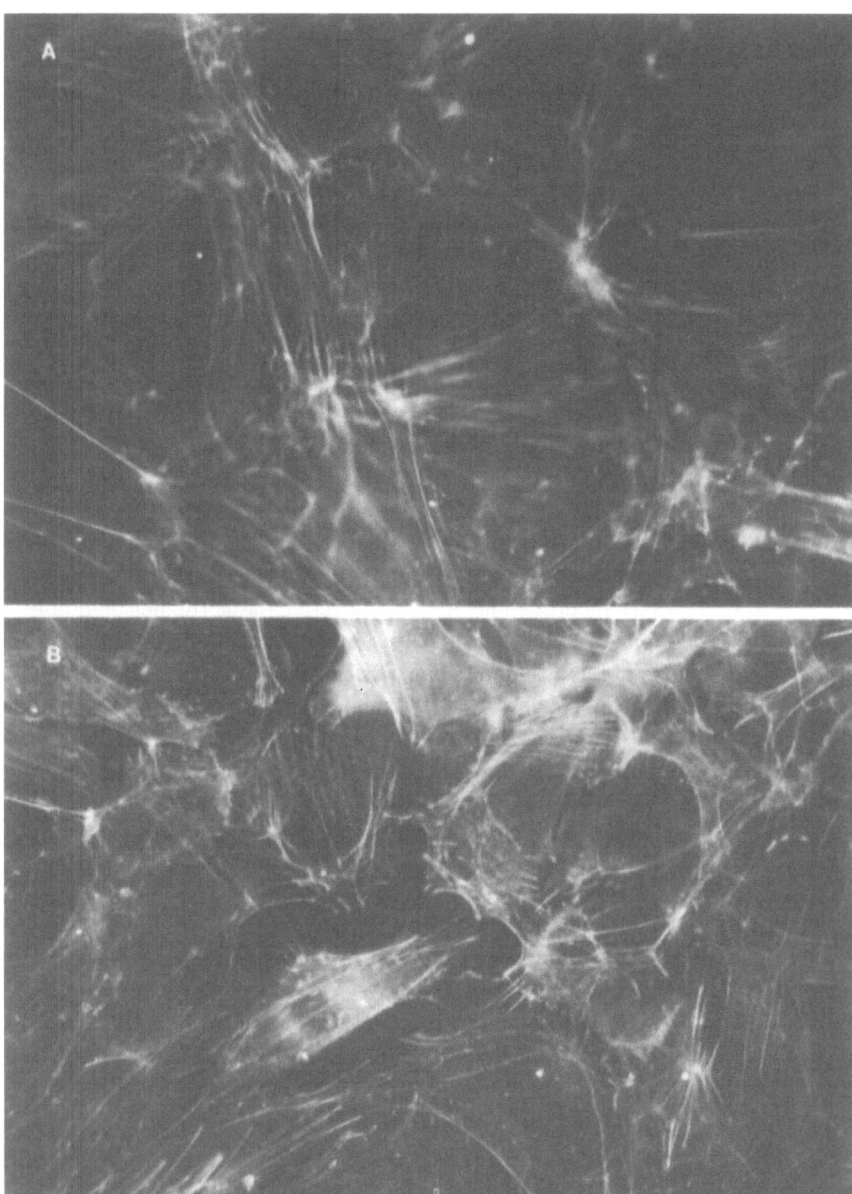

Figure 1. (A) Endothelial cells from human umbilical vein (HUVEC) grown at 37°C in culture medium supplemented with 20% fetal calf serum (FCS) at 5% CO2, third passage. Details about cell culture are given elsewhere[79]. The structure of the cytoskeleton is visualized by phalloïdine/rhodamin staining of polymerized actin. (B) Confluent monolayer of HUVEC grown as above were incubated for one hour in complete medium with 140 μM hydrogen peroxide, rinsed in Phosphate Buffered Saline and further supplemented with complete medium for 20 hours. The structure of the cytoskeleton is visualized by phalloïdin/rhodamin staining of polymerized actin. Details about peroxide treatment and actin staining are described elsewhere[80].

In the course of this phase of the "response-to-injury" reaction, an "innocent by-stander" cell could also be killed, thus generating messengers which will maintain for another cycle the microinflammatory phenomenon and contribute to the progressive phenotypical modification of tissues and cells. We notice here that in response to damages generated within one single cell by endogenously produced superoxide, a microinflammation can be perpetuated.

This is the third step of the physiological cascade of the oxidative stress and it occurs at connective tissue sites. When microinflammation is perpetuated, one could ask, how is it arrested? As a matter of fact, in our experience, all acute inflammation comes sooner or later to a stop. A possible answer to this question is that the endogenous antioxidant pool is not constant, and that it might increase when required, thus modulating the generation of ROS by the immune cells. Another possibility is that a role in halting microinflammation is played by melatonin, which is synthesised in relatively large amounts during the night and which is known to be an inhibitor of the expression of the 5-lipoxygenase gene[20].

In summary, respiration causes cell death which causes a potentially self-maintaining microinflammatory phenomenon. Because activation and damage are induced in endothelial cells at every cycle of the inflammatory process, vessels become more and more dysfunctional and permissive towards immune cells. Thus the same type of injury (i.e. the death of a single cell) will end up to skin modifications with an efficiency which increases with time. These modifications in the connective tissue (tunnels and holes dug in the dermis by the collagenases) will be sensed by dermal fibroblasts. In order to try and somehow restore the previous organization, dermal fibroblasts will secrete molecules to form elastic fibers. Newly synthesised matrix molecules do not fit, per se, into an ordered structure. As a result, the original geometric framework of the extra-cellular matrix will be changed. Cell-matrix interaction is essential for gene expression. The dynamic interaction between newly synthesised extracellular matrix components and migrating cells is constantly modified by cytokines and subjected to feedback regulation at the transcriptional, translational and post-translational level. Perpetuation of such a phenomenon over time will eventually result in entangled collagen fibers, enhanced deposition of elastin and other ECM modifications. The disordered structure observed upon chronic UV irradiation (see below) might be the consequence of one particular arrangement of newly synthesised molecules.

This is a hypothetical description of what can be expected for skin in the absence of external stress. We shall see below, how ultraviolet radiation, gravitational forces, advanced glycation end products, smoke, nutrients and hormonal status will generate the same sequence of steps peculiar to the physiological cascade of the oxidative stress.

3. PHENOMENOLOGY OF AGED SKIN

Old skin is characterised by thinning of the epidermis[21], flattening of the dermal-epidermal junction[22], loss of sub-epidermal vascularization[23], molecular modifications of the connective tissue, such as increase in collagen content in rat dermis[24], threefold increase in elastin content and decrease in collagen content in human dermis[25], changes in the mechanical properties of the skin such as loss of elasticity, loss of cutaneous extensibility, decrease of transepidermal water loss[26,27] and so forth. All these modifications, accompanied by generalised wrinkling, liver spots, dry skin and seborrheic keratose, communicate to the outer world, and to the other individuals, that the owner of that skin is undergoing the irreversible process of aging. This is possibly one of the reasons which render the study of the ageing of the skin so interesting, both scientifically and economically.

When studying a phenomenon, one of the goals is to identify some of the causative agents. As we said before we identified internal, external and external-metabolised agents, which we are going to describe below.

3.1. External Effectors of Skin Ageing: UV Radiation

UV radiation alters the most exposed areas of human skin. The study of actinic ageing (photo-ageing) has been mainly concerned with the analysis of skin from rodents, humans and miniature swine subjected to repeated exposures to UV radiation at different wavelengths according to different protocols. Actinic ageing is characterised by dermal thickening accompanied by loss of elastic properties, and can be easily studied by analysing the histologic properties of skin samples originating from sun-exposed and non-exposed skin areas. In volunteers of European habits, for instance, the upper part of the neck and the buttocks are two anatomical zones ideally suited for taking biopsies and conducting a comparative evaluation.

The relative amounts of collagen III in mouse skin chronically exposed for ten to thirty six weeks to UVB was found to increase by 20%. The amount of elastin has been found to increase by 50% in hairless mice upon chronic exposure (twelve exposures) to UVB or UVA. We recall here that UVA is characterised by wavelengths between 320 and 400 nm, while UVB has wavelengths comprised between 280 and 320 nm, and UVC is defined as the region of ultraviolet waves having wavelength below 280 nm. Sunlight on the surface of the earth contains UVA and UVB, and no UVC. In another study of hairless mice irradiated for ten weeks, UVB radiation was found to increase by 20% the content of glycosaminoglycan, to nearly triple the content of heparin, and to affect in an even more dramatic way their rate of synthesis. More details and references are given in our recent review on photo-ageing of the skin[28].

In miniature pig, whose skin is believed to closely resemble human skin, chronic irradiation with a solar simulator (thrice weekly for 24 months) results in dramatic disorganisation and loss of orientation of collagen fibers, decrease of immunostainable collagen I, increase of immunostainable collagen III, and increase of elastin[29]. In humans, repeated exposures to a solar simulator (seventeen to twenty exposures) resulted in the deposition of amorphous material around the small vessels on the upper dermis, whereas repeated exposures to UVA resulted in the formation of "Zebra bodies", crossbanded filamentous aggregations in the amorphous masses surrounding the blood vessels. Moreover, in the papillary dermis, blood vessels display widely open endothelial gaps and extravasation of blood cells[30]. Elastin was found within the elastotic material in biopsies from three patients with cutis rhomboidalis[31]. Wide attention has been paid to the respective role of UVA and of UVB in the generation of damages which might be related to the induction or to the appearance of photoageing. An exhaustive analysis was carried out in hairless mice[32] by chronically irradiating twenty groups of animals with 5 nm-broad wavebands, thus covering the spectrum between 280 and 380 nm. It was concluded that skinfold thickening, wrinkling and tumor formation were mainly due to UVB, whereas skin sagging was the consequence of UVA.

What mechanism can be invoked for understanding the actinic ageing of the skin? Let us remember that UV radiation provokes infiltration of neutrophils[33,34]. Let us also recall that, upon exposure to UV A or UV B or UV C, the common physiological feature is the apparent vascular injury[35] and the thickening of the stratum corneum. That is to say that the main tissue responses to UV radiation are vascular response and epidermal hyper-plasia. The vascular response accounts for the inflammatory response of the injured tissue and the epidermal hyperplasia is a sign of the tissue remodelling, pertinent to the wound healing process.

These two phenomena operate via cytokines secreted by keratinocytes, fibroblasts, mast cells and endothelial cells: for instance, in response to UV radiation, keratinocytes release the immuno-inflammatory triade TNF-α, Il-1 and INF-γ[36]. To these signals, others could be added because of the possible UV-induced death of resident cells, such as fi-broblasts. So, one could surmise that the cascade of phenomena which shall take place af-ter UV irradiation may be similar to the general process of chronological ageing outlined above.

The adhesion of leukocytes to the endothelial cells of the dermal microvasculature is one of the first events induced by the cytokines secreted during the inflammatory reaction. The cytokine-controlled inflammatory reaction operates via the induction of a state of vas-cular reactivity on one side, and via the activation of immune cells in blood (neutrophils, monocytes, macrophages, lymphocytes) and in tissue (mast cells) on the other side. These activated immune cells operate via reactive oxygen species.

UV irradiation provokes direct damage to membrane lipids with the consequence of the peroxidation cascade. UV irradiation also provokes damages in the DNA and the con-sequent repair activity. This activity is accompanied by the reaction of poly-ADP ribosyla-tion which depletes the cell of the major electron transporter, NAD[7]. As a consequence, the oxidative balance of the cell is impaired and an oxidative stress is generated. Now, upon UV-induced direct or mediated oxidative stress, resident cells may die (exactly as it is expected to be the case for a "non-exogenously-aggressed" cell having spontaneously accumulated excessive damage to its mitochondrial DNA). Oxygen radical-induced per-oxidation of membrane lipids, caused by irradiation, may also contribute to increase mem-brane-phospholipase activity which, in turn, increases the amount of substrate available for prostaglandin formation[37]. Consequently, prostaglandins and leukotrines will be re-leased, tissue mast cells will be activated and secrete inflammatory agonists such as hista-mine and TNF-α. Upon adhesion and diapedesis of neutrophils, active cellular migration from the blood stream inside the vessels to the perivascular matrix will occur. Once in the dermis, neutrophils are able to secrete toxic myeloperoxidases which generate singlet oxy-gen and contribute to the removal of cellular debris by lysis and cleaning of necrotic tissue by macrophages[38,39].

In mammalian cells and in human skin, radiation has been shown to lower the level of endogenous antioxidant enzymes[40,41] present in the skin in order to reduce highly reac-tive oxygen species. Upon the stimulation of tissue damage mediators, induced by UV ra-diation, vascular endothelium will be submitted to oxidative stress of neutrophil origin and will be the target of macrophage cytokines. Thus UV radiation, inflammatory reaction and the resulting oxidative stress may develop to a self-maintaining phenomenon. A potentiat-ing loop is represented by mitochondria-derived ROS functioning as signal transducers of TNF-α induced gene expression[42]. UV-radiation itself induces direct endothelial oxidative stress[43], and the concomitant upregulation of cytokines. In particular, many of their target genes in endothelial cells are under the control of the nuclear transcription factor (NF-kB) known to be activated by UV radiation and ROS[43,44,45]. c-fos and c-jun activation regu-lates the expression of downstream genes in cells exposed to UV radiation[46,47]. Upon UV irradiation, extracellular matrix also influences gene expression in endothelial cells[48].

Functional impairment of skin cells (fibroblasts, keratinocytes, melanocytes, mast cells, Langerhans cells), cross-linking of collagen and other macromolecules in the con-

nective tissue, oxidation of sulphydryl groups generating disulphide bridges, and liberation of proteases, collagenases and elastases are all phenomena induced by UV-generated
ROS[49]. In particular UV radiation stimulates interstitial collagenase of the skin, resulting
in collagen degradation [50].

Changes in the ternary structure of long-lived proteins, such as collagen and fibronectin can in turn produce long-term changes in gene expression. Thus the direct
stimulation of collagenase synthesis by human skin fibroblasts and keratinocytes following UV radiation may contribute to the stimulation of wound-healing-like phenomena
relevant for photo-ageing.

In summary, following UV radiation, the three steps of the physiological cascade of
the oxidative stress do indeed occur. UV radiation not only directly modifies the elasticity
of the tissue, but also induces the synthesis of adhesion molecules such as VCAM-I[51], thus
enhancing a microinflammatory response which, as we have seen above, can be triggered
by cell death or other injuries. Mast cells resident in the normal dermis are concerned by
the proteolytic disassembly of the extra-cellular matrix. Renewal and restructuring of
ECM by activated myofibroblasts will not faithfully restore the original tissue[28]. Under
the control of different growth factors, the tissue will be gradually replaced by a differently organized one, characterized by unique modifications of the various components of
the ECM.

The new geometric framework will modify the repertoire of the molecules and receptors involved in cell-cell and cell-surface interaction, thus imprinting the tissue with its
post-UV functional peculiarities, consistent with the experimental evidence showing en-
tanglement of collagen fibers and loss of elasticity which are the physical properties of
photoaged skin.

3.2. Other External Effectors of Ageing: Wounds and Infections

As we have seen, cell death originating from damages generated during mitochondrial respiration or UV radiation triggers a potentially self-maintaining three step sequence of events which involves immune cells. We pointed out above that tissue repair
after UV injury is reminiscent of wound healing. So traumatisms provoking wounds in the
skin, which have been the object of deep investigations, have served as models for skin
photoageing. Scars are a typical example of newly formed ECM which does not present
with the characterics of a young and healthy connective tissue. In a similar way, bacterial infections accompanying small traumatisms such as the stinging by the thorn of a rose
or by a mosquito, are -in our model- external effectors of ageing. They do indeed trigger
the three steps of the physiological cascade of the oxidative stress, the goal of which is to
free the organism from the intruders.

3.3. Other External Effectors of Ageing: Gravitational and Tractional Forces

Most natural structures, and among them biological structures, depend upon tensile
forces for their integrity. The malleability and adaptability of most organic forms require a
structural organization which might well be independent of gravity (i. e. which does not
need gravity to be maintained). Cells may be described as tensegrity structures since they
generate their own tensional forces and exhibit architectural integrity[52]. This notwith-

standing, cells and tissues are submitted to gravitational forces and to other tensional forces as in the case of accidental pinching or hitting, or because of the physiologically determined pulsating pressure of blood against the arterial wall. Strains imposed by cultural habits can also play a role in skin deformation (ageing) as in the case of the lips or the neck of women in some African tribes. In the course of time, continuously applied external tractional forces provoke the decay of internal tensile forces within the extracellular matrix and cells, thus modifying the physical properties of tissues. A natural model of this long term process is represented by the aortic aneurysm. Vascular dilatation is the consequence of the solicitation of the pulsatile forces on the vascular wall. Associated to this deformation, active recruitment of inflammatory cells into adventitia and the subsequent elaboration of metalloproteinases can be observed. This active remodelling process, expected to strengthen the tissue, will paradoxically contribute to the rapid growth and rupture of the aneurysm[53].

This is to say that tensional modifications of tissue which occur with time have not to be considered as a passive event. On the contrary, they might be (and in some instances are) initiators of the self-entertaining cascade of inflammation.

3.4. Internal Effectors of Ageing: Hormonal Status

Skin is an endocrine-competent organ, capable of producing, converting and reacting to hormones. During the process of tissue repair, stimulation of proteoglycan synthesis by skin fibroblasts and plasminogen activator gene expression are regulated by 3,5,3′ triiodothyronine[54]. Another example is provided by sex hormones which influence the synthesis of some components of the extracellular matrix. While estrogen inhibits the synthesis of dermatan sulfate, androgens do not play any role on its metabolism in the dermis. Since we consider skin ageing, it might be of particular interest to know that estrogens act as immunostimulators and that androgenic anabolic steroids act as immunosuppressors[55]. Skin vascular reactivity and regional blood flow are both dependent on the hormonal status of the subject[56]. Furthermore it has been reported that ageing is associated with endothelial dysfunction in healthy men years before the onset of the age-related decline in women[57]. Last but not least, local blood flow is not only a matter of hormone-dependent endothelial regulation: it also depends on trophic interactions between perivascular nerves and endothelial cells. Hormones and nerve transmitters play a role in this regulation. For instance, arterial distensibility is reduced upon chronic sino-aortic denervation (affecting the parasympathetic system). On the other hand, in the case of long term orthosympathectomy a reduction of the density and of the content of elastin has been observed within the arterial wall[58,59]. Brain hormones are known to regulate blood flow in the case of primary behaviours such as fear reaction and sexual stimulation.

Brain hormones may also directly regulate skin functions. For instance, several neuropeptides are found in the skin[60] and receptors for the hypothalamus-derived growth hormone (GH) have been characterized in skin fibroblasts[61]. GH also contributes to establishing the thickness and the mechanical strength of the skin by increasing collagen synthesis[62].

Last but not least, it has been shown that estradiol enhances leukocyte binding to TNF-α stimulated endothelial cells via an increase of adhesion molecules ICAM-1 and VCAM-1 and E-selectin[63]. Moreover, the steroidogenic capacity of adrenocortical cells is affected by the diminished antioxidant defence system characteristic of ageing[64]. All these lines of evidence indicate that the internal hormonal status is programmed to support, whenever required, a microinflammatory reaction.

3.5. External-Metabolised Effectors of Ageing: Nutrition

If we shift to nutrition as possible causative element of ageing, we shall notice that gut hormones are highly involved in skin function: insulin is the only endocrine hormone absolutely required to support keratinocyte proliferation[65]. In the case of hereditary or iatrogenically induced insufficiency of insulin, glucose levels in the blood are actually raised with a consequent increase of the glucose concentration within the dermis. This will accelerate the rate of formation of advanced glycation end products (AGE)[66] with the consequent formation of crosslinks among ECM molecules. AGE are known to induce the synthesis of VCAM-1 and to trigger the microinflammatory process[51]. This fact may develop vascular dysfunction and the expression of other inducible adhesion molecules. As a consequence, the vasculature becomes a selective target for activated immune cells prone to secrete ROS and migrate in the adjacent tissue, thus triggering the onset of the three steps of the physiological cascade of the oxidative stress.

How is the vascular wall affected by food intake? Enhanced ICAM-1 and VCAM-1 expression have been shown in human atherosclerotic lesions[67,68]. It has also been suggested that a causative link exists between the expression of VCAM-1 and the early phase of experimental hypercholesterolemia-induced atherosclerotic lesions[69]. As a matter of fact, at the site of the vascular wall 50% of the lipid catabolism is performed[70]. In order to internalise lipids, endothelial cells have to peroxidize them. Excess lipid will be excreted from the endothelium in the perivascular region where it will form lipidic bodies. How do lipid peroxides provoke endothelial injury? Smooth muscle cells and macrophages active in digesting lipid bodies become foam cells. This is one of the effects of lipid peroxides which, together with collagen neosynthesis, make vascular wall to age and are considered to initiate atherogenesis. Differently peroxidized lipoproteins differ in their efficacy in inducing NO• synthesis in endothelial cells: minimally oxidized do not, maximally oxidized lipoprotein do, thus starting the three steps of the physiological cascade of the oxidative stress[71].

3.6. Other External-Metabolised Effectors of Ageing: Cigarette Smoke

Cigarette smoke consists of thousands of compounds, many of which are as yet unidentified. Many of these can be expected to react with biomolecular species in the respiratory tract. Exposure of plasma to cigarette smoke has been shown to cause depletion of antioxidants[72]. It is difficult to distinguish whether the oxidative stress generated by cigarette smoke in vivo is due to direct reactions of smoke components or whether it is ought to the ensuing inflammatory response. Accumulation and activation of neutrophils and macrophages contribute to lung and plasma indexes of oxidative injury during smoke inhalation. In particular NO• has been shown to directly activate cyclooxygenases. This suggests that NO• may exacerbate inflammatory conditions via the production of pro-inflammatory prostaglandins[73].

On the other hand it is also possible that constituents in cigarette smoke, such as NO•, induce other biochemical effects in both the respiratory tract and the systemic circulation. Cigarette smoke NO• can modulate several enzymes by rapidly interacting with tyrosine and tryptophan radicals, which are functional components of several enzymes[74]. Although the time-dependent formation of different reactants in cigarette smoke is poorly understood, the formation of nitrated and oxidized forms of tyrosine by cigarette smoke suggests that free radical reactions of nitrogen oxides and other ROS may play a role in cigarette smoke-induced protein modifications. These types of modifications may result in altered activity not only of critical enzymes, but also of membrane receptors and transport proteins, eventually interfering with cell-signalling pathways. Curiously enough, with the sole exception of glutathione, antioxidants do not inhibit protein modifications associated with cigarette smoke[72]. So, if only because it introduces molecular modifications, cigarette smoke can be recognized as an effector of ageing. But there is more to that.

Cigarette smoke-related diseases start with the production of ROS by recruitment and activation of phagocytes in the lung. Damage to the epithelial cells of the respiratory tract will increase its permeability towards carcinogens and other agents. Some of these may cause oxidative damages to lipoproteins, especially LDL, resulting in enhanced uptake of LDL by macrophages, thus accelerating one of the key steps of atherogenesis. Cigarette smoke induces adhesion-promoting processes between leukocytes and the endothelium[75]. As we have seen before, this phenomenon can damage the endothelium by activating neutrophil- and monocyte-induced expression of selectins. These facts illustrate how cigarette smoke may induce an oxidative stress on both the respiratory tract and the circulatory system. Because of this, and because of the role we attributed to the blood vessels in the skin-ageing process, we consider that cigarette smoke plays a "strategic" role in the triggering of the three steps of the physiological cascade of the oxidative stress. Epidemiological studies seem to confirm a role of cigarette smoke in skin ageing[76].

4. CONCLUSION

The response to environmental traumatisms such as UV radiation consists, in the human body, in a tissue reaction which is able to perpetuate independently of the actual presence of the triggering factor (in this case, photons). We have proposed that a microinfammatory reaction is expected to occur as a consequence of physiological damage to mitochondrial DNA within a cell even in the theoretical absence of external stress. Besides environmental traumatisms and cell death because of mitochondrial impairment, we have pointed out that the hormonal and the nutritional status of the individual do influence the degree of the transformation of connective tissue with time. In skin, the end result of such tissue reactions is "repaired" connective tissue, which displays some, but not all, of the physical and biochemical characteristics of the ancient one. This means that the phenotypical changes in the extracellular matrix will be the first observable phenomenon. Interestingly enough, in patients affected by chronic urticaria characterized by the inflammatory infiltrate of neutrophils, cutis laxa has been observed: elastolysis is present between the collagen bundles and around the vessels. In lax skin, the entire dermis is affected by a marked decrease in the number of elastic fibers, particularly conspicuous near the inflammatory cells[77]. Incidentally, this finding and the mechanism of ageing proposed in this chapter might explain or at least suggest why animals lacking a vascular system such as some crustaceans and insects do not age, in the sense that they do not acquire the aged phenotype slowly but undergo in a short time the brutal transition from life to death, phenomenon which has puzzled many a good evolutionist and geneticist.

A first consequence of the general outline of ageing as proposed here is that protection of the skin against external effectors of ageing can be pharmacologically accomplished. Sunscreens shall avoid primary photon-induced damages. Antioxidants shall scavenge the few primary photon-induced free radicals but particularly modulate the overdimensioned inflammatory reaction which follows the physico-chemical aggressions to which we are daily exposed. For today's knowledge and technology, chemical treatment of the skin able to maintain cytokines to minimal functional levels seems to be too difficult to be performed safely. On the other hand, proper administration of melatonin might help in stopping the self-maintaining process of inflammation, as in the case of the UV induced erythema which is efficiently reduced by a topical application of melatonin after the exposure to radiation[78].

A second consequence of this model of ageing is to have realised that the rate of ageing of the skin could ideally be reduced if it were possible to avoid external and external-metabolised effectors of ageing, and to reorient newly synthesised elastic fibers. This attempt to characterise skin ageing is relevant because the aspect of the skin plays a major role in the self image of people. Disabled or ill individuals may get a major psychological benefit from the improvement of the properties of their skin. Independently of this aspect, it is important to point out that an effective anti-ageing treatment for the skin should yield major financial profits to those able to put it at work because of the vastness of the market interested in this cosmetic approach of ageing. It is not unexpected that the ageing-promoting three steps of the physiological cascade of the oxidative stress also occur in organs other than skin. An efficient anti-ageing treatment should be able to interfere with the organ-specific pathway of the inflammatory reaction.

REFERENCES

1. Humbert, W. and Pevet, P. (1994) The decrease of pineal melatonin with age. Ann. N.Y. Acad. Sci. 719, 43–63.
2. Giacomoni, P.U. (1992) Ageing and cellular defence mechanisms. Ann. N.Y. Acad. Sci. 663, 1–3.
3. Black, H.S. (1989) Role of reactive oxygen species in inflammatory process. In: Hensby, C and Lowe, N.J. (eds) Nonsteroidal Anti-inflammatory drugs. Pharmacology of the Skin, vol 2, pp. 1–20. Karger, Basel.
4. Kooy, N.W. and Royall, J.A. (1994) Agonist-induced peroxynitrite production from endothelial cells. Arch. Biochem. Biophys. 310, 352–359.
5. Halliwell, B. (1993)The role of oxygen radicals in human disease, with particular reference to the vascular system.Haemostasis 23 Suppl 1P, 118–126.
6. see for instance, Molecular basis of ageing: mitochondrial degeneration and oxidative damage (H. Joenje ed.) Special Issue of Mutation Research 275 pages113–414 (1992)
7. Balard, B. and Giacomoni, P.U. (1989) Nicotinamide Adenosine Dinucleotide levels in dimethylsulfate treated or UV irradiated mouse epidermis. Mutat. Res. 219, 71–79.
8. Brunet, S. and Giacomoni, P.U. (1990) Heat shock mRNA in mouse epidermis after UV irradiation. Mutat. Res. 219, 217–224.
9. Audic, A. and Giacomoni, P.U. (1993) DNA nicking by ultraviolet radiation is enhanced in the presence of iron and of oxygen. Photochem. Photobiol. 57, 508–512,
10. Malorni,W., Donelli, G., Straface, E., Santini, M.T., Paradisi, S. and Giacomoni, P.U. (1994) Both UVA and UVB induce cytoskeletal dependent surface blebbing in epidermoid cells. J. Photochem. Photobiol. B:Biol. 26, 265–270.
11. Straface, E., Santini, M.T., Donelli, G., Giacomoni, P.U. and Malorni, W. (1995) Vitamin E prevents UVB induced cell blebbing and cell death in A 431 epidermoid cells. Int. J. Rad. Biol. 68, 579–587.
12. Malorni, W., Straface, E., Donelli, G. and Giacomoni, P.U. (1966) UV induced cytoskeletal damage, surface blebbing and apoptosis in cultured human keratinocytes are hindered by α-tocopherol. Eur. J. Dermatol. 6, 414–420.
13. Tachon, P. and Giacomoni, P.U. (1988) Histidine, a clastogenic factor. In: Light in Biology and Medicine (Douglas, Moan and Dall'Acqua Eds) Vol I pages 211–217 Plenum Publishing Corporation.
14. Marrot, L. and Giacomoni, P.U. (1992) Enhancement of oxidative DNA degradation by histidine. The role of stereochemical parameters. Mutat. Res. 275, 69–79.
15. Muiras, M.L., Tachon, P. and Giacomoni, P.U. (1993) Modulation of DNA breakage induced via the Fenton reaction. Mutat. Res. 295, 47–54.
16. Ouzou. S., Deflandre, A. and Giacomoni, P.U. (1994) Protonation of the imidazole ring prevents the modulation by histidine of oxidative DNA degradation. Mutat. Res. 316, 9–16.

17. Cantoni, O., Guidarelli, A., Sestili, P. Giacomoni, P.U. and Cattabeni, F. (1994) L-histidine mediated enhancement of hydrogen peroxide-induced cytotoxicity: relationship between DNA sngle/double strand breakage and cell killing. Pharmacol. Res. 29, 169–178.

18. Lorant D.E., Patel, K.D., McIntire, T.M., McEver, R.P., Prescott, S.M. and Zimmermann, G.A. (1991) Co-expression of GMP 140 and PAF by endothelium stimulated by histamine or thrombin: a juxtacrine system for adhesion and activation of neutrophils. J. Cell. Biol. 115, 223–229.

19. Kovacs, E.G. and Di Pietro, L.A.Y (1994) Fibrogenic cytokines and connective tissue production. FASEB J. 8, 854–861.

20. Carlberg, C and Wiesenberg, I. (1995) The orphan receptor family RZR/ROR, melatonin and 5-lipoxygenase: an unexpected relationship. J. Pineal Res. 18, 171–178.

21. Lavker, R.M., Zheng, P. and Dong, G. (1987) Aged skin: a study by light, transmission electron, and scanning electron microscopy. J. Invest. Dermatol. 88, 44s-51s.

22. Gilchrest, B.A. (1982) Age-associated changes in the skin. J. Am. Geriat. Soc. 30, 139–143.

23. Kligman, A. M. (1979) Perspectives and problems in cutaneous gerontology. J. Invest. Dermatol 73, 39–46.

24. Lapière, C.M. (1988) Aging of fibroblasts, collagen and the dermis. In: Cutaneous Aging. (A.M. Kligman and Y. Takase, Eds.) University of Tokyo Press.

25. Pearce, R.H. and Grimmer, B.A. (1972) Age and the chemical constitution of normal human Dermis J. Invest. Dermatol. 58, 347–361.

26. Daly, C.H. and Odland, G.F. (1979) Age-related changes in the mechanical properties of human skin. J. Invest. Dermatol. 73, 84–87.

27. Leveque, J.L., Corcuff, P., de Rigal, J. and Agache, P. (1984) In vivo Studies of the Evolution of Physical Properties of the Human Skin with age. Int. J. Dermatol. 23, 322–329.

28. Giacomoni, P.U. and D'Alessio, P. (1996) Open Questions in Photobiology IV: Photoageing of the skin. J. Photochem. Photobiol.B:Biol. 33, 267–272.

29. Fourtanier, A. and Berrebi, C. (1989) Miniature Pig as an Animal Model to Study Photoageing. Photochem. Photobiol. 50, 771–784.

30. Kumakiri, M., Hashimoto, K. and Willis, I. (1977) Biologic Changes Due to Long-wave Ultraviolet Irradiation on Human Skin: Ultrastructural Study. J. Invest. Dermatol. 69, 392–400.

31. Chen, V.L., Fleischmajer, R., Schwartz, E., Palaia, M. and Timpl, R. (1986) Immunochemistry of Elastotic Material in Sun-Damaged Skin. J. Invest. Dermatol. 87, 334–337.

32. Bissett, D.L., Hannon, D.P. and Orr, T.V. (1989) Wavelength dependence of histo-logical, physical and visible changes in chronically UV-irradiated hairless mouse skin. Photochem. Photobiol. 50, 763–769.

33. Lavker, R.M., Gerberick, G.F., Veres, D., Irwin, C.J. and Kaidbey, K.H. (1995) Cumulative effects from repeated exposures to suberythemal doses of UVB and UVA in human skin. J. Am. Acad. Dermatol. 32, 53–62.

34. Hawk, J.L.M., Murphy, G.M. and Holden, C.A. (1988) The presence of neutrophils in human cutaneous ultraviolet-B inflammation. Brit. J. Dermatol. 118, 27–30.

35. Stern, W.K. (1972) Anatomic Localization of the Response to Ultraviolet Radiation in Human Skin. Dermatol. 145, 361–370.

36. Barker, J.N.W.N., Mitra, R.S., Griffiths, C.E.M., Dixit, V.M. and Nickoloff, B.J. (1991) Keratinocytes as initiators of inflammation. Lancet 337, 211–214.

37. Hruza, L.L. and Pentland, A.P. (1993) Mechanism of UV induced inflammation J. Invest. Dermatol. 100, 353-413.

38. Riley, P.A. (1994) Free Radicals in Biology: oxidative stress and the effects of ionizing radiation. Int. J. Rad. Biol. 65, 27–33.

39. Sarkisov, D.S. and Pal'tsyn, A.A. (1992) New data on the leukocyte functional morphology in pyo-septic processes. Arkh. Pathol. 54, 3–8.

40. Hannigam, B.M., Richardson, S.A. and McKenna, P.G. (1992) DNA damage in mammalian cell lines with different antioxidant level and DNA repair capacities. EXS 62P, 247–250.

41. Colin, C., Bouissouira, B., Bernard, D., Moyal, D. and Nguyen, Q.L. (1994) Non invasive methods of evaluation of oxydative stress induced by low doses of ultraviolet in humans. 18th International IFSCC Venezia Italy Oct 3–6, 1994, A 105, pages 50–72.

42. Schultze-Osthoff, K., Beyaert, R., Vandevoorde, V., Haegemann,G. and Fiers,W. (1993) Depletion of the mitochondrial electron transport abrogates the cytotoxic and gene inductive effects of TNF. EMBO J. 12, 3095–3104.

43. Bauerle, P.A. (1991) The inducible transcription activator NF: regulation by distinct protein subunits. Biochim. Biophys. Acta 1072, 63–80.

44. Gerritsen, M.E. and Bloor, C.M. (1993) Endothelial cell gene expression in response to injury. FASEB J. 7, 523–532.

45. Collins, T. (1993) Endothelial NF-kB and the initiation of the atherosclerotic lesion. Lab. Invest. 68, 499–507.

46. Weichselbaum, R.R., Hallahan, D., Fuks, Z. and Kufe, D. (1994) Radiation induction of immediate-early genes: effectors of the radiation stress response. Int. J. Radiat. Oncol. Biol. Phys. 30, 229–234.

47. Brunet S. and Giacomoni, P.U. (1990) Specific mRNAs accumulate in long wavelength UV-irradiated mouse epidermis. J. Photochem. Photobiol. B:Biol 6, 431–441.

48. Madri, J.A. and Pratt, B.M. (1988) Angiogenesis. In: The molecular and cellular biology of wound repair (Clark, R. and Henson, P. eds) Plenum, New York.

49. Dalle Carbonare, M. and Pathak, M.A. (1992) Skin photosensitizing agents and the role of reactive oxygen species in photoageing. J. Photochem. Photobiol. B. Biol. 14, 105–124.

50. Wlaschek, M., Briviba, K., Stricklin, G.P., Sies, H.and Scharffetter-Kochanek, K. (1993) Singlet oxygen may mediate the ultraviolet A-induced synthesis of interstitial collagenase. J. Invest. Dermatol. 104, 194–198.

51. Schmidt, A.M., Hori, O., Chen, J.X., Li, J.F., Crandall, J., Zhang, J., Cao, R., Yan, S.D., Brett, J. and Stern, D. (1995) Advanced glycation endproducts interacting with their endothelial receptor induce expression of V-CAM 1 in cultured human endothelial cells and in mice. J. Clin. Invest. 96, 1395–1403.

52. Ingber D.E. and Jamiesson, J.D. (1985) Cells as tensegrity structures: architectural regulation of histodif-ferentiation by physical forces transduced over basement membrane. In: Gene expression during normal and malignant differentiation. pp. 13–33 Academic Press Inc. London.

53. Freestone, T., Turner, R.J., Coady, A., Higman, D.J., Greenhalgh, R.M. and Powell, J.T. (1995) Inflamma-tion and matrix metalloproteinases in the enlarging abdominal aneurism. Arterioscl. Thromb. Vasc. Biol. 15, 1145–1151.

54. Grando, S.A. (1993) Physiology of endocrine skin interrelations. J. Am. Acad. Dermatol. 28, 981–982.

55. Masi, A.T., Feigenbaum, S.L., Chatterton, R.T. and Cutolo, M. (1995) Integrated hormonal-immunological-vascular (H-I-V triad) systems interactions in the rheumatic diseases. Clin. Exp. Rheumatol. 13, 203–216.

56. Bartelink, M.L., De Wit, A., Wollersheim, H., Theeuwes, A. and Thien, T. (1993) Skin vascular reactivity in healthy subjects: influence of hormonal status. J. Appl. Physiol. 74, 727–732.

57. Celemajer, D.S., Sorensen, K.E., Spiegelhalter, D.J. Georgakopoulos, D. Robinson, J. and Deanfield, J.E. (1994) Aging is associated with endothelial dysfunction in healthy men years before the age related decline in women. J. Am. Coll. Cardiol. 24, 471–476.

58. Lacolley, P., Bezie, Y., Girerd, X., Challande, P., Benetos, A., Boutouyrie, P., Ghodsi, N., Lucet, B., Azoui, R. and Laurent, S. (1995) Aortic distensibility and structural changes in sinoaortic-denervated rats. Hyper-tension 26, 337–340.

59. Lacolley, P., Glaser, E., Challande, P., Boutouyrie, P., Mignot, J.P., Duriez, M., Levy, B., Safar, M. and Laurent, S. (1995) Structural changes and in situ aortic pressure-diameter relationship in long-term chemi-cal-sympathectomized rats. Am. J. Physiol. 269 (Heart Circ. Physiol. 38). H407-H416.

60. Lotti, T, Hautmann, G. and Panconesi, E. (1995) Neuropeptides in skin. J. Am. Acad. Dermatol. 33, 482–496.

61. Lobie, P.E., Breipohl, W. and Lincoln, D.T. (1990) Localization of the growth hormone receptor/binding protein in skin. J. Endocrinol. 126, 467–472.

62. Jorgensen,P.H., Andreassen, T.T. andJorgensen, K.D. (1989) Growth hormone influences collagen deposi-tion and mechanical strength of intact rat skin: a dose-response study. Acta Endocrinol. 120, 767–772.

63. Cid, M.C., Kleinman, H.K., Grant, D.S., Schnaper, H.W., Fauci, A.S. and Hoffman, G.S. (1994) Estradiol enhances leukocyte binding to tumor necrosis factor (TNF)-stimulated endothelial cells via an increase in TNF-induced adhesion molecules E-selectin, Intercellular adhesion molecule Type 1 and vascular cell ad-hesion molecule type 1. J. Clin. Invest. 93, 17–25.

64. Azhar, S., Cao, L. and Reaven, E. (1995) Alteration of the adrenal antioxidant defense system during aging in rats. J. Clin. Invest. 96, 1414–1424.

65. Krane, J.F., Murphy D.P. and Carter, D.M. (1991) Synergistic effects of epidermal growth factor (EGF) and insuline like growth factorI/somatomedin C (IGF-I) on keratinocyte proliferation may be mediated by IGF-I transmodulation of the EGF receptor. J. Invest. Dermatol. 96, 419–424.

66. Brownlee, M., Cerami, A. and Vlassara, H. (1988) Advanced glycosylation end-products in tissue and the biochemical basis of diabetic complications. N. Engl. J. Med. 318, 1315–1320.

67. Poston, R.N., Haskard, D.O. Coucher, J.R., Gall, N.P. and Johnson-Tidey R.R. (1992) Expression of Inter-cellular adhesion molecule-1 in atherosclerotic plaques. Am. J. Pathol. 140, 665–673.

68. O'Brien K.D., Allen, M.D., McDonald T.O., Chait, A., Harlan, J.M., Fishbein, D., McCarty, J., Ferguson, M., Hudkins, K., Benjamin, C.D., Lobb, R. and Alpers, C.E. (1993) Vascular cell adhesion molecule-1 is expressed in human coronary atherosclerotic plaques. Implications for the mode of progression of ad-vanced coronary atherosclerosis. J. Clin. Invest. 92, 945–951.

69. Li, H., Cybulsky, M., Gimbrone, M. and Libby, P. (1993) An atherogenic diet rapidly induces VCAM-I, a cytokine-regulatable mononuclear leukocyte adhesion molecule, in rabbit aortic endothelium. Arterioscl. Thromb. 13, 197–204.

70. Sholley, M., Cudas, S., Regelson, W., Franson, R. and Kalimi, M. (1990) Dehydroepiandrosterone alters the morphology and phospholipid content of cultured human endothelial cells. In: Dehydroepiandrosterone (DHEA) Walter De Gruyter and Co., Berlin, New York. pp. 387–395.

71. Chirico, S., Naseem, K.M., Jeremy, J.Y.,Bunce, T.D. and Bruckdorfer, K.R. (1994) The role of minimally-modified LDL in atherosclerosis. J. Free.Rad. Biol. Med. 2, L 28.

72. Eiserich, J.P., van der Vliet, A., Handelman, G.J., Halliwell, B and Cross, C.E. (1995) Dietary antioxydants and cigarette smoke-induced biomolecular damage: a complex interaction. Am. J. Clin. Nutr. 62, 1490S-1500S.

73. Salvemini, D., Misko, T.P., Masferre, J.L., Seibert, K., Currie, M.G. and Needleman, P. (1993) Nitric oxide activates cyclooxygenase enzymes. Proc. Natl. Acad. Sci. USA 90, 7240–7244.

74. Eiserich, J.P., Butler, J., van der Vliet, A., Cross, C.E. and Halliwell, B. (1995) Nitric oxide rapidly scavenges tyrosine and tryptophane radicals. Biochem. J. 310, 745–749.

75. Lehr, H.A. (1993) Adhesion promoting effects of cigarette smoke on leukocytes and endothelial cells. Ann. N.Y. Acad. Sci. 686, 112–119.

76. Kadunce, D.P., Burr, R., Gress, R., Kanner, R., Lyon, J. and Zone, J.J. (1995) Cigarette smoking: risk factor for premature facial wrinkling. Ann. Intern. Med. 114, 840–844.

77. Chun, S.I., and Yoon, J. (1995) Acquired cutis laxa associated with chronic urticaria. J. Am. Acad. Dermatol. 33, 896–899.

78. Bangha, E., Elsner, P. and Kistler, G.S. (1995) Efficacy of topical melatonin (N-acetyl 5-methoxy-tryptamine) in the suppression of UVB-induced erythema. Dermatology 191, 176–180.

79. Jaffe, E.A. (1984) Culture and identification of large vessel endothelial cells. In: Biology of Endothelial Cells (E.A. Jaffe, ed.), pp. 1–13, Martinus Neijhoff, Boston.

80. D'Alessio, P., Moutet, M., Marsac, C. and Chaudière, J. (1996) Modulation pharmacologique des altérations du cytosquelette endothélial induite par le péroxyde d'hydrogène et par le TNF-α. Comptes Rendus de la Société de Biologie 190, 289–297.

12

CELL MAINTENANCE AND STRESS RESPONSE IN AGEING AND LONGEVITY

Thomas B. L. Kirkwood,[1] Clare Adams,[1] Linda Gibbons,[1]
Caroline D. Hewitt,[1] Pankaj Kapahi,[1] Axel Kowald,[1,2] Gareth Leeming,[1]
Gordon J. Lithgow,[1] Kareen Martin,[1] Christopher S. Potten,[3] and
Daryl P. Shanley[1]

[1] Biological Gerontology Group, University of Manchester
  Stopford Building, Oxford Road, Manchester M13 9PT, United Kingdom
[2] Institute for Biotechnology, Technical University Berlin
  Gustav Meyer Allee 25, 13355 Berlin, Germany
[3] Paterson Institute for Cancer Research, Christie Hospital and
  Holt Radium Institute
  Manchester M20 9BX, United Kingdom

1. INTRODUCTION

Evolution theory suggests that ageing is caused by a life-long accumulation of random damage in somatic cells and tissues, due to an evolved limitation in the levels of key maintenance functions[1]. This idea, termed the disposable soma theory, predicts a central role for cell maintenance and stress response mechanisms in regulating the duration of life. There is potentially a large number of such mechanisms and individual theories have focused, in particular, on the roles of free radicals and oxidative damage, aberrant proteins, defective mitochondria, and somatic mutations. The disposable soma theory identifies a common ground for these 'stochastic' theories of ageing. Furthermore, since the same selection forces act to optimise the genetic control of diverse maintenance functions, the disposable soma theory predicts that multiple stochastic causes of ageing will operate together. Central to this concept is the fact that maintenance systems inevitably involve some costs. To the extent that these costs reduce the resources available for growth and reproduction, they will have a negative effect on Darwinian fitness. Thus, it is predicted that selection will ensure sufficient levels of somatic maintenance to keep the organism in sound condition during the normal expectation of life in the wild environment, but investments in maintenance at higher levels will be selected against. An example would be selection to optimise the levels of stress proteins mediating removal and/or refolding of denatured molecules within the cell; the corresponding targets of selection would be the genes that regulate stress protein expression.

Molecular Gerontology, edited by Rattan and Toussaint
Plenum Press, New York, 1996

An important corollary of the principle of optimising the investments in cell maintenance and stress response systems is that the actual optimum levels in a given species will
depend on ecological circumstances and in particular on the prevailing level of environmental mortality[2]. A species subject to high environmental mortality is expected to invest
rather little in maintenance and a lot in replacement (reproduction), whereas a species subject to low environmental mortality may benefit by doing the reverse.

A further implication of the disposable soma theory is the need to consider integrated approaches to the study of cell maintenance. If selection does indeed optimise the
individual cell maintenance and stress response systems to levels which permit gradual accumulation of defects, then it must be expected that these processes will also involve interaction among the different types of damage that can affect the cell. This requires that
we consider the various cell maintenance and stress response systems as components of an
overall network. The concept of a network of maintenance functions brings a coherence to
the study of individual reactions and processes which contribute to the overall process of
senescence, but it also underlines some of the difficulties. The chief considerations are (i)
to understand how each component process affects the function and viability of the network as a whole, and (ii) to disentangle the interactions between the different processes
and levels of functioning. The need for integration also extends beyond the intracellular
mechanisms and must ultimately include an understanding of how the biochemical
changes affecting the functions of individual cells within organs and tissues contribute to
loss of homeostasis and pathophysiological changes in the organism as a whole. For example, in mitotic tissues it is important to understand how intracellular damage may affect
the population dynamics of cell proliferation, death and replacement.

This chapter summarises current research within the Biological Gerontology Group
at the University of Manchester, which is aimed at investigating the role of cell maintenance and stress response systems in ageing and longevity. The work described covers experimental studies—comparative cell biology, transgenics, age changes in stem cells,
molecular genetics of lifespan—and theoretical studies—biochemical networks, optimal
resource allocation. Much of the work relates to ongoing rather than complete projects;
therefore, the description is often simply a status report to indicate current directions.

2. NETWORK MODEL OF AGEING

In order to examine the implications of the network concept for intracellular mechanisms of ageing, we have developed a series of theoretical models of candidate molecular
mechanisms of ageing and studied their potential for interaction[3,4]. The most recent ver-
sion of the network model[4] incorporates the following features: (i) the possible accumulation of defective mitochondria, (ii) the effects of aberrant proteins in protein synthesis,
(iii) the damaging actions of oxygen free radicals, and the protective role of antioxidant
enzymes like superoxide dismutase, and (iv) the turnover of proteins by proteolytic scavengers. This model has been called the MARS model (for mitochondria, aberrant proteins,
radicals, scavengers).

The MARS model has been formulated as a detailed representation in mathematical
form of the main elements of the relevant biochemical processes. Inevitably there has to
be some compromise between realism and analytical tractability and the model does not
include the full complexity of the system, e.g. the antioxidant enzyme network is presently
represented by a single generic antioxidant species. Nevertheless, the model captures the
essential details of the relevant mechanisms and can, in future, be elaborated further. Pa-

rameter values have been estimated wherever possible from published biochemical data, and the number of free parameters has been kept to a minimum. The MARS model has been coded as a FORTRAN program and simulated by computer.

The MARS model has been shown successfully to predict many of the observations and experimental findings reported from ageing cells and organisms. These include (i) a sharp rise with age in the fraction of inactive proteins, explaining the frequently observed loss of specific enzyme activity; (ii) the development with age of only a slight rise in the fraction of erroneous proteins, explaining why a decline in enzyme specificity is only rarely observed; (iii) a significant increase with age in protein half-life; (iv) a decrease with age in the mitochondrial population; (v) an increase with age in the fraction of defective mitochondria; (vi) an increase with age in the average rate of free radical production per mitochondrion; and (vii) a fall with age in the average level of ATP generation per mitochondrion.

A clear demonstration of the importance of this kind of integrative approach to modelling mechanisms of cell senescence was given by the fact that in the MARS model the predicted effects on measurable properties of the system, e.g. the levels of defective mitochondria, were in some cases qualitatively and quantitively different from predictions in earlier models that considered only single systems. Obviously the MARS model can be extended further, e.g. to include the contribution from nuclear somatic mutations[5], and it cannot substitute for experimental investigation. It can, however, significantly assist in the planning and interpretation of experimental studies by indicating the kinds of dynamic interactions that need to be taken into account.

3. COMPARATIVE STUDIES OF CELL MAINTENANCE

An important prediction from the disposable soma theory is that the levels of cell maintenance and stress response systems should show a general correlation with species' lifespans. Previous work has indicated that DNA repair activity (excision repair of UV-induced pyrimidine dimers) and levels of poly-ADP ribose polymerase (PARP) conform with this expectation[5,6]. To test these correlations more generally we are establishing cell cultures for compehensive analysis of comparative cell maintenance. Primary cells have been cultured from a range of mammalian species—currently mouse, rat, rabbit, opossum, marmoset, sheep, pig and human—and the species range is being extended. To minimise effects of possible confounding variables biopsy procedures have been standardised as far as possible. Skin biopsies were obtained from the inside forelimb of animals within the first 5% of lifespan. The human fibroblasts were obtained from normal infant foreskin samples. Primary cells were allowed to grow from the biopsy sample in culture medium and then sub-cultured using standard procedures.

Growth curves were calculated for each species using data obtained from cell counts by Coulter Counter at each cell passage, taking account of seeding efficiency measurements. Cells were cultured to their limits of proliferation or, where spontaneous immortalisation occurred (mouse cells), to the point of crisis. Cumulative population doublings (CPD) were determined and compared across the species. Results confirm previous reports that the replicative capacity of fibroblasts in culture is positively correlated with species longevity[7].

Preliminary studies revealing a positive correlation with lifespan have been made of resistance to stress, in the form of a transient heat stress, and other forms of stress are also under investigation.

4. TRANSGENIC MOUSE MODEL

The production of transgenic animal models with altered levels of cell maintenance and stress response systems is a powerful technique to investigate the causative role of specific types of cell damage in the ageing process. Aberrant proteins have long been thought to play a role in ageing and may arise through a variety of routes (synthetic errors, damage, denaturation etc)[8]. To evaluate the contribution of aberrant proteins in ageing it is necessary to determine whether the altered proteins which arise are the cause or effect of the changes seen during senescence. Classical biological and biochemical studies have attempted to address this question but the interpretation of experimental data has been difficult[8,9]. With the advent of transgenic mouse technology it is now feasible to produce animals with an increased frequency of error proteins in order to examine the role of aberrant proteins and protein turnover during ageing.

We are developing a model in which error rates during protein synthesis are increased by the expression of missense tRNAs. The missense tRNA genes have been constructed from a human serine tRNA which has been mutated at the anticodon loop to produce tRNAs which are unaltered in their amino acid specificity but which base-pair with an "incorrect" codon, thus causing the mRNA to be effectively mistranslated. The wild-type tRNA gene is also present so the resulting frequency of misinsertion will be a function of the relative concentrations of the missense and wild-type tRNAs. The missense tRNA gene is injected into a pronucleus of a fertilized mouse egg *in vitro* and the eggs implanted into a surrogate female 24 hours later. Using this method, the transgene may become randomly incorporated into the genomic DNA. Detection of positive transgenic animals is performed on isolated genomic DNA using the polymerase chain reaction (PCR) with short (20mer) oligonucleotides which are up and downstream of the mutated gene. This results in an amplicon that spans the whole of the tRNA gene plus short flanking sequences either side. Expression of the missense gene can be detected using reverse transcriptase PCR (RT-PCR) on DNase-treated RNA and specific short oligonucleotides for each missense tRNA gene constructed.

Successful development of this model will allow us to examine the effects of increased error levels on the rate of ageing, both generally in terms of survivorship (increase in maximum lifespan) and more specifically in terms of age-related organ and cell changes. In the non-transgenic senescence-accelerated mouse (SAM) the characteristics of ageing and pathological phenotypes observed include a moderate to severe degree of loss of physical activity, hair loss and lack of glossiness, skin coarseness, periophthalmic lesions and shortened lifespan. Early observations of phenotype in some of our transgenic founders, in which expression of the missense tRNAs has been detected, indicate similar types of macroscopic alteration. Further work is in progress to confirm these observations. Likely responses to increased protein error rates include a change in the rate of protein turnover as well as a spectrum of effects on metabolic pathways resulting from the random introduction of incorrect amino acids during protein synthesis. Production of age characteristics and/or pathological phenotypes in the transgenic system will enable us to determine whether aberrant protein production is a good model for age-related disorders and particularly cytoskeletal disorders.

5. STEM CELLS AND AGEING

The presence of a variety of cell types in an organism opens the possibility of different mechanisms of ageing. Thus, several explanations of ageing at the cellular level could be suggested, from a gradual loss of irreplaceable cells, such as neurones, to the limited division potential of dividing cells, such as fibroblasts. Age-related changes of progenitor or stem cells may be of particular importance because any alteration in the numbers or functional competence of stem cells might affect tissue homeostasis. Stem cells are, however, difficult to identify since manipulating them experimentally may alter their state, e.g. causing them to differentiate.

An attractive system for the study of stem cell functions is found in the crypts of the small intestine[10]. The epithelium of the small intestine comprises a large number of finger-like projections, or villi, which have at their base flask-shaped structures called the crypts. The crypts constitute the proliferative compartment and the villi the functional component of the tissue. An average crypt contains 250 cells, of which 150 are proliferative. The stem cells are located at the bottom of the crypt, at the 4th or 5th position. They can divide to give 4 different cell types: the Paneth cells, the enteroendocrine cells, the goblet cells and the columnar enterocytes. The cells proliferate and differentiate in the crypt. The duration of the cell cycle is about 12 hours for most cells but 24 hours for stem cells. Cells leave the crypts at a rate of 12 cells/hour, migrate along the villus and reach the tip 2 days later. At this point, they are shed into the lumen. Because of this well understood developmental lineage and because the stem cells are rather precisely localised within the crypts, the small intestine represents a good model to study the behaviour of the stem cells.

The aims of this study were to use the small intestine crypt as a model to investigate age-related changes in the properties of stem cells in young and senescent male ICRFa mice. This strain was chosen for its lack of specific age-associated pathology. Three approaches have been followed. Firstly, age-related changes in the proximal and distal region of the small intestine were investigated using mice aged from 6 to 30 months old. The parameters measured were the frequency of spontaneous apoptosis, the frequency of mitosis and of cells undergoing DNA synthesis, the differentiation rate (number of goblet cells), the numbers of crypts/mm and villi/mm, the crypts/villi ratio, and the sizes of crypts and villi (height, width, area). Secondly, the ability of stem cells to undergo apoptosis in response to damage was studied by irradiating the whole body of young and old animals with low doses of [137]Cs. Apoptosis is thought to be important for eliminating stem cells which have accumulated DNA damage and therefore this experiment examined the possibility of age-related change in an important tissue maintenance function. Thirdly, the ability of stem cells to regenerate a crypt after irradiation at high doses was compared in young and old animals by determining the number of surviving crypts and their size (height, width, area) four days after the treatment.

The first striking difference was observed in the distal region of the small intestine. The villi were larger and longer in senescent mice (30 months) compared to young mice (6 months), and the cellularity of the lamina propria was reduced. Bloated villi were already seen in the animals at 15 months and the histological changes developed progressively with age. The differences were most pronounced in the distal end of the intestine. The crypts were deeper in old animals and the number of crypts/mm and villi/mm decreased with age. The frequencies of spontaneous apoptosis, mitosis and tritiated thymidine uptake were similar in all age groups, as also were the numbers of goblet cells.

In the low-dose irradiation experiment, a significantly higher level of apoptosis was found in the oldest group (30 months) in both the distal and proximal regions after a 1Gy radiation dose and in the distal region after 8Gy. The apoptotic cells were located about the 4th cell position, corresponding to the stem cell location. In the high-dose experiment, the surviving crypts were fewer and also smaller in old animals (30 months).

Thus, the functions of stem cells appeared to be altered in small intestine crypts of old mice, particularly in the distal region. This is demonstrated by the histological changes seen in the shape of the villi and by the reponses to induced damage. The stem cells in old mice undergo apoptosis at a higher rate and their ability to regenerate a crypt seems impaired. Further experiments will investigate the number of clonogenic cells per crypt, the level of expression of apoptosis-associated genes such as p53 or bcl-2 as well as the proliferation rate in old and young mice.

6. GENETICS OF LIFESPAN AND STRESS RESPONSE

Study of the genetics of lifespan and its association with cell maintenance and stress response systems is an important avenue of research, for which short-lived animal models have obvious advantages. *Caenorhabditis elegans* is proving to be a valuable system for investigating aspects of ageing, particularly the genetic determination of lifespan[11]. As with budding yeast and filamentous fungi, single-gene mutations have been identified which dramatically increase both mean and maximum lifespan (this is referred to as the Age phenotype).

C. *elegans* is a small (1.2 mm) free living, soil-dwelling worm which has been intensely studied as a developmental model and is currently the subject of a comprehensive genome sequencing project. Populations exist primarily as hermaphrodites (XX), with males (XO) occurring only rarely in a ratio of approximately 1:500. Hermaphrodites reproduce by self-fertilisation. The life cycle is three days in length with four larval stages (L1 to L4). The hermaphrodite lives for approximately 20 days at 20°C and the male usually lives approximately 18 days. The isolation of genetic variants with extended lifespan is made possible by an absence of heterosis for life history traits. All of the known single-gene mutations which confer an increase in metazoan lifespan have been identified in *C. elegans*. Some combinations of mutants lead to a 4-fold extension in mean and maximum lifespan[12]. At present, mutation of any one of nine genes is known to confer Age. The first Age locus was defined by the *age-1(hx546)* mutation. This mutation confers an increase of 65% in mean lifespan and 110% in maximum lifespan, due to a decrease in the acceleration of mortality rate with age[13]. *age-1(hx546)* was isolated from a screen for long-lived strains following chemical mutagenesis and displayed normal development and near-normal fertility. Other Age mutations are associated with altered developmental schedule[12].

We do not yet know how many genes can be mutated to confer the Age phenotype, nor whether there are multiple mechanisms leading to Age in the nematode. There are two types of evidence that suggest that all currently known Age mutations extend lifespan in the same way. The first evidence comes from classical genetic studies which indicate that all of the Age genes tested genetically interact; specifically, all the Age mutations can be suppressed by mutation of a downstream gene called *daf-16*.

The second type of evidence comes from phenotypic analysis of Age strains. All of the Age mutations are associated with other phenotypes, in particular stress-resistance. The correlation between stress resistance and lifespan extension suggests a mechanistic relationship. All the Age strains tested display increased thermotolerance (Itt), increased resistance to oxidative stress and increased resistance to ultraviolet radiation. Age mutations are also associated with over-expression of antioxidant enzyme genes (Cu/Zn superoxide dismutase and catalase)[14,15] and molecular chaperone genes (heat shock protein 16)[16]. It has been proposed that lifespan is extended in Age strains due to an increase in the efficiency of free-radical scavenging and protein conformational maintenance. A number of observations are consistent with the idea that Age mutations extend lifespan as a consequence of over-expression of stress-response genes. For example, environmental manipulations known to induce stress-response genes, particularly non-lethal heat shocks, lead to both an increase in stress-resistance and to increases in lifespan[16].

It thus appears that wild type Age genes are regulators (directly or indirectly) of stress-response pathways. By continuing to identify and describe genes which determine *C. elegans* lifespan, we would hope to identify physiological processes which contribute to ageing in more complex systems such as mammals.

7. DIETARY RESTRICTION

It is widely established that laboratory rodents (mice, rats) fed on calorie-restricted diets show increased mean and maximum lifespans[17]. It has been suggested that this is an adaptive response involving a temporary increase in cell maintenance levels, in order that the animal can survive a period of famine with a reduced rate of senescence, thereby permitting maximum reproductive potential to be preserved for when normal feeding is resumed[18,19].

Using life history modelling, a mathematical model of resource allocation has been developed to predict the relative importance of investment in somatic cell maintenance and stress response systems against the production of offspring throughout the life cycle under different conditions of resource abundance. Senescence is modelled as a change in condition, reflecting accumulation of somatic damage, that depends on the amount of resources allocated to reproduction. Each individual has phenotypic plasticity to optimise its use of available resources so as to maximise its Darwinian fitness. Preliminary results indicate that the life extension observed with dietary restriction may indeed be associated with an evolutionary response to fluctuating resource levels.

8. CONCLUSION

It is evident that the network of cell maintenance and stress response systems are important determinants of ageing and longevity. Integrated approaches to unravelling the components and interactions of this network will significantly increase our knowledge and understanding of intrinsic ageing and its relationship to pathogenesis of age-associated diseases.

ACKNOWLEDGMENT

We thank the Cancer Research Campaign, Dunhill Medical Trust, Medical Research Council, Research Into Ageing, and the Wellcome Trust for financial support.

REFERENCES

1. Kirkwood, T.B.L. and Franceschi, C. (1992) Is aging as complex as it would appear? New perspectives in aging research. Ann. New York Acad. Sci., 663, 412–417.

2. Kirkwood, T.B.L. (1992) Comparative life spans of species: why do species have the life spans they do? Am. J. Clin. Nutr., 55, 1191–1195.

3. Kowald, A. and Kirkwood, T.B.L. (1994) Towards a network theory of ageing: a model combining the free radical theory and the protein error theory. J. Theor. Biol., 168, 75–94.

4. Kowald, A. and Kirkwood, T.B.L. (1996) A network theory of ageing: the interactions of defective mito-chondria, aberrant proteins, free radicals and scavengers in the ageing process. Mutation Res., 316, 209–236.

5. Kirkwood, T.B.L. (1989) DNA, mutations and aging. Mutation Res., 219, 1–7.

6. Bürkle, A., Muller, M., Wolf, I. and Kupper, J.H. (1994) Poly (ADP-ribose) polymerase activity in intact and permeabilized leukocytes from mammalian species of different longevity. Mol. Cell. Biochem., 138, 85–90.

7. Rohme, D. (1981) Evidence for a relationship between longevity of mammalian species and lifespans of normal fibroblasts in vitro and erythrocytes in vivo. Proc. Natl. Acad. Sci. USA, 78, 5009–5013.

8. Rosenberger, R.F. (1991) Senescence and the accumulation of abnormal proteins. Mutation Res. 256 255–262

9. Kirkwood, T.B.L., Holliday, R. and Rosenberger, R.F. (1984) Stability of the cellular translation translation process. Int. Rev. Cytol., 92, 93–132.

10. Potten, C.S. and Loeffler, M. (1990) Stem cells: attributes, cycles, spirals, pitfalls and uncertainties. Lessons for and from the Crypt. Development 110: 1001–1020

11. Lithgow, G.J. (1996) Molecular genetics of Caenorhabditis elegans aging. In: Handbook of the Biology of Aging, 5th edn (eds Schneider, E.L. and Rowe, J.W). San Diego, Academic Press, 55–73.

12. Lakowski, B. and Hekimi, S. (1996) Determination of life-span in Caenorhabditis elegans by four clock genes. Science, 272, 1010–1013.

13. Johnson, T.E. (1990) Increased life-span of age-1 mutants in Caenorhabditis elegans and lower Gompertz rate of aging. Science, 249, 908–912.

14. Larsen, P.L. (1993) Aging and resistance to oxidative damage in Caenorhabditis elegans. Proc. Natl. Acad. Sci. USA, 90, 8905–8909.

15. Vanfleteren, J.R. (1993) Oxidative stress and aging in Caenorhabditis elegans. Biochem. J., 292, 605–608.

16. Lithgow, G.J., White, T.M., Melov, S. and Johnson, T.E. (1995) Thermotolerance and extended life span conferred by single gene mutations and induced by thermal stress. Proc. Natl. Acad. Sci. USA, 92, 7540–7544.

17. Holehan, A.M. and Merry, B.J. (1986) The experimental manipulation of ageing by diet. Biol. Rev. 61:329–68.

18. Holliday, R. (1989) Food, reproduction and longevity: is the extended lifespan of calorie restricted animals an evolutionary adaptation? BioEssays, 10, 125–127.

19. Harrison, D.E. and Archer, J.R. (1989) Natural selection for extended longevity from food restriction. Growth Devel. Aging, 53, 3.

13

FROM MOLECULAR GERONTOLOGY TO HEALTHY OLD AGE

Research Strategies

Olivier Toussaint and Suresh I. S. Rattan

Laboratory of Cellular Biochemistry and Biology
The University of Namur
Rue de Bruxelles, 61, B-5000 Namur, Belgium; and
Laboratory of Cellular Ageing
Department of Molecular and Structural Biology
Aarhus University
DK-8000 Aarhus-C, Denmark.

1. INTRODUCTION

Ageing is neither a monolithic nor a monocausal phenomenon. It has many facets and almost all the experimental data suggest that ageing is not controlled by a single mechanism. The diversity of the forms and variations in which age-related alterations are manifested indicate that ageing is stochastic in nature. Yet the apparent practical limit to maximum lifespan within a species, along with the evidence from studies on twins and from linkage studies performed on Drosophila which show that longevity is at least partly heritable, suggest that there is genetic regulation of ageing and longevity[1,2]. Recent studies on human centenarians have shown a correlation between their exceptional longevity and certain genotypes, e.g. for the apolipoprotein (APO-E) gene and the angiotensin converting enzyme (ACE) gene[3]. The genetic aspects of ageing and longevity on the one hand and the stochastic nature of age-related changes on the other are easily brought together if ageing is seen as the failure of maintenance and loss of homeostasis due to interactions between genetic mechanisms of defence (longevity assurance processes) and stochastic causes of damage and perturbations[1,2].

Almost all theories of ageing imply directly or indirectly that the progressive failure of homeostatic mechanisms is crucial for the process of ageing and the origin of age-related diseases. For example, the build-up of oxidative damage in macromolecules, the accumulation of abnormal, erroneous and defective proteins, defects in the signal transduction pathways, deficiency of the immune system, progressive loss of order in biological systems and several other similar hypotheses point towards the failure of maintenance at all levels of organization as a crucial determinant of ageing and lifespan[1,2]. Since the ultimate aim of gerontological research is to achieve a healthy old age, unravelling the fundamental mechanisms of ageing is a pre-requisite for developing appropriate means of increasing mobility, activity, creativity and independence in the elderly[4].

2. NECESSITY OF STUDYING MOLECULAR MECHANISMS OF CELLULAR AGEING

Most pathologies find their origin in cellular processes and age-related diseases are no exception to this rule. The age-associated diseases such as cardiovascular and cere-brovascular diseases, cancer, diabetes, osteoarthritis, osteoporosis, Alzheimer's disease, Parkinson's disease, loss of renal function and several other diseases are mainly due to dysfunctions of the cellular metabolic processes. These dysfunctions are tightly linked to-gether and in turn can lead to variations of the extra-cellular relationships of the involved cell type. Such variations can occur in the nature, intensity and kinetics of interactions not only between several cell types but also within a cell type and in the non-cellular environ-ment like the extra-cellular matrix.

Some of the major cell types involved in these processes are endothelial cells, epithelial cells, fibroblasts, bone cells, astrocytes, neurons, smooth muscular cells and the cells of the immune system. These cell types are subject to a general process of ageing which however will affect differently various cellular components and metabolic pathways. Therefore, for a rational approach to understand, prevent or cure age-related pathologies, we must first under-stand how different cell types age in an organ and how their interactions give rise to various pathologies. Given the diversity of the interactions between a given cell type and its environ-ment and between the cellular components themselves, the study of cellular ageing must be an inter-disciplinary research with participation of molecular biologists, geneticists, biochem-ists, physiologists and pathologists.

Significant progress has been made in our understanding of the phenomenology and of some of the mechanistic aspects of cellular ageing (see articles by Derventzi et al., Stathakos et al. and Slagboom and Knook in this book). However, it remains to be eluci-dated how various counting mechanisms, for example telomere loss and the loss of meth-ylated cytosines, regulate proliferative senescence and how the dysregulation of cellular responsiveness to external and internal signals brings about various alterations in cellular structure and function including the selective repression and expression of genes.

Furthermore, it is also important to find out how environmental events affect differ-ent aspects of cellular ageing. Various experimental models exist which have shown that unique or successive sublethal stress can lead to stress-induced growth arrest of cells (see article by Toussaint and Remacle in this book). Most of cellular response to stress involve the induction of repair and elimination systems. The efficiency of most repair systems af-ter stress has been described to decrease with ageing, including the induction of heat shock protein synthesis, induction of superoxide dismutase and the induction of various DNA repair systems. It has also been stressed that the kinetics and intensity of response of a given cellular protective system to a given stress are different from the kinetics and in-tensity of the induction of a single component of a protective system.

An age-related decrease in the inducibility of protective systems in stressful condi-tions could explain why cell survival to thermal stress, oxidative stress, antibiotics, phor-bol esters, radiations, burns, arthrosis, cutaneous injuries, and exposure to paraquat is lower in old cells. All the components of protective systems which are induced after or during stress are down regulated once the repair and elimination of damage is satisfactory. However, it is possible that a stress will have an irreversible effect on gene expression. Therefore it is possible that, after repeated stress, further changes can appear in the cellular genetic expression which are not specific to cellular ageing, but to other irreversible processes corresponding to 'pathological' cellular states appearing after repeated stress or after constant low-stress conditions. There are two ways to test this hypothesis. The first is to study the mechanisms of cellular regulation and the second is to study the gene expression occurring in normal or stress-induced ageing. The genes specific to stressed cells would then be compared to the genes specifically found in age-related pathological situations. This approach could be complementary to studying changes in gene expression found in vivo in age-related pathological conditions due to congenital predisposition.

3. SEARCHING FOR GERONTOGENES

There may or may not be a genetic programme that determines the exact time of death, but there appears to be an evolutionary constraint in terms of maximum achievable lifespan within a species. The view that the lifespan of an organism is intrinsically limited and lies largely within a species-specific range necessarily implies certain notions of genetic elements of regulation. In this context, the term gerontogene refers to any such genetic elements that are involved in ageing. The idea of gerontogenes does not contradict the non-adaptive nature of ageing, rather, it reasserts the importance of the genetic mechanisms of somatic maintenance in assuring germ-line continuity. Since the existence of genes for programmed self-destruction is on the whole discounted on the basis of evolutionary arguments against the notion of adaptive nature of ageing[5, 6], the concept of gerontogenes is closely linked with the genes involved in homeostasis and longevity assurance, instead of certain special genes for ageing.

The term gerontogenes does not refer to a tangible physical reality of real genes for ageing but refers to an emergent functional property of a number of genes which influence ageing. For this purpose, the term "virtual" gerontogenes has been suggested[7]. In science, the term "virtual" is used for entities that it is helpful to regard as being present although they have no physical existence. The paradigm of such an entity is the virtual image of optics. Another example is the currently popular "virtual reality", which fulfils the same definition.

The concept of virtual genes therefore refers to the emergent property of several genes whose functions are tightly coupled and whose combined action and interaction resemble the effect of one gene. Treating such a group as a virtual gene is a useful conceptual tool while the search continues for the genetic elements of regulation of complex biological processes, such as ageing. Although differentiation and development provide good examples of highly complex systems involving a large number of genes, it may be inappropriate to apply the concept of virtual genes to them, because these processes are under direct genetic control and have evolved as a result of natural selection. This situation is unlike ageing, where no natural selection for any specific genes is envisaged. Therefore, the concept of virtual genes is appropriate only for phenomena such as ageing, in which a genetic involvement is expected without direct genetic control open to natural selection.

The idea that gerontogenes are virtual implies that every time a gene is discovered which appears to have a role in the process of ageing, it will, on sequencing and identification, turn out to be a familiar normal gene with a defined function. Its role as a gerontogene can only be realised in the context of its emergent property in relation to several other genes that influence its activity and interactivity. Such genes cannot be hidden or cryptic, because their identities can in principle become well known. Individually, the functions of such genes can in principle be clearly established. Yet, as a result of concerted action and interaction, the combined effect of these genes resembles that of a "gene for ageing" although these were not specially designed or naturally selected for causing ageing. This idea of virtual gerontogenes is in line with the evolutionary explanation of the ageing process as being an emergent phenomenon caused by the lack of eternal maintenance and repair instead of being an active process caused by evolutionary adaptation.

Obviously, not every gene is potentially a virtual gerontogene. However, potentially every gene can affect the survival of an organism. Therefore, a distinction must be made between immediate survival or death on the one hand and the process of ageing on the other. The inactivation of any essential gene will result in the death of an organism without having anything to do with the process of ageing. The set of possible virtual gerontogenes can be narrowed down to sets of genes involved in the maintenance and repair of the cellular and sub-cellular components.

Evidence for the hypothesis that candidate virtual gerontogenes operate through one or more of the mechanisms of somatic maintenance and repair comes from experiments performed to retard ageing and to increase the lifespan of organisms. For example, anti-ageing and life-prolonging effects of calorie restriction are seen to be accompanied by the stimulation of various maintenance mechanisms. These include increased efficiency of DNA repair, increased fidelity of genetic information transfer, more efficient protein synthesis, more efficient protein degradation, more effective cell replacement and regeneration, improved cellular responsiveness, fortification of the immune system, and enhanced protection from free-radical- and oxidation-induced damage[3, 7]. Similarly, anti-ageing effects of a dipeptide, carnosine on human diploid fibroblasts[8] and a cytokinin, kinetin on human fibroblasts and on insects[9, 10] also appear to be due to the effect of these chemicals on maintaining the efficiency of defence mechanisms, including efficient protein synthesis and turnover and the removal of oxidative damage.

Genetic selection of Drosophila for longer lifespan also appears to work mainly through an increase in the efficiency of maintenance mechanisms, such as antioxidation potential (see article by Brack et al., in this book). An increase in lifespan of transgenic Drosophila containing extra copies of Cu-Zn superoxide dismutase (SOD) and catalase genes is due primarily to enhanced defences against oxidative damage[11]. The identification of long-lived mutants of the nematode Caenorhabditis elegans, involving various genes[12] may provide other examples of virtual gerontogenes because in almost all these cases increased lifespan is accompanied by an increased resistance to oxidative damage, an increase in the activities of SOD and catalase enzymes, and an increase in thermotolerance[13, 14].

Attempts at identifying determinants of longevity by finding a correlation between maximum lifespan of a species and other biological characteristics have also shown that it is the efficiency of various defence mechanisms that correlates best with longevity. Some of the well-known maintenance mechanisms whose activity levels and efficiencies are directly correlated with species lifespan include DNA repair[15–17], cell proliferative capacity[18] and antioxidative potential[19]. Molecular studies using a comparative approach, including the use of transgenic organisms, will be useful in identifying the genes that are most important in this respect, and can be expected to form the basis of appropriate strategies for future gerontological research.

4. MODULATING AGEING AND LIFESPAN

The modulation of ageing and lifespan has occupied human mind since its inception. Throughout human history, search for means to prevent or retard ageing has followed three main lines: (i) cleansing from impurities and wastes; (ii) nutritional supplements, including the use of medicinal plants; and (iii) replacement therapy. The immense popularity of various spas and water therapies even today, is an example of the first type of anti-ageing approach. The claims made for various herbal and other medicinal plant products, such as ginseng, ginkgo biloba and garlic, as nutritional supplements and anti-ageing drugs have received some support, however preliminary, from laboratory and/or clinical tests[3]. Replacement therapy, especially hormonal replacement therapy as an anti-ageing treatment has been used and misused for quite some time. The logic behind this approach has been that since the levels of various hormones decrease during ageing, supplying these hormones from outside may compensate for this loss and rejuvenate the system. Various therapeutic procedures have been followed to replace the lost hormones, such as gland transplantation, secretory-cell injections and hormone injections. Whereas many of these approaches, such as monkey testicle transplants in the 1950s, have been frauds[20], others have been refined to some extent and are still in use. For example, the subcutaneous injection of recombinant biosynthetic human growth hormone for six months in elderly men resulted in some increase in lean body mass, a decrease in adipose-tissue mass, and a slight increase in vertebral bone density[21]. Similarly, there have been some claims regarding the anti-ageing effects of dehydroepiandrosterone (DHEA), which is the primary precursor of sex steroids[22].

Recently, the so-called sleep hormone melatonin has received much attention for its role as an anti-ageing hormone[23, 24]. In a small-scale study, melatonin has been shown to delay ageing and prolong the lifespan of C57BL mice when they were fed orally with this hormone from early age[25]. In the same study it was also shown that grafting of pineals from young mice into the thymus of syngenic old mice increased their survival and inhibited age-related structural changes in thymus[25]. Extensive studies need to be done on the effects of melatonin and the pineal gland on the survival and ageing of a wide range of species before the usefulness of melatonin for human beings can be guaranteed[26, 27]. Furthermore, several questions remain unanswered regarding the wider applicability of such approaches because of our present lack of understanding of the regulation of synthesis of various hormones, their modes of action, metabolism and interrelations with other hormones.

Experimentally, the most effective anti-ageing and life-prolonging strategy has proved to be calorie restriction. Although most of the studies on dietary restriction and ageing have been performed on rats and mice, results have started to emerge from studies on long-term dietary restriction of non-human primates, rhesus monkeys[28, 29]. A large number of studies have established that calorie restriction delays ageing and prolongs the lifespan of various animals[30]. Most importantly, the anti-ageing effects of dietary restriction operate mainly by improving the efficiency of various mechanisms of maintenance and repair such as DNA repair, fidelity of genetic information transfer, rate of protein synthesis and degradation, cell replacement and regeneration, cellular responsiveness, the immune system, and protection from free-radical- and oxidation-induced damage[3]. However, at this stage, it is very difficult to say what kind of voluntary dietary-restriction regimes will have similar anti-ageing and life-prolonging effects in human beings, considering that there are significant differences in the biology and, perhaps more importantly, the sociology of human beings as compared with other animals.

Some other attempts to modulate ageing and lifespan include antioxidant supplementation in insects and rodents, and supplementation of cell-culture-media with growth factors, hormones and other compounds[3]. Attempts to develop gene therapeutic approaches to delay the signs of ageing and prolong lifespan have had only limited success until now. Most of these studies have been performed on fruitflies only. One of the first such studies to report lifespan extension by gene manipulation employed protein elongation factor EF-1α gene for making transgenic Drosophila which lived longer at higher temperature. However, this effect was seen only under specific conditions of heat-shock promoter regulation of EF-1α gene and not under normal conditions of fruitfly survival (see article by Brack et al., in this book). The other successful attempt at delaying ageing and prolonging the lifespan of Drosophila is described by Orr and Sohal[11], who made long-lived transgenic lines which overexpressed SOD and catalase genes.

5. FUTURE RESEARCH STRATEGIES

As discussed throughout this book, individually no tissue, organ or system becomes functionally exhausted, even in very old organisms, yet it is their combined interaction , interdependence and constant remodelling that determines the survival of the whole. The same logic needs to be applied for studies at the molecular and genetic levels, particularly with respect to developing gene-therapy to delay ageing. The search for an all-encompassing ageing gene(s) is unlikely to be successful. Estimates of the number of genes which could influence ageing and lifespan of mammals run up to a few hundred out of about one hundred thousand genes, and their allelic variants[31]. Therefore, manipulating any single gene or a few genes that show some effects on ageing and lifespan will help to identify genes that might qualify as being a part of the virtual gerontogene family. Of course, such studies will be valuable in finding ways to "fine-tune" the network and to prevent the onset of various age-related diseases and impairments by maintaining the efficiency of homeostatic processes. In the short term, such studies will also result in the development of a variety of so-called anti-ageing products, by concentrating on individual members of the gerontogene family. In contrast, direct gene therapy directed at the overall ageing process seems to hold little promise.

There are various modes of maintenance and longevity assurance processes at various levels of biological organization. These mechanisms include the processes of cellular and sub-cellular repair, cell division, cell replacement, neuronal and hormonal responsiveness, immune response, detoxification, free radical scavenging, stress-protein synthesis, macromolecular turnover and maintaining the fidelity of genetic information transfer. In order to unravel the molecular basis of ageing and modulate the process, it is proposed that the most promising research strategies will incorporate an analysis of the formation and functioning of maintenance and repair networks (Fig. 1). Some of the mechanisms expected to be crucial in this regard are those involved in maintaining: (i) the structural and functional integrity of the nuclear and mitochondrial genome; (ii) the accuracy and speed of transfer of genetic information from genes to gene products; (iii) the turnover of defective and abnormal macromolecules; and (iv) the efficiency of intracellular and extracellular communication and responsiveness.

Much is known about each of the above-mentioned categories of maintenance mechanisms. The application of modern, more sensitive methods can further establish in detail what happens to these processes during ageing. However, the ultimate aim of biogerontological research is to understand why these changes occur, how they affect various

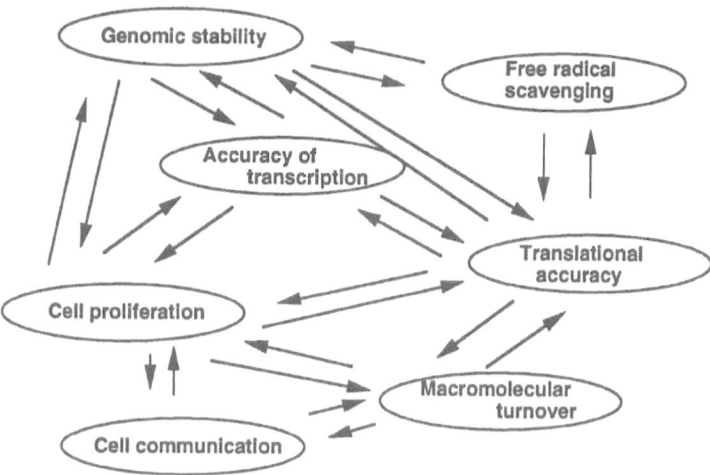

Figure 1. A section of the highly complex interactive network of maintenance and repair mechanisms. which forms the basis of an organism's property of homeostasis.

other constituents of the network, and how these can be modulated in order to maintain the healthy span of life. The attainment of this goal will require the development of experimental approaches in which the interactions between mechanisms of maintenance at various levels is studied and the reasons for their failure are elucidated.

The following lines of research can form the basis of a promising strategy to understand ageing:

1. Studying the extent of maintenance and repair of the genes involved in maintaining the stability of the nuclear and mitochondrial genome.
2. Studying the efficiency of transcription of these genes and post-transcriptional processing of their transcripts during ageing.
3. Studying the accuracy and efficiency of translation of these genes and analysing the specificity, stability and turnover of their gene products, including post-translational modifications.
4. Experimental modulation of various types of maintenance mechanisms (for example, responsiveness to stress, efficiency of signal transduction pathways, regulators of cell cycle progression etc.) and studying its effects on other levels such as gene stability and gene product synthesis and turnover.
5. Searching for natural or induced mutants (including in transgenic and knockout organisms) with altered levels of maintenance and repair of the crucial genes. This includes studies on non-human model systems, on human patients with premature ageing syndromes and on human centenarians in order to find genetic markers of lifespan potential.
6. Searching for age-specific and age-related disease-specific biomarkers for diagnostic purposes and for monitoring the effects of potential therapeutic agents.

The above approaches can narrow down the number of genes and the critical processes which are central to forming an optimum network for survival. Elucidating the nature and components of such a network will open up new possibilities of interfering with the gerontogene network and fine tune the system for both successful ageing and for discovering the real maximum limit to human lifespan.

Finally, there is an urgent need of bringing about significant changes in our research strategies and priorities if the goal "from molecular gerontology to healthy old age" is to be achieved. To quote Robin Holliday[4], ".......Firstly, there must be recognition, especially in clinical circles, of the central importance of gerontology for an overall understanding of age-related disease. Secondly, there must be recognition that much of modern cell biology, genetics and molecular biology is related in one way or another to cell and tissue homeostasis and maintenance, and therefore also relates to ageing."

REFERENCES

1. Holliday, R. (1995) Understanding Ageing. Cambridge University Press, Cambridge.
2. Rattan, S.I.S. (1995) Ageing – a biological perspective. Molec. Aspects Med., 16, 439–508.
3. Schächter, F., Faure-Delanef, L., Guénot, F., Rouger, H., Froguel, P., Lesueur-Ginot, L. and Cohen, D. (1994) Genetic associations with human longevity at the APOE and ACE loci. Nature Genet., 6, 29–32.
4. Holliday, R. (1996) The urgency of research on ageing. BioEssays, 18, 89–90.
5. Rose, M.R. (1991) Evolutionary Biology of Aging. Oxford University Press, New York.
6. Kirkwood, T.B.L. (1992) Biological origins of ageing. In: Oxford Textbook of Geriatric Medicine. (Evans, J.G. and Williams, T.F., Eds.), pp. 35–40, Oxford University Press, Oxford.
7. Rattan, S.I.S. (1995) Gerontogenes: real or virtual? FASEB J., 9, 284–286.
8. McFarland, G.A. and Holliday, R. (1994) Retardation of the senescence of cultured human diploid fibroblasts by carnosine. Exp. Cell Res., 212, 167–175.
9. Rattan, S.I.S. and Clark, B.F.C. (1994) Kinetin delays the onset of ageing characteristics in human fibroblasts. Biochem. Biophys. Res. Commun., 201, 665–672.
10. Sharma, S.P., Kaur, P. and Rattan, S.I.S. (1995) Plant growth hormone kinetin delays ageing, prolongs the lifespan and slows down development of the fruitfly Zaprionus paravittiger. Biochem. Biophys. Res. Commun., 216, 1067–1071.
11. Orr, W.C. and Sohal, R.S. (1994) Extension of life-span by overexpression of superoxide dismutase and catalase in Drosophila melanogaster. Science, 263, 1128–1130.
12. Lakowski, B. and Hekimi, S. (1996) Determination of life-span in Caenorhabditis elegans by four clock genes. Science, 272, 1010–1013.
13. Larsen, P.L. (1993) Aging and resistance to oxidative damage in Caenorhabditis elegans. Proc. Natl. Acad. Sci. USA, 90, 8905–8909.
14. Lithgow, G.J., White, T.M., Melov, S. and Johnson, T.E. (1995) Thermotolerance and extended life-span conferred by single-gene mutations and induced by thermal stress. Proc. Natl. Acad. Sci. USA, 92, 7540–7544.
15. Grube, K. and Bürkle, A. (1992) Poly(ADP-ribose) polymerase activity in mononuclear leukocytes of 13 mammalian species correlates with species specific life span. Proc. Natl. Acad. Sci. USA, 89, 11759–11763.
16. Whitehead, I. and Grigliatti, T.A. (1993) A correlation between DNA repair capacity and longevity in adult Drosophila melanogaster. J. Gerontol., 48, B124-B132.
17. Bohr, V.A. and Anson, R.M. (1995) DNA damage, mutation and fine structure repair in aging. Mutat. Res., 338, 25–34.
18. Rattan, S.I.S. and Stacey, G.N. (1994) The uses of diploid cell strains in research in aging. In: Cell & Tissue Culture: Laboratory Procedures. (Griffiths, J.B., Doyle, A. and Newell, D.G., Eds.), pp. 6D:2.1–2.12, John Wiley, Chichester, UK.
19. Sohal, R.S., Agarwal, S., Dubey, A. and Orr, W.C. (1993) Protein oxidative damage is associated with life expectancy of houseflies. Proc. Natl. Acad. Sci. USA, 90, 7255–7259.
20. Hamilton, D. (1986) The Monkey Gland Affair. Chatto and Windus, London.
21. Rudman, D., Feller, A.G., Nagraj, H.S., Gergans, G.A., Lalitha, P.Y., Goldberg, A.F., Schlenker, R.A., Cohn, L., Rudman, I.W. and Mattson, D.E. (1990) Effects of human growth hormone in men over 60 years old. New Eng. J. Med., 323, 1–6.
22. Bellino, F.L., Daynes, R.A., Hornsby, P.J., Lavrin, D.H. and Nestler, J.E. (1995) Dehydroepiandrosterone (DHEA) and Aging. Annals of the New York Academy of Sciences, vol. 774., New York.

23. Reiter, R.J. (1995) The pineal gland and melatonin, in relation to aging: a summary of the theories and of the data. Exp. Gerontol., 30, 199–212.

24. Kumar, V. (1996) Melatonin: a master hormone and a candidate for universal panacea. Ind. J. Exp. Biol., 34, 391–402.

25. Pierpaoli, W. and Regelson, W. (1994) Pineal control of aging: effect of melatonin and pineal grafting on aging mice. Proc. Natl. Acad. Sci. USA, 91, 787–791.

26. Huether, G. (1996) Melatonin as an antiaging drug: between facts and fantasy. Gerontol., 42, 87–96.

27. Reppert, S.M. and Weaver, D.R. (1995) Melatonin madness. Cell, 83, 1059–1062.

28. Bodkin, N.L., Ortmeyer, H.K. and Hansen, B.C. (1995) Long-term dietary restriction in older-aged rhesus monkeys: effects on insulin resistance. J. Gerontol. Biol. Sci., 50A, B142-B147.

29. Lane, M.A., Baer, D.J., Rumpler, W.V., Weindruch, R., Ingram, D.K., Tilmont, E.M., Cutler, R.G. and Roth, G. (1996) Calorie restriction lowers body temperature in rhesus monkeys, consistent with a postulated anti-aging mechanism in rodents. Proc. Natl. Acad. Sci. USA, 93, 4159–4164.

30. Masoro, E.J. (1995) Dietary restriction. Exp. Gerontol., 30, 291–298.

31. Martin, G.M. (1992) Biological mechanisms of ageing. In: Oxford Textbook of Geriatric Medicine. (Evans, J.G. and Williams, T.F., Eds.), pp. 41–48, Oxford Univ. Press, Oxford.

INDEX

Aberrant proteins, 194, 196
Abnormal proteins, 67
ACE, see angiotensin converting enzyme
Acetyl-L-carnitine, 99
Actinic ageing, 183
Adhesion molecules, 135, 185
Adrenocortical cells, 186
Advanced glycation end products (AGE), 8, 65, 182, 187
Aedes aegypti, 169
Age mutations, 198, 199
AGE, see advanced glycation end products
Age-1 mutation, 198
Age-associated pathology, 140
Age-related disease, 134, 140, 141
Ageing *(also: Aging)*
    definition, 1, 25, 37, 53, 75, 177, 179
    successful, 3
Altered gene expression, 17–20, 116
Altered proteins, 67
Alzheimer's disease, 27, 98, 202
Aminoglycosides, 61, 62
Amyloidogenesis, 134
Angiotensin converting enzyme (ACE), 20, 201
Antibiotics, 61, 62, 202
Anticodon loop, 196
Antigenic stimuli, 137
Antioxidant enzymes, 166, 169, 184, 194
Antioxidants, 97–99, 152, 166, 169, 187
Antisense RNA, 170
Aortic aneurysm, 185
AP-1 transcription factor, 138
APO C loci, 143
Apolipoprotein E, 20, 201
Apoptosis, 28, 119, 120, 131, 135, 139, 197
Arachidonic acid, 180
Ascorbate, 178
Astrocytes, 202
Ataxia telangiectasia, 10
Atherogenesis, 187
Atherosclerosis, 10, 133, 140–142
Atherosclerotic lesions, 187

ATP synthesis, 39–41, 180, 194
ATP turnover, 91
Autoimmunity, 136; *see also* centenarians, autoantibodies

B chronic lymphocytic leukemia, 134–136
B lymphocytes, 134, 141
β-galactosidase reporter gene, 161, 171
Bacteria, 141
Bcl-2, 143, 198
BCLL, 134; *see* B chronic lymphocytic leukemia
Biomarkers, 207
Biopsies, 195
Blastoderm stage, 154
Blebs, 179
Bone cells, 202
Brain, 94
Burns, 202

c-fos, 138, 184
c-jun, 138, 184
C57BL mice, 205
Caenorhabditis elegans, 153, 198, 199
Calcium influx, 96
Cancer, 15, 133, 142, 202, 204
Capillaries, 179
Carbonyl, 64
Cardiovascular disease, 134, 143, 202
Carnosine, 204
Catalase, 166, 167, 169, 204, 206
Cataract, 134
CD 3+, 133, 136
CD 4+, 133, 136
CD 5, 135
CD 8+, 133, 136
CD 16+, 141
CD 16-, 141
CD 19+, 135
CD 28+, 138
CD 57+, 141
CD 57-, 141
Cdk-inhibitors, 17, 18